FIELD - PROGRAMMABLE GATE ARRAY

Edited by **George Dekoulis**

Field - Programmable Gate Array
http://dx.doi.org/10.5772/63664
Edited by George Dekoulis

Contributors

Ajay Singh, Miguel Angel Martínez Prado, Juvenal Rodríguez, Carlos Miguel Torres Hernández, Gilberto Herrera, Diana Carolina Toledo Pérez, Cristian Anghel, Cristian Stanciu, Constantin Paleologu, Alexandru Amaricai, Oana Boncalo, Nikhil Marriwala, O.P Sahu, Anil Vohra, Francesco Grancagnolo, Gianluigi CHIARELLO, Claudio CHIRI, Giuseppe COCCIOLO, Alessandro CORVAGLIA, Marco PANAREO, Aurora PEPINO, Giovanni TASSIELLI, Raffaele Giordano, Wen-Jyi Hwang, Chien-Min Ou, Steffen Mauch, Johann Reger, Jose Torres, Raimundo Garcia, Jesus Soret, Julio Martos, Nordin Aranzabal, Abraham Menendez, Pedro A. Martinez, Adrian Suarez, Gabriel Torrens

Published by InTech

Janeza Trdine 9, 51000 Rijeka, Croatia

Publishing Process Manager Edi Lipovic
Technical Editor SPi Global
Cover InTech Design team

First published May, 2017
Printed in Croatia
Legal deposit, Croatia: National and University Library in Zagreb

Additional hard copies can be obtained from orders@intechopen.com

Field - Programmable Gate Array, Edited by George Dekoulis
p. cm.
Print ISBN 978-953-51-3207-3
Online ISBN 978-953-51-3208-0

For EU product safety concerns:
IN TECH d.o.o., Prolaz Marije Krucifikse Kozulić 3, 51000 Rijeka, Croatia,
info@intechopen.com or visit our website at intechopen.com.

Meet the editor

Professor George Dekoulis received a first class BEng (Hons) in Communications Engineering from De Montfort University in Leicester, UK, in 2001. He joined the Space and Planetary Physics Group at Lancaster University in 2001. He has received several research awards from STFC, UK and EPSRC, UK and the "IET Hudswell International Research Scholarship". He was awarded a Ph.D in 2007 in Space Engineering and Communications. He is currently a professor in Aeronautical and Space Engineering at Aerospace Engineering Institute, Cyprus. He was a professor in Electrical and Electronics Engineering at Middle East Technical University, Turkey. He has worked as Research Associate in Space Engineering and Technical Support Engineer on Space Systems at Lancaster University, and as R & D Aerospace Systems Design Engineer for Electromech, UK. He is researching and developing Aerospace Engineering Systems using state-of-the-art all-digital technologies.

Contents

Preface

This edited volume is a collection of reviewed and relevant research chapters, concerning the developments within the "Field-Programmable Gate Array" field of study. The book includes scholarly contributions by various authors and edited by a group of experts pertinent to semiconductors. Each contribution comes as a separate chapter completed within itself but directly related to the book's topics and objectives.

The book has 11 chapters.

The target audience comprises scholars and specialists in the field.

InTechOpen

Chapter 1

Efficient Hardware Architecture for Correlation-Based Spike Detection and Unsupervised Clustering

Chien-Min Ou and Wen-Jyi Hwang

Additional information is available at the end of the chapter

http://dx.doi.org/10.5772/66105

Abstract

This chapter presents a novel hardware architecture for correlation-based spike detection and unsupervised clustering. The architecture is able to utilize the information extracted from the results of spike clustering for efficient spike detection. The architecture supports the fast computation for the normalized correlation and OSORT operations. The normalized correlation is used for template matching for accurate spike detection. The OSORT algorithm is adopted for unsupervised classification of the detected spikes. The mean of spikes of each cluster produced by the OSORT algorithm is used as the templates for subsequent detection. The architecture adopts postnormalization technique for reducing the area costs. Modified OSORT operations are also proposed for facilitating unsupervised clustering by hardware. The proposed architecture is implemented by field programmable gate array (FPGA) for performance evaluation. In addition to attaining high detection and classification accuracy for spike sorting, experimental results reveal that the proposed architecture is an efficient design providing low area cost and high throughput for real-time offline spike sorting applications.

Keywords: spike sorting, spike detection, spike clustering, field programmable gate array, brain machine interface

1. Introduction

There is an increasing demand in data-acquisition systems for neurophysiology to record simultaneously from many channels over long time periods [1]. These experiments accumulate large amounts of data, which would be processed by spike sorting systems [2] for analyzing the activities of neurons. A typical spike sorting system usually involves complicated spike detection and classification operations for separating spikes from background noise and clustering the detected spikes. Large amount of spike-trains would impose heavy computation load for a software spike sorting system, resulting in long processing time.

One approach to reduce the computation time is to implement a spike sorting system by hardware. A number of spike sorting systems based on field programmable gate array (FPGA) [3] have been proposed. In [4], a spike classification architecture using probabilistic neural networks [5] is proposed. Although high speed-up over its software counterpart is observed, the system does not provide spike detection. In addition, the architecture requires the user to input the number of clusters. Therefore, it does not support a fully unsupervised hardware classification. Similarly, the architecture proposed in [6] also focuses on the classification. The generalized Hebbian algorithm (GHA) [7] is implemented by FPGA in this work for feature extraction. The features produced by the architecture are then clustered by the k-means algorithm. The architecture does not include the hardware implementation of the k-means algorithm. In addition, the number of clusters still needs to be prespecified in the k-means algorithm. The architecture in [8] is able to carry out the feature extraction and clustering in hardware. In the architecture, the feature extraction and clustering are based on GHA and fuzzy c-means [9] algorithm, respectively. Therefore, similar to the architecture in [6], fully unsupervised classification is difficult for the architecture in [8] because the number of clusters still needs to be known beforehand.

An alternative FPGA-based hardware architecture [10] is to adopt OSORT algorithm [11] for spike-sorting. The OSORT algorithm is able to perform clustering without the prior knowledge of the number of clusters. Unsupervised classification therefore can be carried out. The architecture also includes a spike detection circuit based on the nonlinear energy operators (NEOs) [12], which is an energy-based detection algorithm. Similar to other energy-based algorithms [13, 14], the NEO algorithm is simple and effective. However, although the energy-based algorithms can operate in conjunction with a number of automatic threshold algorithms [12–14], proper selection of threshold values for these algorithms may still be difficult when noise becomes large. Therefore, their performance may deteriorate rapidly as noise energy increases.

The template-based spike detection algorithm may be suited for the detection of spikes from the source sequences with high noise level. A matched filter [15, 16] is a typical technique based on templates. To detect the presence of the templates, the filter correlates known templates with the input spike trains. This operation can also be viewed as a likelihood ratio detection (LRT) [17]. A drawback of the matched filter is that the templates are required. Although the adaptive generation of templates is possible [18], only a single template is produced for spike trains. However, because spike trains in general are formed from two or more neurons, a single template may not be sufficient for the detection of all the spikes generated by the neurons. The template-based algorithm presented by [19] has been found to be effective for the detection of noisy spikes. It adopts the OSORT algorithm for automatic template generation. Normalized correlation operations then carry out the spike detection using the templates produced by the OSORT algorithm. Nevertheless, in the algorithm, both the template generation and correlation computation have high computational complexities.

The software implementation of the algorithm may not be suitable for fast analysis of spike trains.

The objective of this chapter is to present a novel FPGA-based hardware architecture for efficient spike detection and clustering. The architecture supports both the normalized correlation operations for spike detection and the OSORT operations for spike clustering. Moreover, the clustering results produced by the OSORT algorithm are used as the templates for spike detection. The architecture is the hardware implementation of the algorithm in [19]. It then has the advantages of accurate spike detection, fully unsupervised spike clustering, and fast computation.

We have implemented a spike sorting system on a network-on-chip (NOC) platform for the evaluation of the proposed architecture. The platform is based on FPGA. It consists of a soft-core processor [20] and the proposed architecture. The proposed spike sorting architecture is used as a hardware accelerator of the soft-core processor for spike sorting. The simulator developed in [21] is adopted to generate extracellular recordings. In this paper, comparisons with the existing software and hardware implementations are made. Experimental results show that the proposed architecture attains a high speed-up over its software counterpart for spike sorting. It also has a lower area cost over existing hardware architectures. Experimental results reveal that the proposed architecture is an effective alternative for real-time spike sorting with accurate detection and clustering.

2. The algorithm

Before presenting the hardware architectures for spike sorting, we first review the spike detection and clustering algorithms adopted by this work. Detailed discussions of these algorithms can be found in [19].

2.1. Normalized correlation

Consider a spike train $\mathbf{X}$, where the mth sample of $\mathbf{X}$ is denoted by $x[m]$. Moreover, the mth segment of the spike train $\mathbf{X}$ is denoted by $\mathrm{x}_m = [x[m], x[m-1], \ldots, x[m-N+1]]^T$, where N is the length of the segment. Suppose the spike train is processed by a matched filter with template $\mathrm{t} = [t[0], \ldots, t[N-1]]^T$. Let $\overline{\mathbf{x}}_m$ and $\overline{\mathbf{t}}$ be the normalized version of $\mathbf{x}_m$ and $\mathbf{t}$, respectively. That is,

$$\overline{\mathbf{x}}_m = \frac{\mathbf{x}_m}{\left\|\mathbf{x}_m\right\|}, \ \overline{\mathbf{t}} = \frac{\mathbf{t}}{\left\|\mathbf{t}\right\|}. \qquad (1)$$

The normalized output at m, denoted by, $\bar{y}[m]$, is computed from

$$\bar{y}[m] = \sum_{k=0}^{N-1} \bar{x}[m-k]\bar{t}[k] = \bar{\mathbf{x}}_m^T \bar{\mathbf{t}} \tag{2}$$

This is the inner product of segment $\bar{\mathbf{x}}_m$ and template $\bar{\mathbf{t}}$, which indicates the normalized correlation between these two vectors. The segment $\mathbf{x}_m$ is detected as a spike when $\bar{y}[m]$ is larger than a prespecified threshold η. Let $d(\bar{\mathbf{x}}_m, \bar{\mathbf{t}})$be the squared distance between $\bar{\mathbf{x}}_m$and $\bar{\mathbf{t}}$. It can be shown that

$$d(\bar{\mathbf{x}}_m, \bar{\mathbf{t}}) = 2 - 2\bar{\mathbf{x}}_m^T \bar{\mathbf{t}}. \tag{3}$$

Because $d(\bar{\mathbf{x}}_m, \bar{\mathbf{t}}) \geq 0$,

$$\bar{\mathbf{x}}_m^T \bar{\mathbf{t}} \leq 1. \tag{4}$$

Our normalized correlation operations are based on $\bar{\mathbf{x}}_m$ and $\bar{\mathbf{t}}$. When $\bar{\mathbf{x}}_m^T \bar{\mathbf{t}} \geq \eta$, then $\mathbf{x}_m$is detected as a spike. From Eq. (4), it follows that

$$\eta \leq 1 \tag{5}$$

The normalized correlation operations shown in Eq. (2) may have high computational complexities. Although the template $\bar{\mathbf{t}}$ can be computed beforehand, the computation of $\bar{\mathbf{x}}_m$ still needs to be carried out online, involving the calculation of$||\mathbf{x}_m||$and $\bar{\mathbf{x}}_m = \frac{\mathbf{x}_m}{||\mathbf{x}_m||}$. Note that N multiplications, N–1 additions, and one squared root operation are required for the computation of $||\mathbf{x}_m||$. Moreover, N divisions are needed for $\bar{\mathbf{x}}_m$. Finally, the inner product of $\bar{\mathbf{x}}_m^T \bar{\mathbf{t}}$ requires N multiplications and N–1 additions. In total, the basic implementation of the normalized correlation operations requires 2N multiplications, (2N–2) additions, N divisions, and 1 squared root operation. When the fast computation is an important concern, the hardware implementation may then be desirable.

2.2. OSORT algorithm

The OSORT algorithm is an unsupervised template-based clustering algorithm for spike sorting. It does not require feature extraction, and the number of clusters c is automatically determined by the algorithm. Define $C_i, i = 1, \ldots, c$, as the clusters of spikes generated by the OSORT algorithm, where $\mathbf{t}_i$, $i = 1, \ldots, c$, is the average value of the spikes belonging to C_i.

Given a current detected spike **s** for clustering, in the OSORT algorithm, the squared distances $d_i = d(\mathbf{s}, \mathbf{t}_i)$ for $i = 1, \ldots, c$ are first computed. The minimum distance d_{i*} is then identified, where $i* = \arg\min_i d_i$. We will assign **s** to C_{i*} when d_{i*} is less than a prespecified threshold τ_1. In this case, because C_{i*} has a new member, its mean value $\mathbf{t}_{i*}$ will also be updated. Otherwise, a new cluster is created, where **s** is its only member. After the updating of $\mathbf{t}_{i*}$, the cluster merging process will be activated. The process involves the computation of the distance between $\mathbf{t}_{i*}$ and $\mathbf{t}_j$, $i* \neq j$. Two clusters $\boldsymbol{C}_{i*}$ and $\boldsymbol{C}_{j*}$ will be merged when $d(\mathbf{t}_{i*}, \mathbf{t}_{j*}) < \tau_2$, where $j* = \arg\min_{j, j \neq i*} d(\mathbf{t}_{i*}, \mathbf{t}_j)$. **Figure 1** summarizes the operations of the OSORT algorithm.

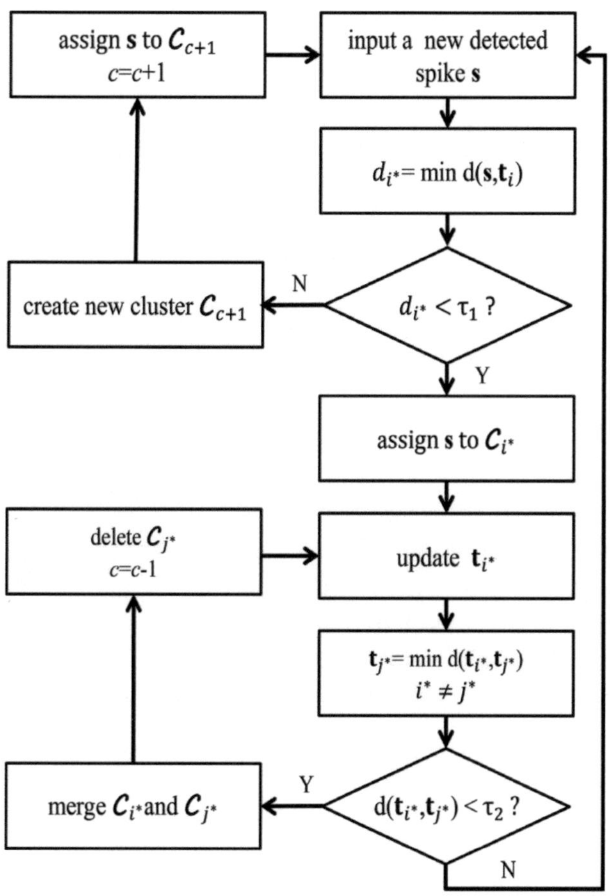

Figure 1. The flowchart of OSORT operations.

2.3. Normalized correlation and OSORT algorithm for spike sorting

By combining the normalized correlation with the OSORT algorithm, an effective spike sorting system for both spike detection and classification can be realized. The system is a feedback system capable of automatic template generation for spike detection and unsupervised clustering for the classification. The block diagram of the system is revealed in **Figure 2**.

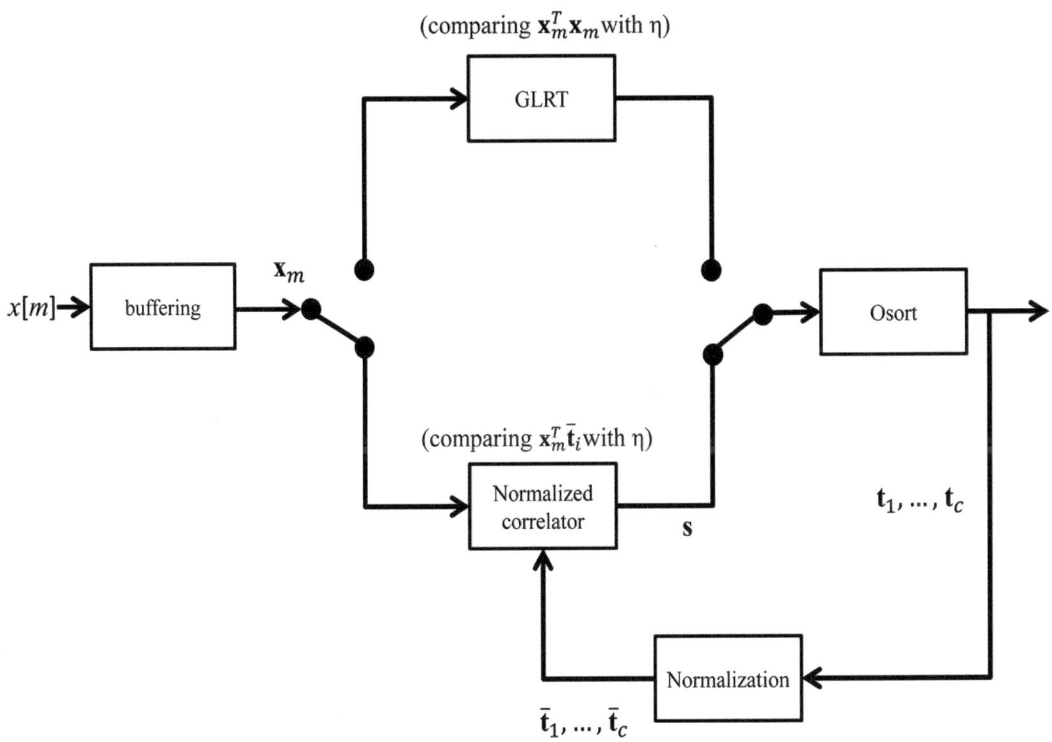

Figure 2. The block diagram of the spike detection/sorting system based on GLRT test, normalized correlator, and OSORT algorithms.

Initially, the clusters and templates produced by the OSORT are not available. As a result, it may be difficult to carry out the normalized correlation for spike detection. One way to solve this problem is to use only the block energy for the detection of spikes. The detected spikes are then clustered by the OSORT algorithm for the generation of initial templates.

After the templates become available, the spike detection is then based on the normalized correlation $\bar{\mathbf{x}}_m^T\bar{\mathbf{t}}$. The input block is detected as a spike when any of the c normalized correlation exceeds the threshold η. Because of the normalized correlation operations, the threshold value is bounded as shown in Eq. (5). Note that the templates for spike detection will be updated regularly so that the variations of input signals can be tracked to improve the spike detection performance.

The hardware architecture for implementing the spike sorting system is depicted in **Figure 3**. The architecture contains two modules and one controller. The first module of the architecture, termed normalized correlator module, is responsible for the spike detection. It is capable of performing both the GLRT and normalized correlation operations. The second module, termed OSORT module, carries out the unsupervised OSORT spike sorting. The global controller coordinates the operations of these two modules. The architecture and detailed operations of the normalized correlator module and OSORT module are presented in the following two sections.

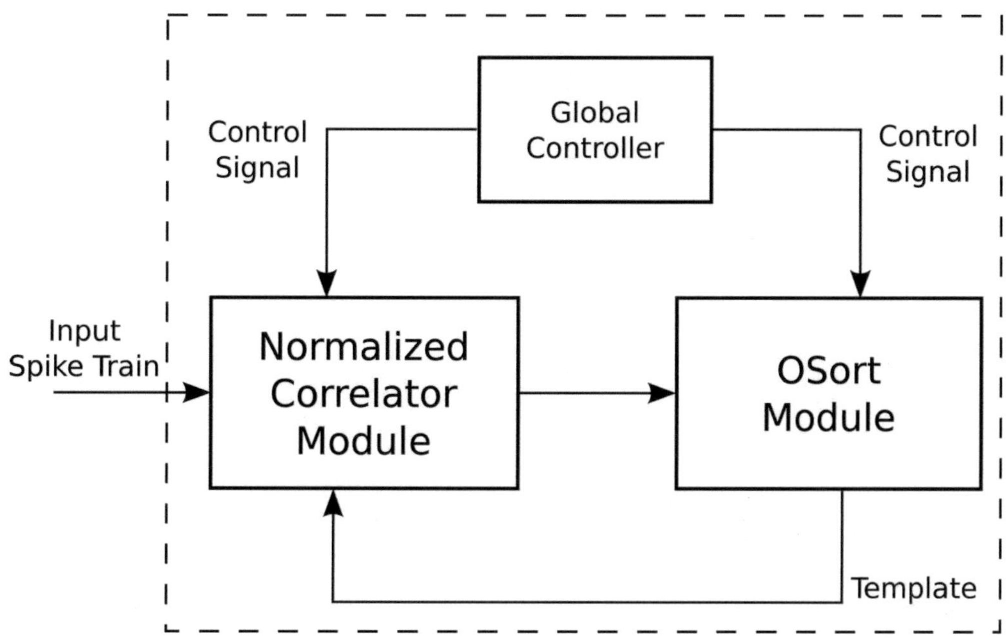

Figure 3. The hardware architecture of the proposed spike sorting system.

3. Architecture of the normalized correlator module

The block diagram of the normalized correlator module is revealed in **Figure 4**. The module supports the filtering, block energy computation, correlation computation, detection, and buffering. The filtering operation is the preprocessing step for the spike detection. It reduces the DC offset and noises. The objective of the block energy computation is to find $||\mathbf{x}_m||^2$, which is then followed by correlation computation for calculating $\overline{\mathbf{x}}_m^T \overline{\mathbf{t}}$. The detection results are then produced by the comparison operations. The detected spikes are stored in the switch buffer, which can be accessed by the OSORT module for subsequent clustering operations. Without loss of generality, the length of spike is set to be $N = 64$ for our discussion.

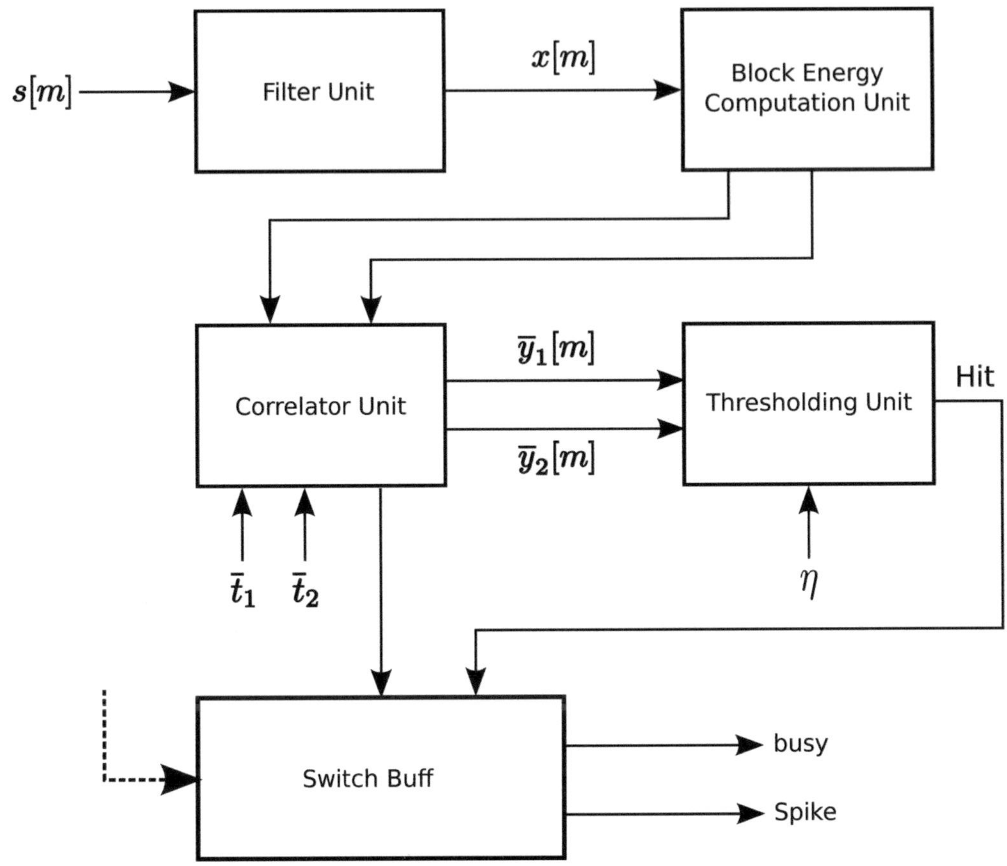

Figure 4. The block diagram of the normalized correlator module.

3.1. Filter unit and block energy computation unit

The filter unit is a hardware implementation of a band-pass Butterworth filter. The filter unit contains multipliers, shift registers, and adders. The architecture is a simple realization of the direct form I of IIR filters. To implement the block energy computation unit, we first note that a basic approach may involve N multiplications for the energy computation, resulting in high-area costs. The proposed approach is based on the fact that

$$\left\|\mathbf{x}_m\right\|^2 = \left\|\mathbf{x}_{m-1}\right\|^2 + x^2\left[m\right] - x^2\left[m-N\right]. \tag{6}$$

Consequently, the calculation of $||\mathbf{x}_m||^2$ needs only two multiplications. This is because $||\mathbf{x}_{m-1}||^2$ (i.e., the block energy of the previous block) is already available. **Figure 5** shows the resulting design, which contains two multiplier, a single N-stage shift register, and two adders. The goal of the shift register is to store the previous samples (i.e., $x[k]$, $k = m - 1,\ldots, m - N$) of

http://dx.doi.org/10.5772/66105

$x[m]$. The shift register therefore is able to offer the sample $x[m - N]$ for the calculation of $x^2[m - N]$. In addition, the shift register can be employed for the correlation computation.

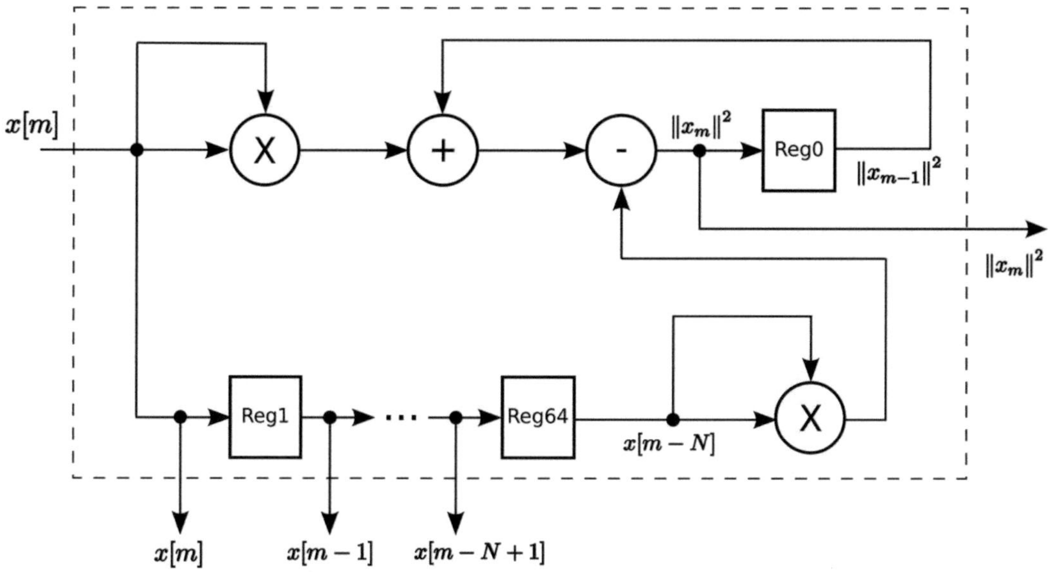

Figure 5. Architecture of block energy computation unit.

3.2. Correlator unit

The goal of the unit is to carry out the normalized correlation $\bar{\mathbf{x}}_m^T\bar{\mathbf{t}}$. Note that the normalized template $\bar{\mathbf{t}}$ can be obtained offline from the OSORT circuit. Therefore, it is only necessary to find $\bar{\mathbf{x}}_m$ online. One simple approach to compute $\bar{\mathbf{x}}_m$ is to divide each sample of $\mathbf{x}_m$ by $||\mathbf{x}_m||$. Because the block $\mathbf{x}_m$ contains N samples, N dividers are required. In the proposed architecture, a novel postnormalization approach is employed, where the inner product $\mathbf{x}_m^T\bar{\mathbf{t}}$ is computed first. Because $\mathbf{x}_m^T\bar{\mathbf{t}}$ is a scalar, we can then use only one divider to compute $\bar{\mathbf{x}}_m^T\bar{\mathbf{t}}$ by dividing $\mathbf{x}_m^T\bar{\mathbf{t}}$ by $||\mathbf{x}_m||$.

Figure 6 shows the architecture of the correlator unit for the case of two templates. The circuit consists of $2N$ multipliers, one squared root circuit, two accumulators, and one divider. Moreover, there are two registers for storing the normalized templates $\bar{\mathbf{t}}_1$ and $\bar{\mathbf{t}}_2$. Recall that the shift register in the block energy computation unit contains the samples of $\mathbf{x}_m$. Based on $\mathbf{x}_m$ and $\bar{\mathbf{t}}_i$, $i = 1, 2,$the computation of each $\mathbf{x}_m^T\bar{\mathbf{t}}_i$, $i = 1, 2$, is carried out in parallel. Moreover, the multiplication results are accumulated in a pipelined fashion. The accumulation results are then scaled by a factor of $1/||\mathbf{x}_m||$. Because the block energy computation unit provides

$||\mathbf{x}_m||^2$, only a squared root circuit and an inverse circuit are needed for the calculation of $1/||\mathbf{x}_m||$, as shown in **Figure 6**.

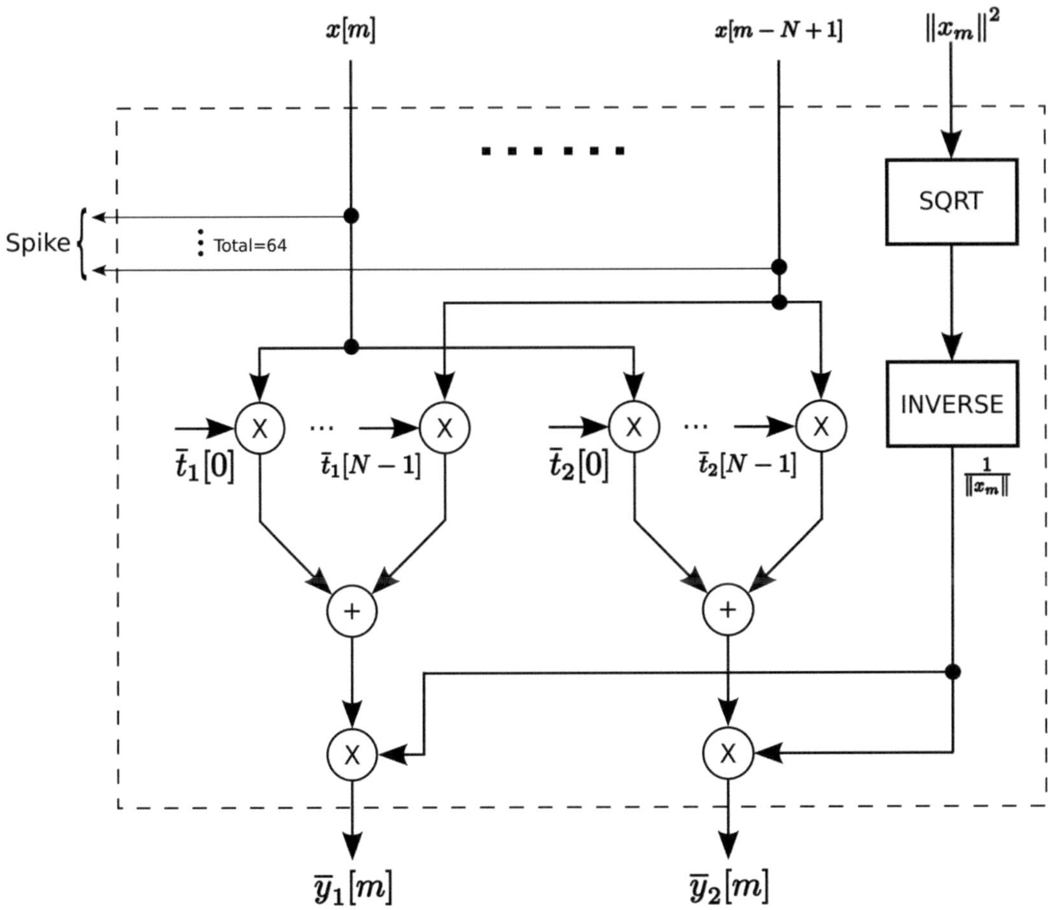

Figure 6. Architecture of correlator unit.

3.3. Threshold unit

Although the operations of the unit can be easily accomplished by a simple comparison circuit, the detection accuracy may be further improved by taking the detection results of the neighboring blocks into consideration. Because the neighboring blocks are overlapping, they may be similar. As a result, the normalized correlation values of the neighboring blocks may also be similar. Therefore, it is likely that an occurrence of a single spike may result in the issues of multiple hits.

To solve this problem, when the normalized correlation value of a block is above the threshold, a hit is not immediately declared. The architecture will then examine the normalized correlation values of the previous blocks. A hit would actually be issued only if k out of K preceding blocks have normalized correlation values above a threshold. In this way, the false alarm rate (FAR) can be effectively lowered. **Figure 7** shows the corresponding architecture, which contains a K-stage shift register storing the comparison results of the K previous blocks. Each stage of the shift register contains only a single-bit information, where 1 indicates that the corresponding block has normalized correlation value above the threshold η, and 0 otherwise. Therefore, if the sum of all the K stages is larger or equal to k, then at least k preceding blocks have normalized correlation value above the threshold. In this case, the architecture issues a hit.

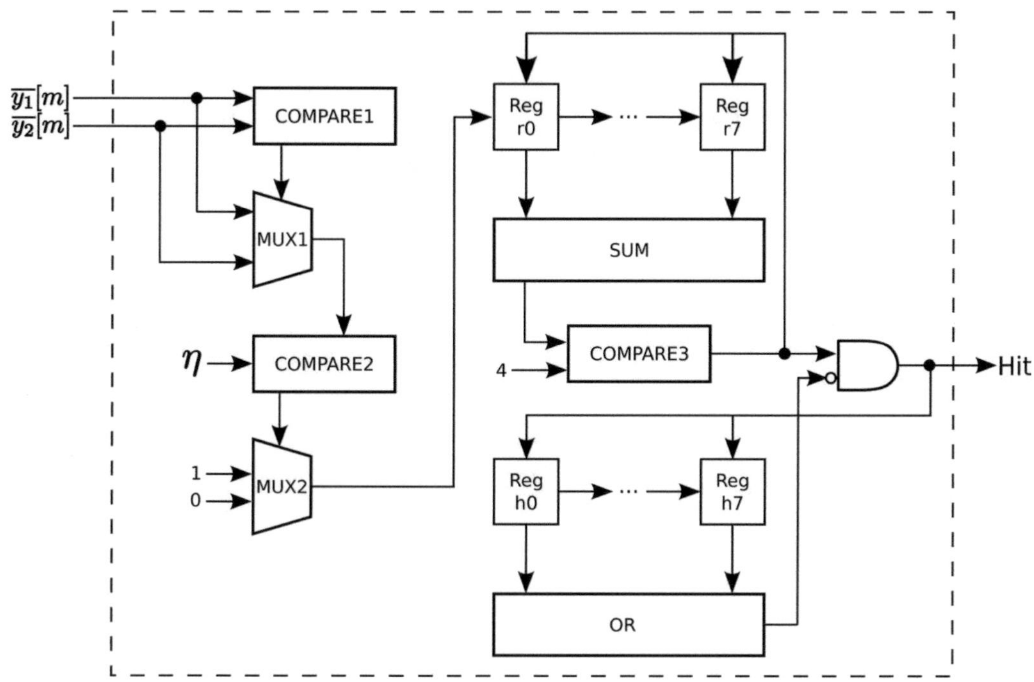

Figure 7. Architecture of threshold unit.

3.4. Switch buffer

The goal of switch buffer is to store the detected spikes for subsequent clustering operations. As shown in **Figure 8**, there are two buffers (denoted as Buffer x and Buffer y) in the circuit. When one of the buffers stores the detected spikes, the other provides the detected spikes to the OSORT module for clustering operations. The switch controller in the circuit is responsible for the determination of the buffer to store the detected spikes. The flowchart of the operations of the switch controller is shown in **Figure 9**. From the flowchart, it can be observed that the controller assigns the detected spikes to a buffer in accordance with the availability of that buffer. A buffer is available when it has empty cells for storing new detected spikes, and is not currently providing spikes to the OSORT module.

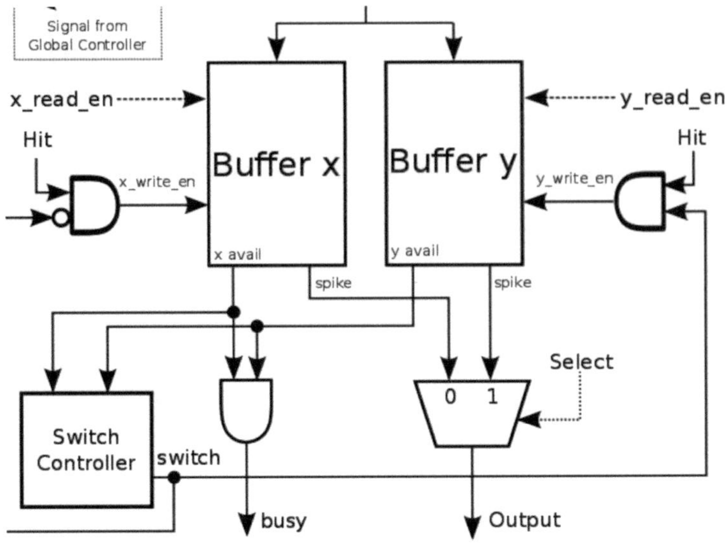

Figure 8. Architecture of switch buffer.

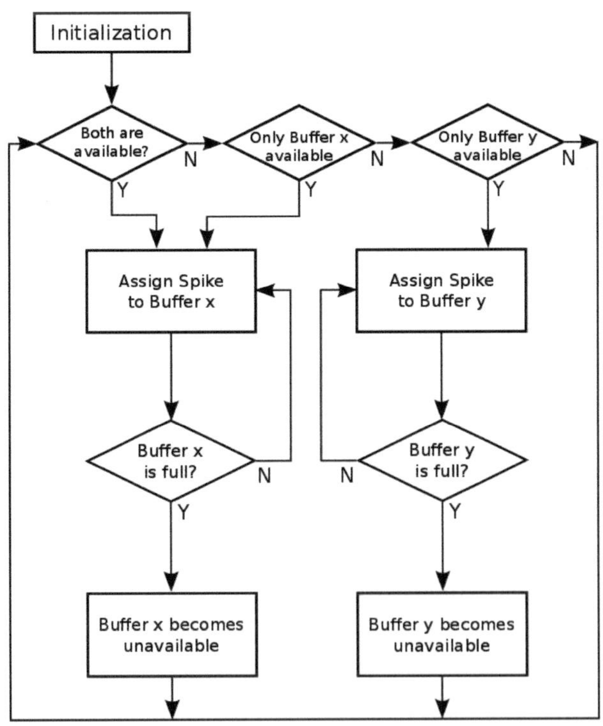

Figure 9. Flowchart of switch controller.

4. The architecture of OSORT module

The OSORT module contains buffers, distance computation unit, mean updating unit, comparator, and controller, as shown in **Figure 10**. The centroid and the size of each cluster are stored in the buffers. The distance computation unit and mean updating unit are responsible for squared distance computation and the updating of centroid of the clusters, respectively. The control unit coordinates different components of the OSORT module for carrying out the unsupervised clustering operations.

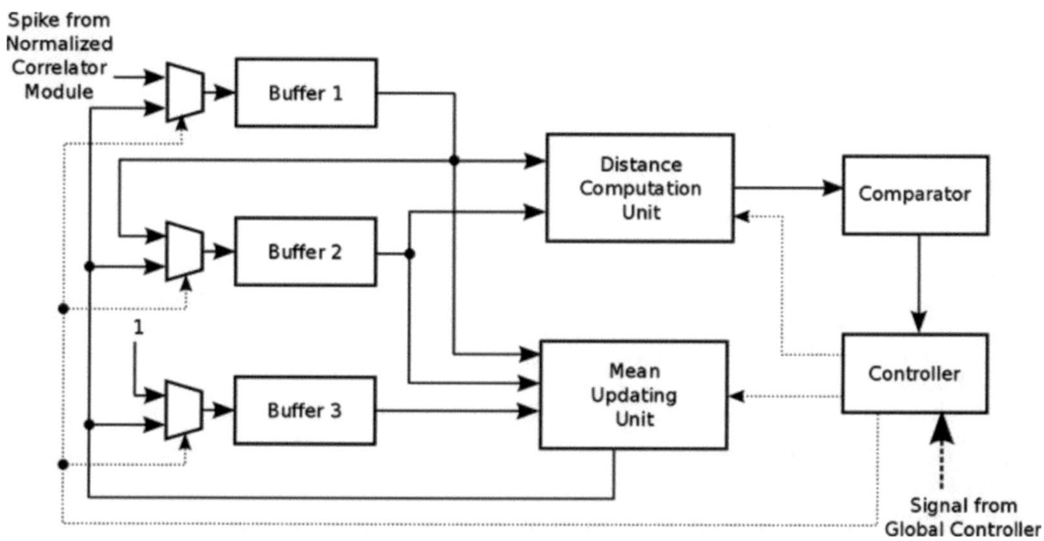

Figure 10. Architecture of OSORT module.

4.1. Buffers

There are three buffers in the OSORT module, which are denoted by Buffer 1, Buffer 2, and Buffer 3, respectively. Buffer 1 holds an input spike detected by the correlators. Buffers 2 and 3 contain the mean value and size of each cluster, respectively. Buffer 1 is a simple N-stage shift register, fetching or delivering one sample of the input spike at a time. As shown in **Figure 11**, Buffer 2 contains QN-stage shift registers. Each shift register holds the mean value of a cluster. Therefore, Q is the upper-bound of the number of clusters. Buffer 2 updates or provides mean values of clusters one at a time. Because the mean value $\mathbf{t}_i$ of a cluster $\mathcal{C}_i$ is the average value of the spikes mapping to that cluster, the mean value also contains N samples. Accessing the mean value of the cluster is also carried out one sample at a time. Buffer 3 records the size of each cluster. There are Q entries in the buffer. The ith entry contains the number of spikes in the cluster $\mathcal{C}_i$.

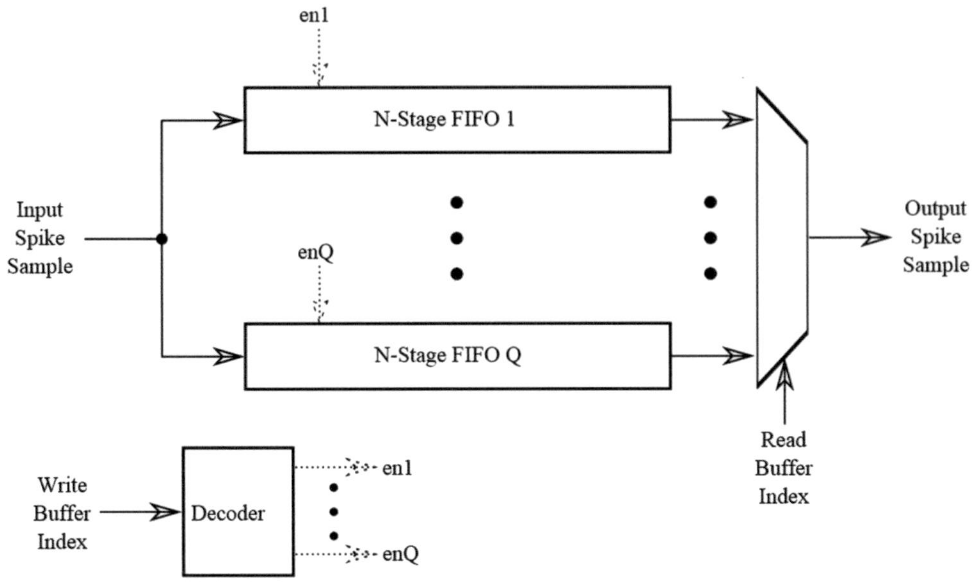

Figure 11. Architecture of Buffer 2.

4.2. Distance computation unit and mean updating unit

Because buffers in the memory unit can be accessed one sample at a time, the circuits in the distance computation unit and mean updating unit provide only sample-wise computations. This is beneficial for reducing the area costs. There are two cases when the distance computation unit needs to be activated. In the first case, a new spike is arrived. To find the cluster for the new spike, the squared distance computation is required. In the second case, it is desired to merge two clusters. The squared distance calculation is needed for finding the closest clusters. The distance computation unit takes the samples fetched from Buffer 1 and Buffer 2 as inputs. The unit finds the squared distance between the spikes stored in Buffer 1 and the mean value of a cluster selected in Buffer 2. Upon the completion of the squared distance computation, the comparison of the new squared distance with the current minimum distance stored in a register of the unit is carried out. If the new squared distance is smaller than the current minimum distance, then it becomes the new current minimum distance. The same squared distance computation and comparison operations will be repeated for until all the mean values in Buffer 2 are searched. This scheme is useful for finding the best matching mean value stored in Buffer 2 to the spike stored in Buffer 1.

The mean updating unit is activated after a new spike is assigned to a cluster, or after two clusters are merged. In these cases, the mean of the updated cluster needs to be computed. Note that the clusters to be updated are determined by the distance computation unit. The mean updating unit is only responsible for the computation of the new mean of the updated clusters. The circuit takes waveforms stored in Buffer 1 and Buffer 2, and the cluster size stored

http://dx.doi.org/10.5772/66105

in Buffer 3 as inputs. The updated mean is the weighted sum of the waveforms obtained from Buffer 1 and Buffer 2, as shown in **Figure 12**. The weights are determined from the cluster sizes from Buffer 3. The updated results are then stored back to Buffer 2 and Buffer 3.

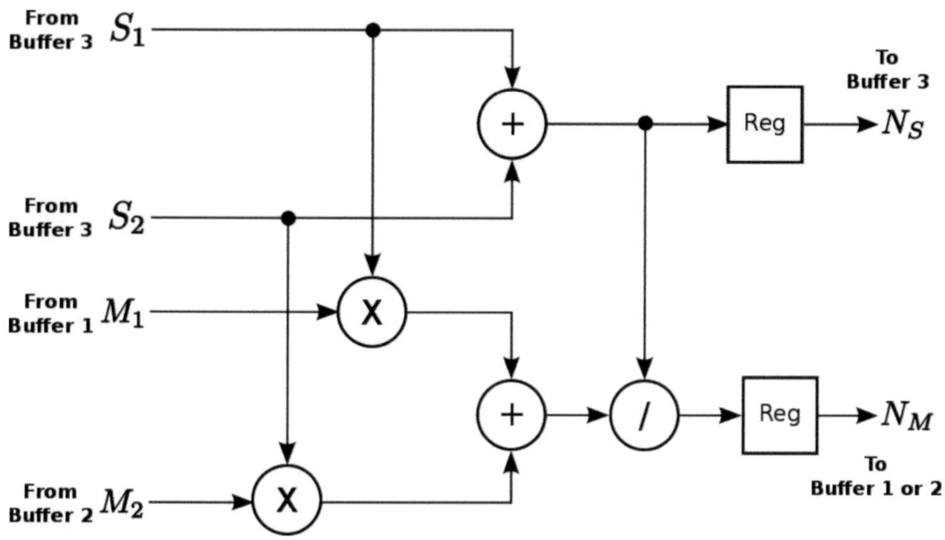

Figure 12. Architecture of mean updating unit.

4.3. Control unit

The control unit activates components of the memory unit and cluster computation unit for the unsupervised clustering. The states of the control unit are summarized in **Table 1**. In addition to the tasks carried out by each state, the activated circuit components associated with each state are also included in the table. **Figure 13** shows the flowchart of the OSORT algorithm in terms of the states defined in **Table 1**.

States	Activated components	Operations
State 1	Buffer 1 Distance computation unit	Fetch a new spike to Buffer 1
State 2	Buffers 1, 2	Find the best matching cluster to the spike in Buffer 1
State 3	Buffers 1, 2, 3	Creating a new cluster
State 4	Buffers 1, 2, 3 Mean updating unit	Update the mean and size of a cluster
State 5	Buffers 2, 3	Remove a cluster

Table 1. States of the control unit in OSORT module.

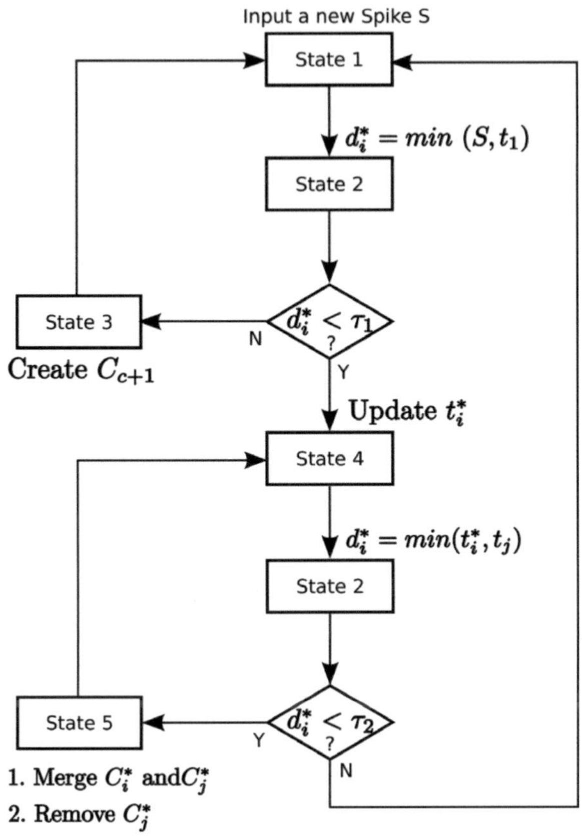

Figure 13. The flowchart of the controller in the OSORT module.

The flowchart in **Figure 13** is consistent with that shown in **Figure 1**. Although the combinations of the states shown in **Table 1** are able to implement the OSORT algorithm, additional modifications may still be desirable to facilitate the hardware implementation. One modification implemented in the controller is to handle the cases when the current number of clusters reaches the upper limit Q, and the creation of new cluster is still desired. In this case, the least recently updated cluster will be replaced by the new cluster. To carry out this modification, an additional field is added to each entry of Buffer 3. The entry indicates the number of updates in the past for the corresponding cluster. This modification can be viewed as an additional function supported in State 3 in **Table 1** for the creation of a new cluster.

5. Global controller and NOC

As depicted in **Figure 3**, the global controller in the proposed spike sorting system coordinates the operations of the normalized correlator module and OSORT module. The major goal of the global controller is to fetch detected spikes from the normalized correlator module and deliver them to the OSORT module. When a buffer in the switch buffer of the normalized correlator module becomes full, the global controller starts to fetch the detected spikes one at a time from the buffer to the OSORT module.

The fetching operations are repeated until all the spikes stored in the buffer are fetched. At this time, the buffer becomes available again for storing the new detected spikes, as shown in **Figure 14**. The proposed hardware spike sorting system is configured as a user component in a NOC system, which is designed by the QSYS platform. In addition to the proposed system, we see from **Figure 15** that the NOC contains the NIOS II processor, a DMA controller, an on-chip RAM, and a hardware timer.

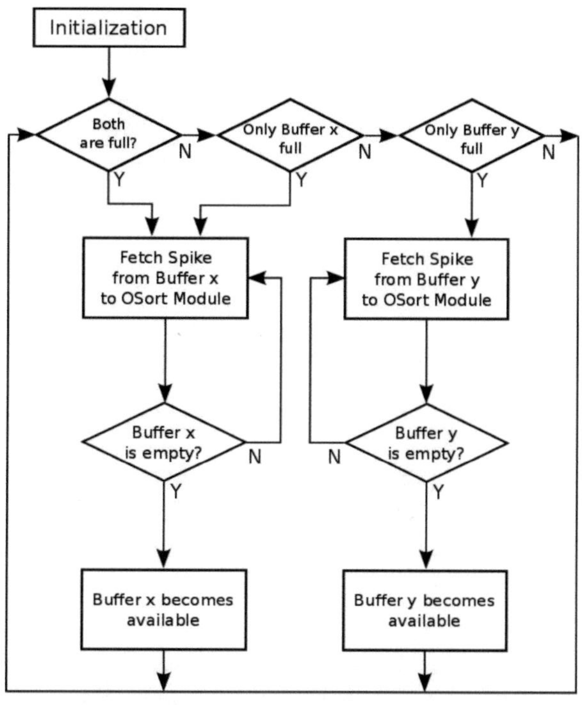

Figure 14. The flowchart of the global controller.

The raw spike trains are stored in the on-chip RAM. The DMA controller is responsible for delivering the spike trains to the proposed hardware system without the intervention of the NIOS II processor. The DMA controller is able to halt the delivery of spike trains automatically when both buffers in the switch buffer of the normalized correlator module are unavailable and/or full. The hardware timer is used to measure the computation speed of the proposed system. The NIOS II processor integrates different components in the NOC. It activates the DMA controllers for the delivery of spike trains. After that, the processor collects the spike sorting results from the OSORT modules of the proposed spike sorting system. The processor is also able to read the information provided by the hardware timer for the measurement of the computation speed of the proposed circuit.

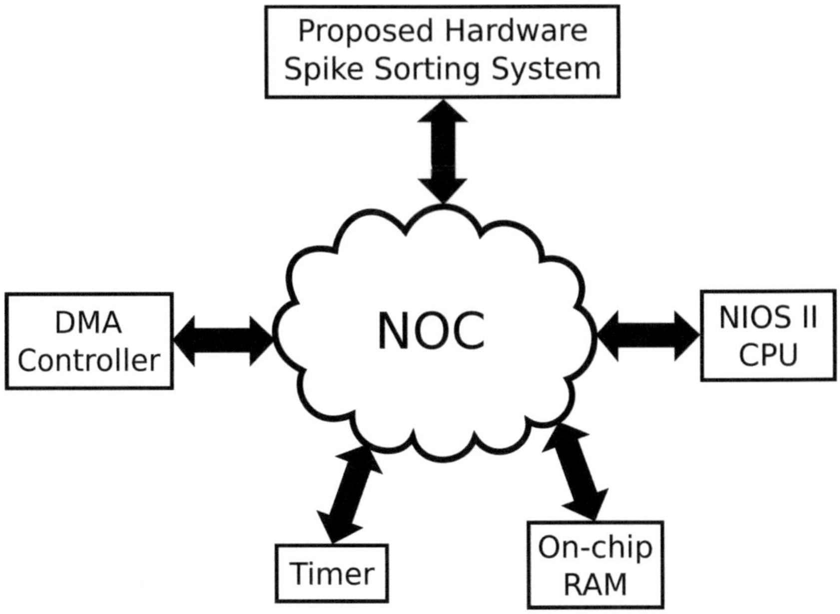

Figure 15. The proposed NOC system for spike sorting.

6. Experimental results

This section presents some experimental results of the proposed architecture. The extracellular recordings for the experiments are based on the simulator developed in [21], where the ground truth about spiking activity can be accessed. Each spike has length 2.67 ms. The sampling rate for the spike recording is 24,000 samples/s. Therefore, there are 64 samples (i.e., $N = 64$) in each spike.

The performance of the proposed architecture for spike detection is first evaluated. The performance evaluation is based on the true positive rate (TPR) and false alarm rate (FAR). The TPR of a detection algorithm is defined as the number of true spikes detected by the algorithm divided by the total number of true spikes. The FAR of a detection algorithm is the number of silent segments, which are falsely detected as spikes by the algorithm, divided by the total number of the segments detected by the algorithm. The TPR and FAR of various detection algorithms are included in **Table 2**. In the experiments, the spike trains are from two neurons. Therefore, there are two templates (i.e., $c = 2$) for the proposed normalized correlator.

Because the normalized correlation is effective for detecting real spikes and ignoring silent segments, we can observe from **Table 2** that the proposed architecture has superior performance over the other algorithms. We use the example shown in **Figure 16** to further demonstrate this fact. In the example, a noisy spike train with SNR = −3 dB is used for the spike detection. **Figure 16** reveals the normalized correlation values $\bar{y}_i[m]$, $i = 1, 2$, for the spike train. From

http://dx.doi.org/10.5772/66105

Figure 16, we see that, because of large noise corruption, it is difficult to locate spikes even by direct eye inspection. However, based on the normalized correlation values provided by the proposed architecture, the location of true spikes can still be effectively identified.

SNR (dB)		Normalized correlator	Noncoherent energy detector [17]	NEO [12]	SWT [22]	Matched filter [16]
10	TPR	93.64%	91.37%	93.10%	94.82%	89.65%
	FAR	0.40%	5.35%	3.57%	6.77%	2.80%
1	TPR	90.04%	88.03%	87.21%	92.43%	82.90%
	FAR	0.92%	6.36%	22.49%	79.36%	3.02%
–3	TPR	82.71%	82.60%	80.53%	86.66%	80.31%
	FAR	1.06%	9.52%	57.87%	82.43%	8.92%

Table 2. TPR and FAR values of various spike detection algorithms.

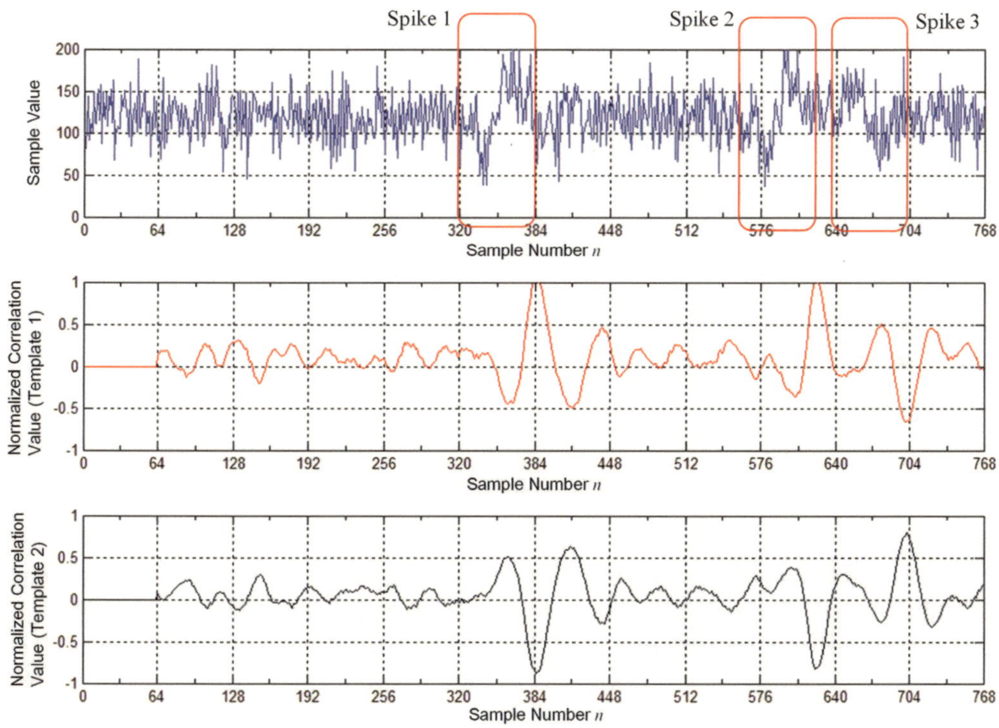

Figure 16. An example of the proposed normalized correlator for noisy spike detection with SNR = –3 dB for $c = 2$ templates.

Next we evaluate the area complexities. Because adders, multipliers, dividers, comparators, and registers are the basic building blocks of the proposed architecture, the area complexities are separated into five types: the number of adders, multipliers, dividers, comparators, and registers. **Tables 3** and **4** show the area complexities of the normalized correlator and OSORT modules, respectively. It can be observed from **Table 3** that, in the normalized correlator module, the correlator unit and switch buffer have larger area complexities. The number of adders, multipliers, and registers grows with the block dimension N and the number of templates c in the correlator unit. Let L be the capacity (i.e., the maximum number of spikes) of each buffer in the switch buffer. The number of registers in the switch buffer therefore is dependent on L and N, as shown in **Table 3**. The area complexities of the other types are of $O(1)$. Therefore, the proposed circuit has low consumption of dividers and comparators. From **Table 4**, we observe that only the area complexities of the buffers in the OSORT module grow with N. The other parts of the OSORT module have fixed area complexities.

	Filter unit	Block energy computation	Correlator	Thresholding unit	Switch buffer	Subtotal
Adders	$O(1)$	$O(1)$	$O(cN)$	$O(1)$	0	$O(cN)$
Multipliers	$O(1)$	$O(1)$	$O(cN)$	$O(1)$	0	$O(cN)$
Dividers	0	0	1	0	0	1
Comparators	0	0	0	$O(1)$	0	$O(1)$
Registers	$O(1)$	$O(N)$	$O(cN)$	$O(1)$	$O(LN)$	$O(cN + LN)$

Table 3. Area complexities of the normalized correlator module.

	Buffer	Distance computation unit	Mean updating unit	Subtotal
Adders	0	$O(1)$	$O(1)$	$O(1)$
Multipliers	0	$O(1)$	$O(1)$	$O(1)$
Dividers	0	0	$O(1)$	$O(1)$
Comparators	0	$O(1)$	0	$O(1)$
Registers	$O(cN)$	$O(1)$	$O(1)$	$O(cN)$

Table 4. Area complexities of the OSORT module.

The proposed architecture has been implemented by FPGA for performance measurement. The target FPGA device for the hardware implementation is Altera STRATIX IV EP4SGX230. The design platform for the experiments is the Altera QUARTUS II with QSYS. **Table 5** shows the hardware utilization of the proposed architecture. There are four different FPGA hardware resources considered: adaptive look-up tables (ALUTs), dedicated logic registers, block memory bits, and DSP blocks. The DSP blocks are dedicated to the implementations of adders, multipliers, dividers, and comparators. The ALUTs, dedicated logic registers, and block memory bits can be used for the implementation of registers, as well as adders, multipliers, dividers, and comparators. It can be observed from **Table 5** that the consumption of DSP blocks of normalized correlator is higher than that of the OSORT module. This is because the normalized correlator requires more number of arithmetic operators. There are 182,400 ALUTs, 182,400 dedicated logic registers, 1288 DSP blocks, and 14,625,792 block memory bits in the target FPGA device. It can be observed from **Table 5** that only limited hardware resources are consumed by the proposed circuit.

	ALUTs	Dedicated logic registers	Block memory bits	DSP blocks
Normalized correlator module	13,966	29,733	0	532
OSORT module	22,604	22,562	320	88
Total	39,355/182,400 (21.57%)	52,517/182,400 (28.79%)	320/14,625,792 (<0.1%)	642/1288 (49.84%)

Table 5. The utilization of FPGA resources of the proposed circuit. The switch buffer capacity for the measurement is $L = 40$.

In addition to consuming low hardware resources, the proposed architecture is able to provide high throughput. **Table 6** reveals the throughput of the proposed architecture for various clock rates and switch buffer size L. The throughput is defined as the number of spike samples which can be processed by the proposed architecture per second. The unit of the throughput in the table therefore is mega samples per second (Msamples/sec). It can be observed from **Table 6** that the throughput grows with L and/or clock rate. In particular, when $L = 32$ and clock rate is 100 MHz, the throughput is 25.04 Msamples/sec. The throughput of its software counterpart running on Intel I7-930 processor at clock rate 2.8 GHz and 16 GB RAM is only 0.69 Msamples/sec. The throughput of the proposed architecture therefore is 36 times higher than that of its software counterpart.

Clock rate	$L = 20$	$L = 40$	$L = 80$
50 MHz	9.22 Msamples/sec	10.25 Msamples/sec	13.31 Msamples/sec
75 MHz	13.00 Msamples/sec	15.34 Msamples/sec	20.00 Msamples/sec
100 MHz	17.80 Msamples/sec	20.25 Msamples/sec	25.04 Msamples/sec

Table 6. The throughput of the proposed circuit for various clock rates and switch buffer capacities L.

7. Concluding remarks

The proposed architecture has been found to be effective for real-time spike sorting. It features high accuracy, low hardware resource consumption, and high throughput. The combination of the normalized correlation and OSORT algorithm is beneficial for accurate spike detection with high TPR and low FAR even for low SNR values. The postnormalization approach adopted by the normalized correlator circuit is also able to reduce the area costs for the normalization operations. In addition, the switch buffer in the correlation circuit can effectively coordinate the operations of spike detection and classification for achieving high throughput. Experimental results reveal that the proposed architecture achieves TPR = 82.71% and FAR = 1.06% for SNR = −3 dB. The ALUT consumption is only 21.57% for the FPGA device STRUTIX IV EP4SGX230. The throughput is 25.04 Msamples/sec for the clock rate 100 MHz. All these facts demonstrate the effectiveness of the proposed architecture.

Acknowledgements

The authors would like to acknowledge the financial support of the Ministry of Science and Technology, Taiwan, under grant MOST 105-2221-E-003-011-MY2.

Author details

Chien-Min Ou[1] and Wen-Jyi Hwang[2*]

*Address all correspondence to: whwang@csie.ntnu.edu.tw

1 Department of Electronics Engineering, Chien-Hsin University of Science and Technology, Taoyuan, Taiwan

2 Department of Computer Science and Information Engineering, National Taiwan Normal University, Taipei, Taiwan

References

[1] Einevoll, G. T.; Franke, F.; Hagen, E.; Pouzat, C.; Harris, K. D. Towards reliable spike-train recordings from thousands of neurons with multielectrodes, Current Opinion in Neurobiology 2012, 22, 11–17.

[2] Gibson, S.; Judy, J. W.; Markovic, D. Spike sorting: the first step in decoding the brain, IEEE Signal Processing Magazine 2012, 29, 124–143.

[3] Hauck, S.; Dehon, A. Reconfigurable Computing: The Theory and Practice of FPGA-Based Computing, Morgan Kaufmann: San Francisco, CA, USA, 2008.

[4] Zhu, X.; Yuan, L.; Wang, D.; Chen, Y. FPGA implementation of a probabilistic neural network for spike sorting. In Proceeding of the IEEE International Conference on Information Engineering and Computer Science, Piscataway, New Jersey, USA. 2010; pp. 26–29.

[5] Dutta, K.; Prakash, N.; Kaushik, S. Probabilistic neural network approach to the classification of demonstrative pronouns for indirect anaphora in Hindi. Expert Systems with Applications 2010, 37, 5607–5613.

[6] Yu, B.; Mak, T.; Li, X.; Xia, F.; Yakovlev, A.; Sun, Y.; Poon, C.S. A reconfigurable Hebbian eigenfilter for neurophysiological spike train analysis. In Proceedings of the IEEE International Conference on Field Programmable Logic and Applications, Piscataway, New Jersey, USA. 2010; pp. 556–561.

[7] Haykin, S. Neural Networks and Learning Machines, 3rd ed.; Pearson: Upper Saddle River, NJ, USA, 2009.

[8] Hwang, W. J.; Lee, W. H.; Lin, S. J.; Lai, S. Y. Efficient architecture for spike sorting in reconfigurable hardware, Sensors 2013, 13, 14860–14887.

[9] Miyamoto, S.; Ichihashi, H.; Honda, K. Algorithms for Fuzzy Clustering, Springer: Berlin/Heidelberg, Germany, 2010.

[10] Gibson, S.; Judy, J. W.;Markovic, D. An FPGA-based platform for accelerated offline spike sorting, Journal of Neuroscience Methods 2013, 215, 1–11.

[11] Rutishauser, U. Online detection and sorting of extracellularly recorded action potentials in human medial temporal lobe recordings, in vivo, Journal of Neuroscience Methods 2006, 154, 204–224.

[12] Mukhopadhyay, S.; Ray, G. C. A new interpretation of nonlinear energy operator and its efficacy in spike detection, IEEE Transactions on Biomedical Engineering 1998, 45, 180–187.

[13] Quiroga, R. Q.; Nadasdy, Z.; Ben-Shaul, Y. Unsupervised spike detection and sorting with wavelets and superparamagnetic clustering, Neural Computation 2004, 16, 1661–1687.

[14] Gibson, S.; Judy, J. W.; Markovic, D. Technology-aware algorithm design for neural spike detection, feature extraction, and dimensionality reduction, IEEE Transactions on Neural Systems and Rehabilitation Engineering 2010, 18, 469–478.

[15] Mtetwa, N.; Smith, L. S. Smoothing and thresholding in neuronal spike detection, Neurocomputing 2006, 69, 1366–1370.

[16] Sato, T.; Suzuki, T.; Mabuchi, K. Fast template matching for spike sorting, Electronics and Communications in Japan 2009, 92, 57–63.

[17] Oweiss, K.; Aghagolzadeh, M. Detection and classification of extracellular ***action potential recordings, Chapter 2 of Statistical Signal Processing for Neuroscience, Academic Press, Tokyo., pp.15–74, 2010.

[18] Kim, S.; McNames, J. Automatic spike detection based on adaptive template matching for extracellular neural recordings, Journal of Neuroscience Methods 2007, 165, 165–174.

[19] Hwang, W. J.; Wang, S. H.; Hsu, Y. T. Spike detection based on normalized correlation with automatic template generation, Sensors 2014, 14, 11049–11069.

[20] NIOS II Processor Reference Handbook; Altera Corporation: San Jose, CA, USA, 2015. Available online: http://www.altera.com/literature/lit-nio2.jsp (accessed on 8 April 2015).

[21] Smith, L. S.; Mtetwa, N. A tool for synthesizing spike trains with realistic interference. Journal of Neuroscience Methods 2007, 159, 170–180.

[22] Kim K.; Kim, S. A wavelet-based method for action potential detection from extracellular neural signal recording with low signal-to-noise ratio, IEEE Transactions on Biomedical Engineering 2003, 50, 999–1011.

Chapter 2

Efficient FPGA Implementation of a CTC Turbo Decoder for WiMAX/LTE Mobile Systems

Cristian Anghel, Cristian Stanciu and
Constantin Paleologu

Additional information is available at the end of the chapter

http://dx.doi.org/10.5772/67017

Abstract

This chapter describes the implementation on field programmable gate array (FPGA) of a turbo decoder for 3GPP long-term evolution (LTE) standard, respectively, for IEEE 802.16-based WiMAX systems. We initially present the serial decoding architectures for the two systems. The same approach is used; although for WiMAX the scheme implements a duo-binary code, while for LTE a binary code is included. The proposed LTE serial decoding scheme is adapted for parallel transformation. Then, considering the LTE high throughput requirements, a parallel decoding solution is proposed. Considering a parallelization with $N = 2^p$ levels, the parallel approach reduces the decoding latency N times versus the serial decoding one. For parallel approach the decoding performance suffers a small degradation, but we propose a solution that almost eliminates this degradation, by performing an overlapped data block split. Moreover, considering the native properties of the LTE quadratic permutation polynomial (QPP) interleaver, we propose a simplified parallel decoder architecture. The novelty of this scheme is that only one interleaver module is used, no matter the value of N, by introducing an even-odd merge sorting network. We propose for it a recursive approach that uses only comparators and subtractors.

Keywords: LTE, WiMAX, turbo decoder, single interleaver, Max LOG MAP, parallel architecture, FPGA

1. Introduction

The channel coding theory was intensively studied during the last decades, but the interest on this topic increased even more following the pioneering work of Berrou et al. on turbo codes [1–3].

In their early existence, the turbo codes proved to obtain great decoding performances, so that they were used in many standards as recommendations. They transformed into a more appealing solution once the processing capacity increased for the field programmable gate array (FPGA) and digital signal processor (DSP). Their implementation complexity was not prohibitive anymore, this allowing them to become mandatory.

In this context, the Third-Generation Partnership Project (3GPP) organization early proposed these novel coding techniques. It should be mentioned that turbo codes were introduced in standard by the first version of Universal Mobile Telecommunications System (UMTS) technology (in 1999). Moreover, the next UMTS releases (the following high-speed packet access) contributed with new and interesting features, while turbo coding remained still unchanged. Furthermore, several modifications were introduced by the long-term evolution (LTE) standard. Even if they were not significant as volume, their importance arose in terms of concept. In this framework, the 3GPP proposed for LTE a new interleaver scheme, while maintaining exactly the same coding structure as in UMTS. Also, the turbo codes were introduced by the Institute of Electrical and Electronics Engineers (IEEE) in 802.16 standards, known as the base for WiMAX systems.

In Ref. [4], an UMTS dedicated turbo decoding binary scheme is developed, whereas for WiMAX systems a similar duo-binary architecture is presented in Refs. [5] and [6]. Thanks to the new LTE/LTE-advanced (LTE-A) interleaver, the decoding performances are improved, as compared to the ones corresponding to the UMTS standard. In addition, the new LTE interleaver comes with native properties suited for a parallel decoding approach inside the algorithm, thus taking advantage on the main idea brought by turbo decoders (i.e., exchanging the extrinsic values between the two decoding units). In Ref. [7], a serial decoding scheme implemented on FPGA is presented. However, parallelization is still required when high throughput is required, as in the particular case of LTE systems using diversity techniques.

In the past years, many interesting parallel decoder schemes were studied by the researchers. In this context, the obtained results are measured on two directions. The direction number 1 is represented by the decoding performance degradation between the parallel and the serial solutions. The direction number 2 is the hardware resources occupied for such parallel decoder implementation. In Ref. [8], a first group of parallel decoding solutions is presented. It is based on the classical maximum a posteriori (MAP) algorithm. This method passes through the trellis twice, first time to compute the forward state metrics (FSM) and the second time to obtain the backward state metrics (BSM) and simultaneously the log likelihood ratios (LLR). Following this approach, several approaches were developed in order to reduce the theoretical latency of the decoding process of $2K$ clock periods for each semi-iteration (where K is the data block length).

In Refs. [9] and [10], a second set of parallel architectures that take advantage of the quadratic permutation polynomial (QPP) interleaver algebraic-geometric properties is described. In these works, efficient hardware implementations of the QPP interleaver are proposed. However, the parallelization factor N still represents the number of used interleavers in the developed architectures.

In Ref. [11], a third approach was reported, which consists in using a folded memory. All the data needed for parallel processing are stored on the same time. On the other hand, the main

http://dx.doi.org/10.5772/67017

challenge of this kind of implementation is to correctly distribute the data to each decoding unit, once a memory location containing all N values is read. In order to solve this issue, an architecture based on two Batcher sorting networks was proposed. However, even in this approach, N interleavers are still needed to generate all the interleaved addresses that input the master network.

In this chapter, we present the optimized implementations for serial architectures for WiMAX and LTE turbo decoding schemes. Then, for LTE systems, we describe a parallel decoding architecture introduced in Refs. [12] and [13], which also relies on a folded memory-based approach. Nevertheless, the main difference as compared to the already existing solutions presented above is that our proposed approach includes only one interleaver. Additionally, with an even-odd merge sorting unit [14, 15], the parallel architecture maintains the same structure as the serial one, the only difference being given by the fact that the soft-input soft-output (SISO) decoding unit is included N times in the scheme. The block memory number and dimensions remain unchanged between the two proposed decoding structures. In terms of decoding performance, the obtained results for the serial and parallel approaches are almost similar. We propose an overlapped data block split that reduces the small degradation introduced by the parallel architecture.

Finally, we present throughput and speed results obtained when targeting a XC5VFX70T [16] chip on Xilinx ML507 [17] board. Moreover, we provide simulation curves for the three considered cases, i.e., serial decoding, parallel decoding and parallel decoding with overlap.

2. The coding scheme

2.1. WiMAX systems

Section 8.4 from 802.16 standard [18] presents the coding scheme on the basis of which the proposed decoder is implemented. **Figure 1** shows the duo-binary encoder. The native coding rate is 1/3. In order to obtain other coding rates, a puncturing block must be used. Accordingly, a depuncturing block must be added to the receiver architecture.

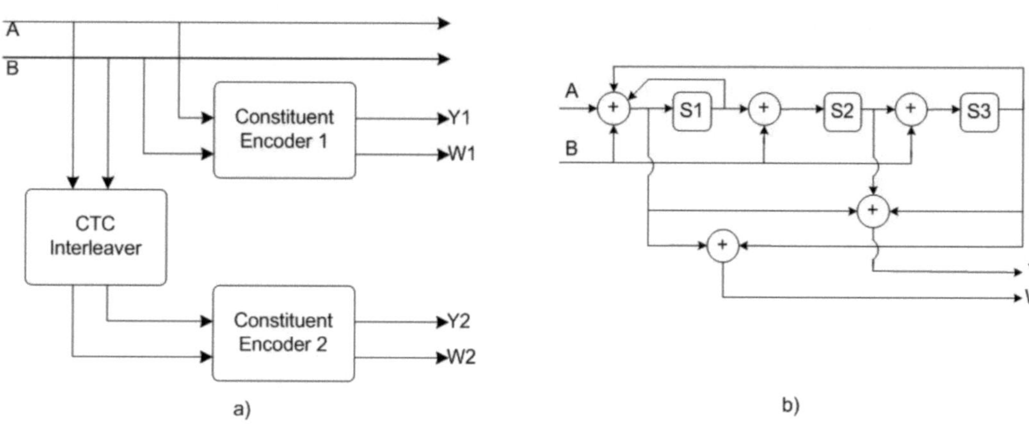

Figure 1. (a) 802-16e turbo coding scheme; (b) constituent encoder.

Let us define the following parameters: coding rate R; block dimension (in pairs of bits, i.e., dibits) K, which is computed independent of a coding rate, as a function of the uncoded block size; the number of iterations L, i.e., the latency *Latency* (in clock periods); information bits rate R_b [Mbps]; and system clock frequency F_{clk} [MHz].

As mentioned in Ref. [6], the main problem of a convolutional turbo code (CTC) decoder implementation is represented by the amount of required hardware resources. Moreover, in order to reach the targeted high data rate, the system clock has to be fast. Equation (1) presents the decoding throughput.

$$R_b = \frac{2K}{\text{Latency}\ T_{clk}} \tag{1}$$

For a fixed latency algorithm, according to Eq. (1), the output throughput is improved when achieving a higher clock frequency. Another way is to reduce latency using a parallel architecture; however, this increases the occupied area and may lead to a smaller clock frequency due to longer routes. Moreover, another direct constraint is the significant memory needed for storing data. This issue also affects the frequency, since a large number of used memory blocks leads to a large resource spread on chip and, obviously, longer routes.

Taking into account the previously mentioned aspects, we can conclude that all the parameters presented above are related, so that a global optimization is not possible. Consequently, we have chosen to balance each direction in order to meet throughput requirements.

2.2. LTE systems

A classic turbo coding scheme is presented in the 3GPP LTE specification, including two constituent encoders and one interleaver module (**Figure 2**). The data block C_k can be observed at the input of the LTE turbo encoder. The K bits from this input data block are transferred at the output, as systematic bits, in the steam X_k. At the same time, the first constituent encoder processes the input data block, resulting the parity bits Z_k, whereas the second constituent encoder processes the interleaved data block C_k', resulting the parity bits Z_k'. Combining the systematic bits and the two streams of parity bits, we obtain the following sequence (at the output of the encoder): $X_1, Z_1, Z'_1, X_2, Z_2, Z'_2, \ldots, X_K, Z_K, Z'_K$.

In order to drive back the constituent encoders to the initial state (at the end of the coding process), the switches from **Figure 2** are moved from position A to position B. Since the final states of the two constituent encoders are not the same (different input data blocks produce different final state), this switching procedure generates tail bits for each encoder. These tail bits are sent together with the systematic and parity bits, thus resulting the following final sequence: $X_{K+1}, Z_{K+1}, X_{K+2}, Z_{K+2}, X_{K+3}, Z_{K+3}, X'_{K+1}, Z'_{K+1}, X'_{K+2}, Z'_{K+2}, X'_{K+3}, Z'_{K+3}$.

As it was previously mentioned and discussed in Ref. [7], the LTE turbo coding scheme introduces a new interleaving structure. Thus, the input sequence is rearranged at the output using:

http://dx.doi.org/10.5772/67017

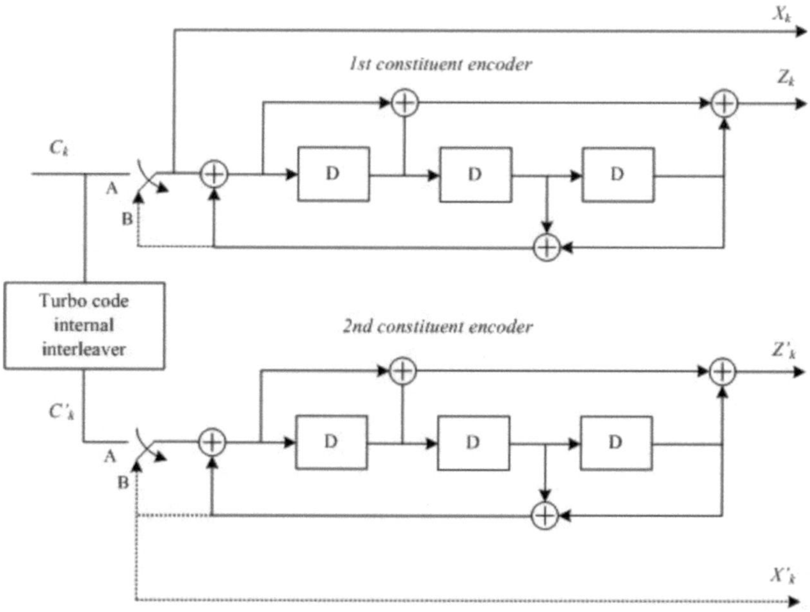

Figure 2. LTE turbo coding scheme.

$$C'_i = C_{\pi(i)}, \quad i = 1, 2, ..., K, \tag{2}$$

where the interliving function π applied over the output index i is defined as

$$\pi(i) = (f_1 \cdot i + f_2 \cdot i^2)\mathrm{mod}K \tag{3}$$

The input block length K and the parameters f_1 and f_2 are provided in Table 5.1.3-3 in Ref. [19].

3. The decoding algorithm

3.1. WiMAX systems

The decoding architecture consists of two decoding units called constituent decoders. Each such unit receives systematic bits (in natural order or interleaved) and parity bits, as shown in **Figure 1**.

The block diagram implements a maximum-logarithmic-maximum A posteriori (Max-Log-MAP) algorithm. For the case of turbo binary codes, the decoder scheme will represent, in the log likelihood ratio (LLR) space, each binary symbol as a single likelihood ratio. But in the situation of turbo duo-binary codes, the decoding unit requires three likelihood ratios in the same space. If we consider the duo-binary pair A_k and B_k, the LLRs may be computed as:

$$\Lambda_{a,b} = (A_k, B_k) = \log \frac{P(A_k = a, B_k = b)}{P(A_k = 0, B_k = 0)} \tag{4}$$

where (*a*,*b*) are (0,1), (1,0), or (1,1). The ratio set is updated by each decoding unit (constituent decoder) for each input pair, using the corresponding LLRs and parity bits, also seen as LLRs. Then, the output LLRs minus the input LLRs provides the extrinsic values. The trellis for a duo-binary code contains eight states, each such state with four inputs and four outputs, as presented in **Figure 3**. Using the systematic and parity pairs LLRs, for each branch, the metric $\gamma_k(S_i \rightarrow S_j)$ is computed, i.e.,

$$\gamma_k(S_i \rightarrow S_j) = \Lambda^i_{a,b}(A_k, B_k) + w\Lambda(W_k) + y\Lambda(Y_k) \tag{5}$$

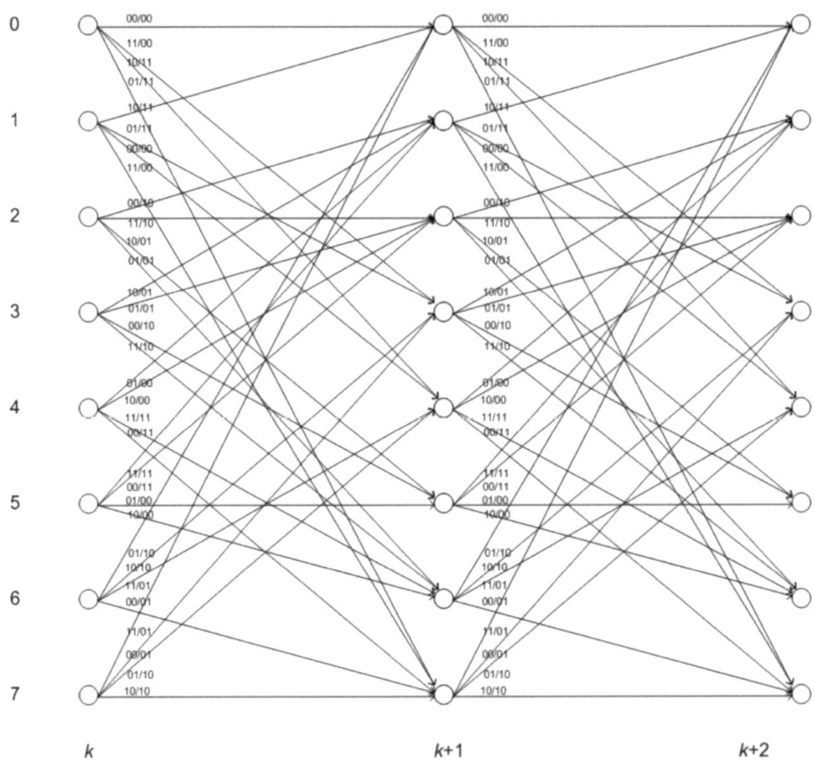

Figure 3. WiMAX decoder trellis.

The constituent decoder (**Figure 4**) performs the corresponding processing forward and backward over the trellis. When moving forward, the decoder computes the unnormalized metric $\alpha'_{k+1}(S_j)$ corresponding to each computed normalized metric $\alpha_k(S_i)$ associated with state S_i, using (**Figure 4**)

$$\alpha'_{k+1}(S_j) = \max_{S_i \rightarrow S_j} \{\alpha_k(S_i) + \gamma_k(S_i \rightarrow S_j)\} \tag{6}$$

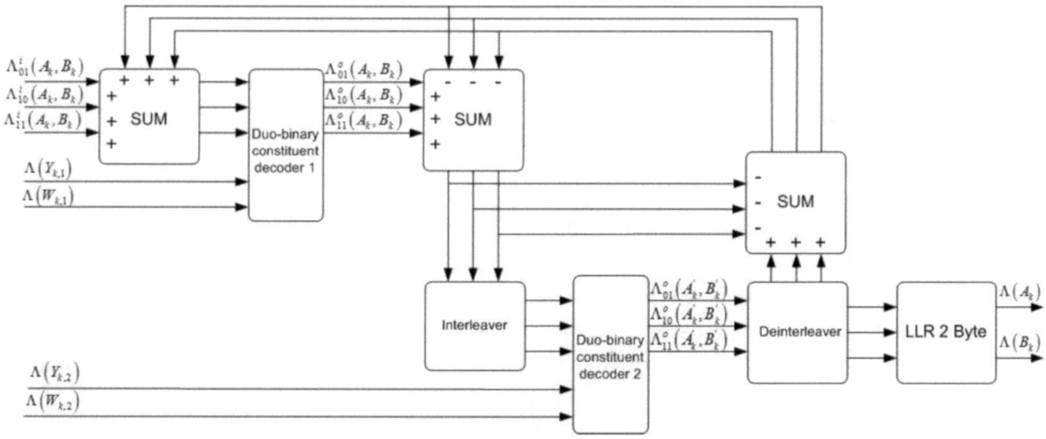

Figure 4. Decoder block scheme.

where the operator "maximum" is executed over all four branches entering the state S_j at the time stamp $k + 1$. Once the metrics for all states are updated at time stamp $k + 1$, the normalization versus the state S_0 value is made by the decoder. Analogously to forward processing, for backward moving, the decoder computes:

$$\beta_k^{'}(S_i) = \max_{S_i \to S_j}\{\beta_{k+1}(S_j) + \gamma_k(S_i \to S_j)\} \tag{7}$$

where the operator "maximum" and the normalization method are similar to Eq. (6).

The initialization with null values is carried out for all the forward and backward metrics at all states. Once the new values are computing and stored, the decoding unit executes the second step in the decoding procedure, i.e., the LLRs computing as in Eq. (4). The decoding unit starts by computing the likelihood ratio for each branch

$$Z_k(S_i \to S_j) = \alpha_k(S_i) + \gamma_k(S_i \to S_j) + \beta_{k+1}(S_j) \tag{8}$$

and continues with the value

$$t_k(a,b) = \max_{S_i \to S_j:(a,b)}\{Z_k\} \tag{9}$$

where the operator "maximum" is computed over all eight branches generated by the pair (a, b). At the end, the output LLR is computed as

$$\Lambda^o_{a,b}(A_k, B_k) = t_k(a,b) - t_k(0,0) \tag{10}$$

The decoding procedure is executed for a decided number of iterations or until a convergence criterion is reached. Then, a final decision is taken over the bits. This is achieved by computing for each bit from the pair (A_k, B_k) the corresponding LLR:

$$\Lambda(A_k) = \max\{\Lambda^o_{1,0}(A_k,B_k), \Lambda^o_{1,1}(A_k,B_k)\} - \max\{\Lambda^o_{0,0}(A_k,B_k), \Lambda^o_{0,1}(A_k,B_k)\} \tag{11}$$

$$\Lambda(B_k) = \max\{\Lambda^o_{0,1}(A_k,B_k), \Lambda^o_{1,1}(A_k,B_k)\} - \max\{\Lambda^o_{0,0}(A_k,B_k), \Lambda^o_{1,0}(A_k,B_k)\}, \tag{12}$$

where $\Lambda^{o}_{0,0}(A_k, B_k) = 0$. Finally, by comparing each LLR with a null threshold, i.e., looking at the sign, the hard decision is made.

3.2. LTE systems

The decoding architecture for the LTE systems is presented in **Figure 5**. The two decoding units called recursive systematic convolutional (RSC) use theoretically the MAP algorithm. The MAP solution, a classical one, ensures the best decoding performances. Unfortunately, at the same time, it is characterized by an increased implementation complexity and also it may include variables with a large dynamic range. These are the reasons why the classical solution with the MAP algorithm is used only as a reference for the expected decoding performance. When it comes to real implementation, new suboptimal algorithms have been studied: Logarithmic MAP (Log MAP) [20], Max Log MAP, Constant Log MAP (Const Log MAP) [21] and Linear Log MAP (Lin Log MAP) [22].

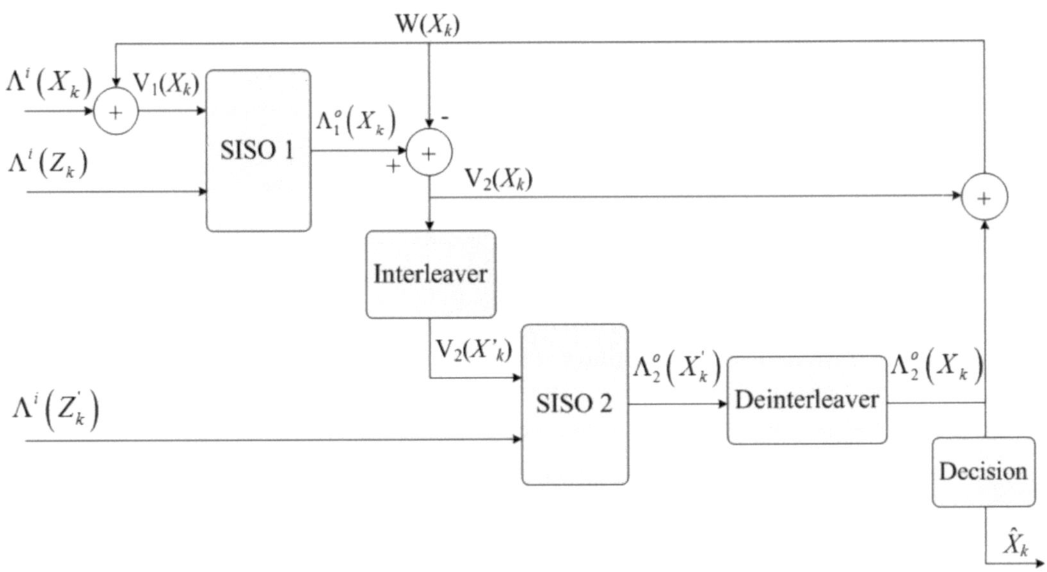

Figure 5. LTE turbo decoder.

For the LTE systems, we consider a decoding architecture based on the Max Log MAP algorithm. This suboptimal algorithm overcomes the problems of implementation complexity and dynamic range by paying the price of lower decoding performance when compared with the MAP algorithm. However, this degradation can be maintained inside some accepted limits. Starting from the Jacobi logarithm, only the first term is used by the Max Log MAP algorithm, i.e.,

$$\max^{*}(x,y) = \ln(e^{x} + e^{y}) = \max(x,y) + \ln(1 + e^{-|y-x|}) \approx \max(x,y)\ . \tag{13}$$

The trellis diagram for the turbo decoding architecture of the LTE systems contains eight states, as presented in **Figure 6**. Each state of the diagram has two inputs and two outputs. The branch metric between the states S_i and S_j is

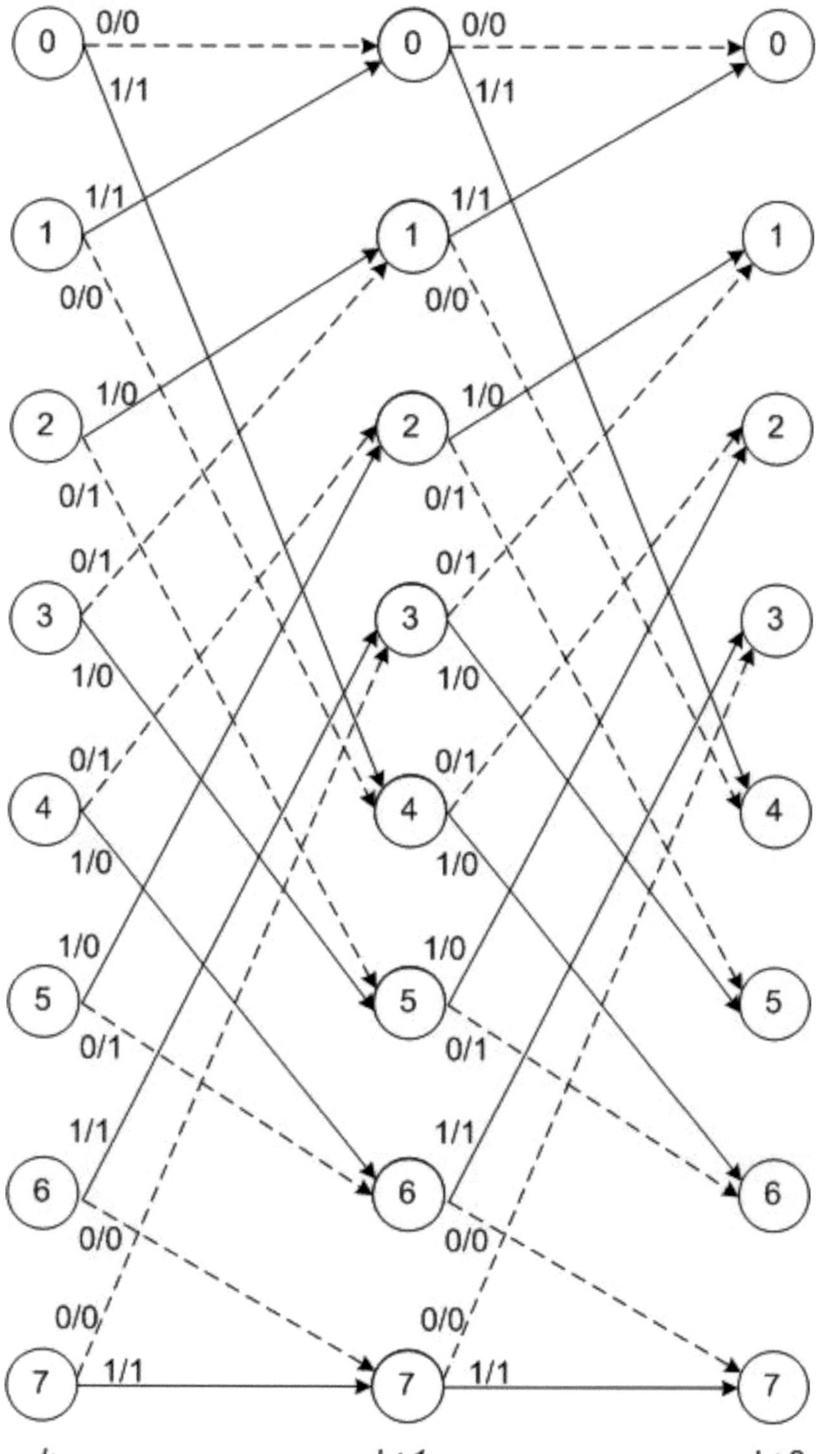

Figure 6. LTE turbo coder trellis.

$$\gamma_{ij} = \mathrm{V}(X_k)X(i,j) + \Lambda^i(Z_k)Z(i,j)\,, \tag{14}$$

where $X(i,j)$ and $Z(i,j)$ are the data, respectively, the parity bits, both associated with one branch and $\Lambda^i(Z_k)$ is the LLR for the input parity bit. For SISO 1 decoding unit, this input LLR is $\Lambda^i(Z_k)$, whereas for SISO 2 it becomes $\Lambda^i(Z_k^{'})$. For SISO 1, $\mathrm{V}(X_k) = \mathrm{V}_1(X_k) = \Lambda^i(X_k) + \mathrm{W}(X_k)$, whereas for SISO 2, $\mathrm{V}(X_k) = \mathrm{V}_2(X_k^{'}) = \mathrm{IL}\{\Lambda_1^o(X_k) + \mathrm{W}(X_k)\}$, where "IL" operator denotes the interleaving procedure. In **Figure 5**, $W(X_k)$ is the *extrinsic information*, whereas $\Lambda_1^o(X_k)$ and $\Lambda_2^o(X_k^{'})$ are the output LLRs generated by the two SISOs.

Looking at the LTE turbo encoder trellis, one can notice that between two states, there are four possible values for the branch metrics:

$$\begin{aligned} \gamma_0 &= 0 \\ \gamma_1 &= \mathrm{V}(X_k) \\ \gamma_2 &= \Lambda^i(Z_k) \\ \gamma_2 &= \mathrm{V}(X_k) + \Lambda^i(Z_k)\,. \end{aligned} \tag{15}$$

The LTE decoding process follows a similar approach as for WiMAX systems, i.e., it moves forward and backward through the trellis.

3.2.1. Backward recursion

The algorithm moves backward over the trellis computing the metrics. The obtained values for each node are stored in a normalized manner. They will be used for the LLR computation once the algorithm will start moving forward through the trellis. We name $\beta_k(S_i)$ the backward metric computed at the kth stage, for the state S_i, where $2 \le k \le K+3$ and $0 \le i \le 7$. For the backward recursion, the initialization $\beta_{K+3}(S_i) = 0, 0 \le i \le 7$ is used at the stage $k = K + 3$. For the rest of the stages $2 \le k \le K + 2$, the computed backward metrics are

$$\hat{\beta}_k(S_i) = \max\{\left(\beta_{k+1}(S_{j1}) + \gamma_{ij1}\right), \left(\beta_{k+1}(S_{j2}) + \gamma_{ij2}\right)\}\,, \tag{16}$$

where S_{j1} and S_{j2} are the two states from stage $k + 1$ connected to the state S_i from stage k and $\hat{\beta}_k(S_i)$ represents the unnormalized metric. Once the unnormalized metric $\hat{\beta}_k(S_0)$ is computed for state S_0, all the backward metrics for states $S_1 \ldots S_7$ are normalized as

$$\beta_k(S_i) = \hat{\beta}_k(S_i) - \hat{\beta}_k(S_0) \tag{17}$$

and then stored in the dedicated memory.

3.2.2. Forward recursion

When the backward recursion is finished, the algorithm moves forward through the trellis in the normal direction. This specific phase of the decoding is similar to the one for Viterbi algorithm. In this case, the storing procedure is needed only for the previous stage metrics, i.e., for computing the current stage k metrics, only the forward metrics from the last stage $k-1$ are needed. We will name $\alpha_k(S_i)$ the forward metric corresponding to state at the stage k, where $0 \le k \le K-1$ and $0 \le i \le 7$. For the forward recursion, the initialization $\alpha_0(S_i) = 0,\ 0 \le i \le 7$ is used at the stage $k = 0$. For the rest of the stages $1 \le k \le K$, the unnormalized forward metrics are computed as

$$\hat{\alpha}_k(S_j) = \max\left\{ \left(\alpha_{k-1}(S_{i1}) + \gamma_{i1j}\right), \left(\alpha_{k-1}(S_{i2}) + \gamma_{i2j}\right) \right\}, \tag{18}$$

where S_{i1} and S_{i2} are the two states from stage $k-1$ connected to the state S_j from stage k. Once the unnormalized metric $\hat{\alpha}_k(S_0)$ is computed for state S_0, all the forward metrics for states $S_1 \ldots S_7$ are normalized as

$$\alpha_k(S_i) = \hat{\alpha}_k(S_i) - \hat{\alpha}_k(S_0)\ . \tag{19}$$

The decoding algorithm can obtain now an LLR estimated for the data bits X_k since it has for each stage k the forward metrics just computed and also the backward metrics stored in the memory. For the first time, this LLR is obtained by computing the likelihood of the connection between the state S_i at stage $k-1$ and the state S_j at stage k as

$$Z_k(S_i \rightarrow S_j) = \alpha_{k-1}(S_i) + \gamma_{ij} + \beta_k(S_j)\ . \tag{20}$$

The likelihood of having a bit equal to 0 (or 1) is when the Jacobi logarithm of all the branch likelihood corresponds to 0 (or 1) and thus:

$$\Lambda^o(X_k) = \max_{(S_i \rightarrow S_j):X_i=1} \{Z_k(S_i \rightarrow S_j)\} - \max_{(S_i \rightarrow S_j):X_i=0} \{Z_k(S_i \rightarrow S_j)\}, \tag{21}$$

where "max" operator is recursively computed over the branches, which have at the input a bit of 1 $\{(S_i \rightarrow S_j) : X_i = 1\}$ or a bit 0 $\{(S_i \rightarrow S_j) : X_i = 0\}$.

4. Proposed serial decoding scheme

4.1. WiMAX systems

One important remark about the decoding algorithm is that the outputs of one constituent decoder represent the inputs for the other constituent decoder. At the same time, knowing that the interleaver and deinterleaver procedures apply over the data blocks (so the complete block is needed) in a nonoverlapping manner will allow the usage of a single constituent decoder. This decoding unit operates time multiplexed and the corresponding proposed scheme is presented in **Figure 7**.

In **Figure 7**, we can identify storing requirements: the memory blocks that store data from one semi-iteration to another and the memory blocks used from one iteration to another. IL stands for the interleaver/deinterleaver procedure, while CONTROL is the management unit, controlling the decoder functionalities. This module provides the addresses used for

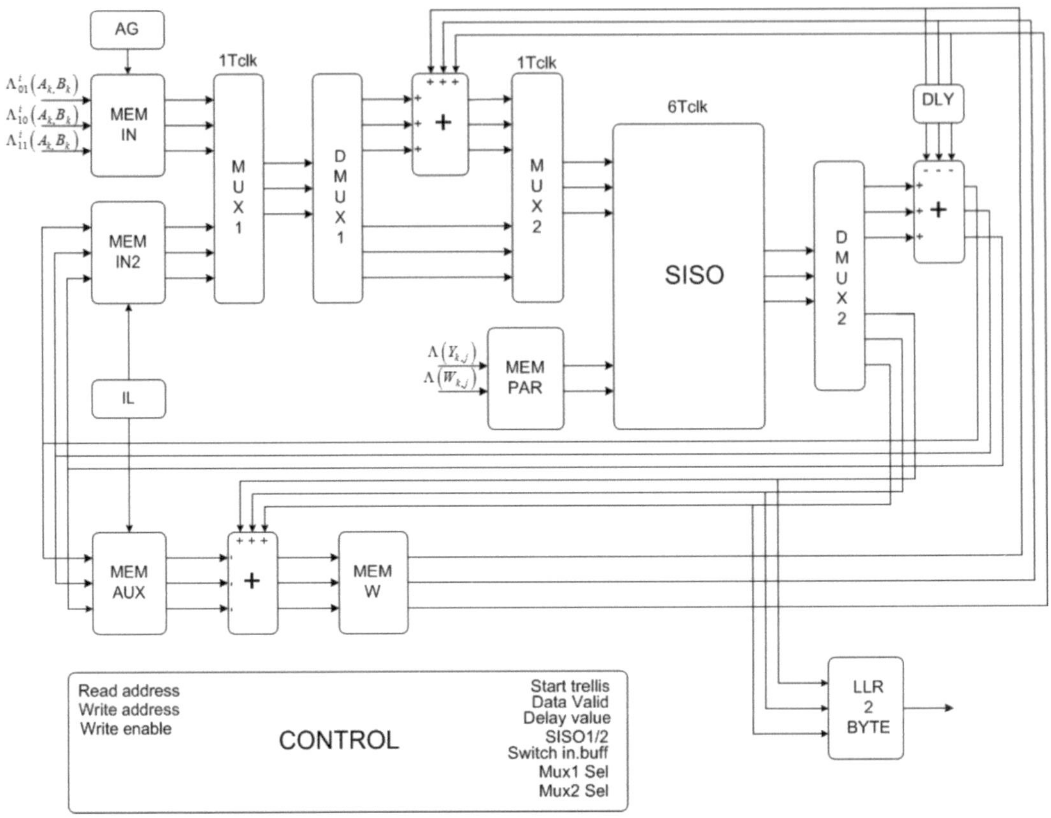

Figure 7. Proposed decoder scheme.

read and write, the signals used to trigger the forward and backward movements through the trellis, the selection for one of the two SISO units and also the control of MUX and DEMUX blocks. The input buffer is also selected since the decoding architecture can accept a new-encoded data block while still processing the previous one. The most important module shown in **Figure 7** is the SISO unit, which is the decoding structure. **Figure 8** depicts the block scheme of this decoding unit. One can observe the unnormalized metric computing modules BETA (backward) and ALPHA (forward) and the module GAMMA that computes the transition metric. This last one ensures also the normalization: the metrics values obtained for state S_0 are subtracted from the metrics values obtained for the states $S_1 \ldots S_7$. The output LLRs are computed inside the L module and normalized inside the NORM module. The MUX-MAX module provides the correct inputs when moving forward or backward through the trellis. It also computes the maximum function. The backward metrics are stored in MEM BETA memory during backwards recursion, their values being read when executing the forward recursion, in order to compute the estimated LLRs.

It is important to mention that some studies have been conducted regarding the normalization function. Trying to increase the system frequency (in order to reduce the decoding latency and

http://dx.doi.org/10.5772/67017

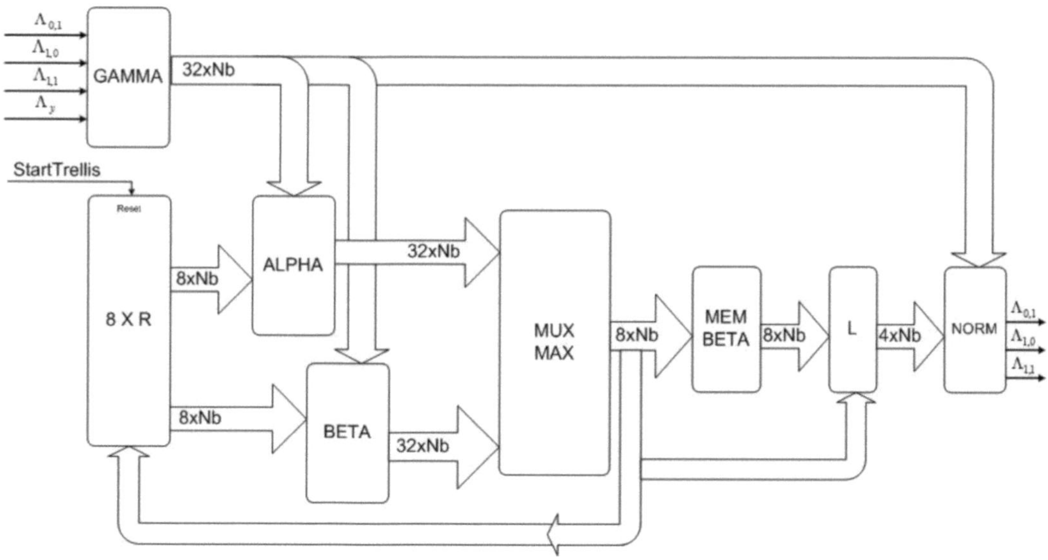

Figure 8. SISO block scheme.

so, to increase the decoded data throughput), one may think of removing the normalization and so to reduce the amount of logic on the critical path. This solution is not applicable because five extra bits would be needed for metrics values. From here more the memory blocks and more the complex arithmetic. Finally, all these will lead to a lower system frequency, so no benefit on this approach. On the other hand, we propose a dedicated approach to implement the metric computation blocks (ALPHA, BETA and GAMMA). Based on the trellis state, we identified the relations for each metric, 32 equations being used for transition metric computation (we remind that for each of the eight trellis states we have four possible transitions). Moreover, only 16 are distinct (the other 16 are the same) and from these 16, some are null. Using this approach, a complexity decrease is obtained.

Figure 9 depicts the timing diagram for the proposed SISO. This corresponds to the scenario with one SISO unit and some MUX and DEMUX blocks replacing the two SISO units from the theoretical decoding architecture (see **Figure 7**).

In **Figure 9**, R/W ($K-1{:}0$) means reading/writing memory from addresses $K-1$ to 0, R/W {IL ($K-1{:}0$)} means reading/writing memory from interleaved addresses $K-1$ to 0 and COMPUTE means that the block is processing the input data.

4.2. LTE systems

The same remark about the two SISO units from **Figure 5** working in a nonoverlapping manner applies for LTE systems as for WiMAX ones. The same approach is used, i.e., the proposed decoding architecture includes only one SISO unit and some MUX and DEMUX blocks. **Figure 10** depicts the block scheme of the proposed decoding architecture.

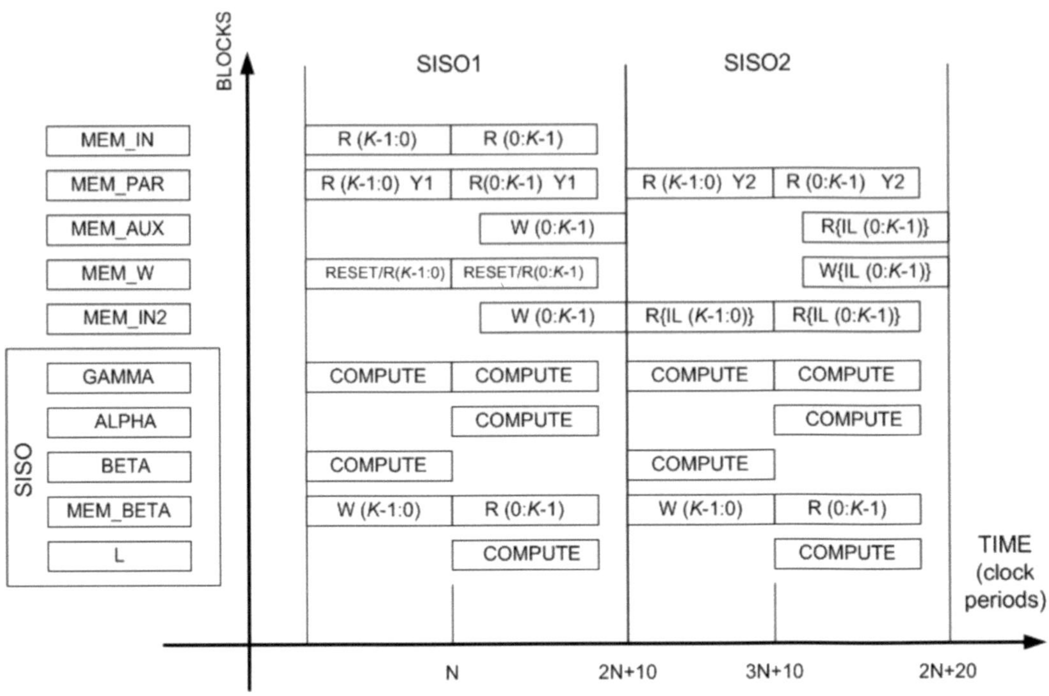

Figure 9. Time utilization for one turbo iteration.

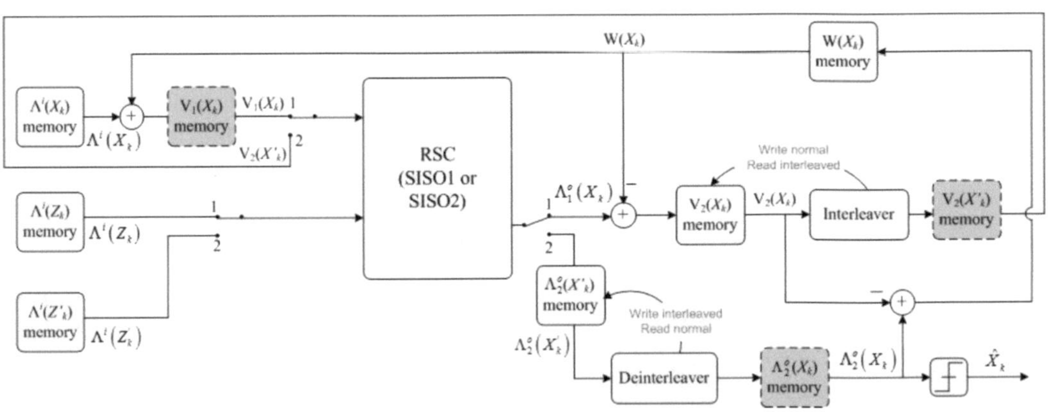

Figure 10. Proposed serial turbo decoder block scheme.

One can observe the memory blocks in **Figure 10**. Some are used to store data between two successive semi-iterations, respectively, between two successive iterations. Others, in dotted-line, are virtual memories used just to clarify the introduced notations. Moreover, the interleaver and deinterleaver modules are distinctively introduced in the scheme, but in fact they are the same. Both include a block memory called ILM (interleaver memory) and an interleaver. The novelty of this approach compared to the previous serial implementation proposed in Ref. [7] is the ILM. This memory will allow a fast transition to a parallel decoding architecture. The input data memories

http://dx.doi.org/10.5772/67017

(on the left side in **Figure 10**) and the ILM are switched buffers, allowing new data to be written while the previous block is still decoded. The ILM is filled with the interleaved addresses; at the same time, the new data are stored in the input memories. The saved addresses are then used as read addresses for the interleaver unit and as write addresses for the deinterleaver unit. Here, we detail the way the architecture from **Figure 10** works. The vectors $V_1(X_k) = \Lambda^i(X_k) + W(X_k)$ and $\Lambda^i(Z_k)$ are read from the corresponding memories by SISO 1. For the first semi-iteration, the memories are read in both directions, in order to ensure the forward and backward movements on the trellis. When this decoding phase is completed, the second semi-iteration starts, SISO 2 reads in both directions the memories storing the vectors $V_2(X_k^{'}) = IL\{V_2(X_k)\} = IL\{\Lambda_1^o(X_k) - W(X_k)\}$ and $\Lambda^i(Z_k^{'})$. IL stands again for the interleaver process.

In detail, SISO 1 reads the input memories and starts the decoding process, outputting the computed LLRs. Having the LLRs available and the extrinsic values, the vector $V_2(X_k)$ is computed and then stored in a normal order in the memory. The ILM content read in the normal order provides the reading addresses for $V_2(X_k)$ memory, emulating the interleaver process. The reordered LLRs $V_2(X'_k)$ are available, the corresponding values for the three tail bits X'_{K+1}, X'_{K+2} and X'_{K+3} being added at the end of this sequence. The same SISO unit acts now as SISO 2, this time reading data inputs from the other memory blocks. The two switching mechanisms from **Figure 10** change the position between these two semi-iterations (when in position 1, $V_1(X_k)$ and $\Lambda^i(Z_k)$ memories are active, while in position 2, $V_2(X'_k)$ and $\Lambda^i(Z_k^{'})$ memories are used).

The SISO unit provides at the end of each semi-iteration K values for the LLRs. The LLRs obtained after the second semi-iteration are stored in the $\Lambda_2^o(X_k^{'})$ memory (the content of ILM, already available for the $V_2(X_k)$ interleaver process, is used also as writing address for $\Lambda_2^o(X_k^{'})$ memory, after a delay is added).

The memories $\Lambda_2^o(X_k^{'})$ and $V_2(X_k)$ are read in the normal order to allow $W(X_k)$ computation; $W(X_k)$ is written in the corresponding memory and at the same time it is used for a new semi-iterations. In other words, the memory for $W(X_k)$ is updated during a semi-iteration. The time diagram for the proposed serial decoding architecture is presented in **Figure 11**, the intervals colored with gray indicating the writing periods for $W(X_k)$ memory. As mentioned in this chapter, the input memories and the ILM (the upper four memory blocks in the image) are switched buffers and they are filled with new data while the previous-coded block passes the last phase of its decoding process. The same notations as shown in **Figure 9** are used.

All the memory blocks in **Figure 10** have 6144 locations, this being the maximum coded data block length defined by the standard. Only the memory blocks with the input data for SISO units have 6144 + 3 locations because they store also the tail bits. All locations contain 10 bits. Using a Matlab simulator in finite precision, it has been observed that six bits are needed for the integer part, in order to cover the dynamic range of the variables and three bits are needed for the fractional part to maintain the decoding performance close to the theoretical one, with a certain accepted level of degradation. The 10th bit is for sign.

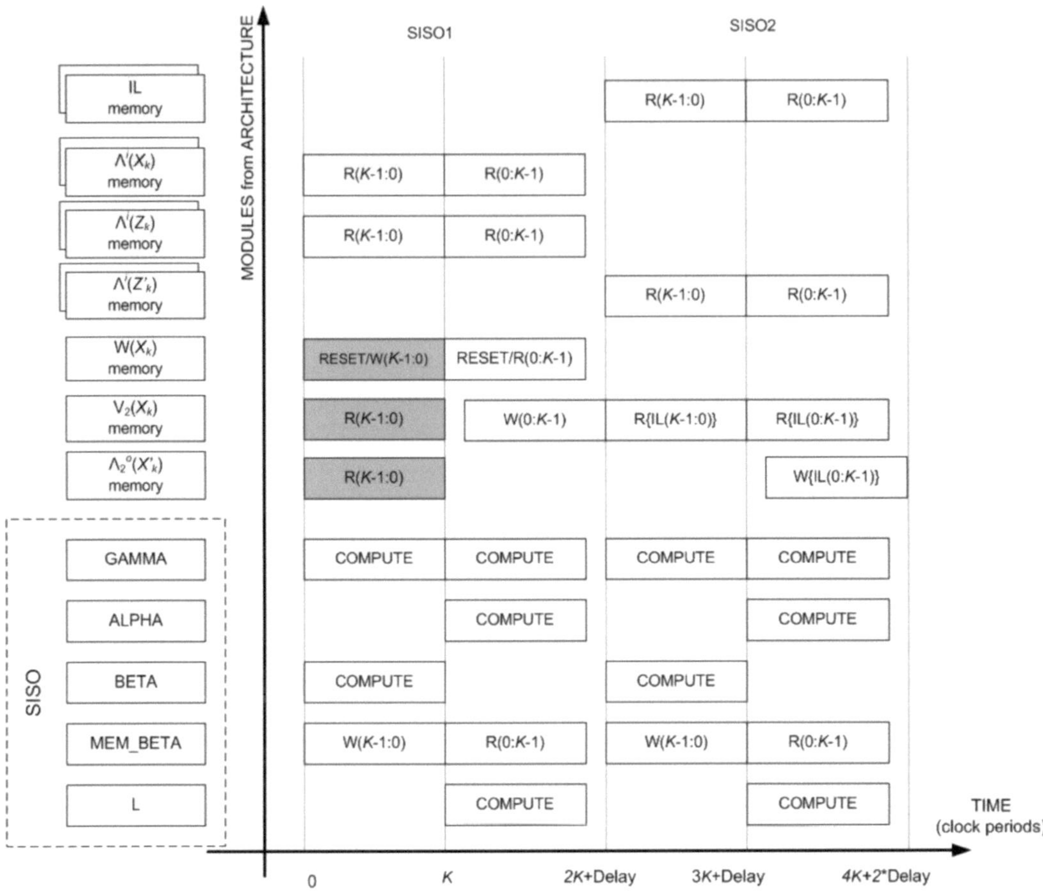

Figure 11. Time diagram for a serial turbo decoder.

The SISO decoding unit is similar to the one depicted in **Figure 8**. ALPHA and BETA modules compute the unnormalized forward metrics and the unnormalized backward metrics, respectively. The GAMMA module computes the transition metrics and executes also the normalization (the metrics for state S_0 are subtracted from the metrics corresponding to states $S_{1,}$ …, S_7). The output LLRs are computed inside the L module and normalized by the NORM module. The selection of the inputs for forward and backward moving on the trellis and also the maximum function are executed by the MUX-MAX module. Finally, the MEM BETA module stores the backward metrics.

The L module produces the output log likelihood ratios. These are then normalized inside the NORM module. The MUX-MAX makes the inputs selection (for forward or backward trellis runs) and implements also the maximum operator. The MEM BETA module keeps the backward metrics corresponding values into the memory.

Using the same approach for both WiMAX and LTE proposed serial decoding architectures, the same remarks apply. So, for the LTE turbo decoder also, the normalization function allows a reduced dynamic range for the variables. Trying to eliminate it, in order to reduce the number of logic levels on the critical path, will not lead to a higher system frequency because again, more memory blocks are required, more complex arithmetic (since variables are expressed on more bits) is used and finally, as an overall consequence, lower clock frequency is reported for the design.

And for ALPHA, BETA and GAMMA modules inside the SISO decoding unit, again the dedicated equations are used to compute the metrics. Sixteen such relations are implemented for transition metric computation (eight states in trellis with two possible transitions each). In fact, only four equations are distinct (as indicated in Eq. (15)]. And from these four equations, one of them is null. This way the computational effort is minimized for this proposed architecture.

The interleaving and deinterleaving procedures implement the same equation. The interleaved index is computed using a modified form of Eq. (3), i.e.,

$$\pi(i) = \{[(f_1 + f_2 \cdot i)\text{mod } K] \cdot i\}\text{mod } K \tag{22}$$

For the interleaving process, the data are written in the memory block in the natural order and then it is read in the interleaved order, while for the deinterleaver process the data are written in the interleaved order and then it is read in the natural order.

The computation in Eq. (22) is executed in three phases. First, the value for $(f_1 + f_2 \cdot i)\text{mod}K$ is obtained. The index i (describing the natural order) multiplies this partial result and the obtained value is passed once again through modulo K block. And as a remark for this computation: the formula is increased with f_2 for consecutive values of index i. So a register adds f_2 for each new index. If the register current value is higher than K, K is subtracted and the result is placed back in the register. This processing requires one system clock period, the results being generated in a continuous manner.

5. Proposed parallel decoding scheme

The serial architecture described in **Figure 10** for LTE systems can be reorganized in a parallel setup, by instantiating the RSC SISO module N times in the structure. We propose a configuration that concatenates the N values associated with the N RSCs and employs a single memory location for all the memories in the scheme. The K locations with 10 bits per location (corresponding to the serial architecture) are replaced by K/N positions with $10N$ bits per position (working for the parallel format).

The most important benefit brought by the proposed serial decoding scheme is the single usage of the interleaver module before the decoding stage. The ILM is updated, each time a new data block enters the decoder, while the previous block is still being decoded. This approach prepares a fast and simple transition to the parallel scheme. Considering that the factor N is known, the ILM will have K/N locations, with N values being written at each location (i.e., the ILM can be prepared for the parallel processing that follows). As mentioned in Ref. [16], a Virtex 5 block memory can be organized from a configuration of 32k locations × 1 bit to a setup of 512 locations × 72 bits. In the costliest scenario (i.e., K = 6144), based on the N values and representing the stored values on 10 bits, the parallel ILM can be employed as:

- 768 locations × 80 bits
- 1536 locations × 40 bits
- 3072 locations × 20 bits
- 6144 locations × 10 bits

Only two BRAMs are used, the same as in the case of serial ILM.

Figure 12 shows the ILM working principle. As one can observe, during the writing procedure, each index i from 0 to $K - 1$ generates a corresponding interleaved value. All the computed values are stored in the ILM, in the same order. We will consider the ILM as a matrix, the rows being the memory locations and the columns being the positions on each location. The first K/N interleaved values are placed on the first column. The second set of K/N values is stored on the second column and the procedure continues. In order to perform the described method, a true dual port BRAM is selected. In **Figure 12**, each time a new value is added on row WA at column WP (near the already existing content at columns till WP-1), the content of row WA + 1 is also read from the memory. In the next clock period, a new value is added at row WA + 1 at column WP (near the already existing content at columns till WP − 1), while reading also the content of row WA + 2. And so on. When the interleaver function is used, the ILM is read in a normal way and the N interleaved values from a row are employed as reading addresses for the $V_2(X_k)$ memory. Furthermore, the new LTE interleaver module (with the QPP algebraic properties) will always place at the same row the N values that should be read in the interleaved order from ILM. The only additional task is a reordering process needed to match the corresponding RSCs. An example is presented in **Figure 13** for the values $K = 40$ and $N = 8$. On the left side, the content of the $V_2(X_k)$ memory is shown. Each column is composed of the outputs generated by one of the N RSC SISOs. On the right side, the content of ILM memory is described. Each minimum value from a line of the ILM represents the line address for the $V_2(X_k)$ memory (see the gray color circle in the illustration). By using a reordering module, each position from the outputted line is directed to its corresponding SISO. For example, position c from the first read line (index 10) is sent to SISO g, whereas position c from the second read line (index 13) is sent to SISO a. The same procedure applies also for the deinterleaving process, only that the write addresses are extracted from ILM, while the reading ones are used in the natural order.

For the reordering module, an even-odd merge sorting network is applied. The corresponding method was introduced by Batcher in Ref. [14] and is part of the sorting network group that includes several sorting approaches. One such example is the bubble sorting, which sorts in a repeated manner the adjacent pairs of elements. Another example is the shell sorting, which groups the input data into an array and then performs the array's column sorting (also in a repeating manner). After each associated iteration, the array becomes one column smaller. A third example is the even-odd transposition sorting, which sorts alternatively the odd-indexed

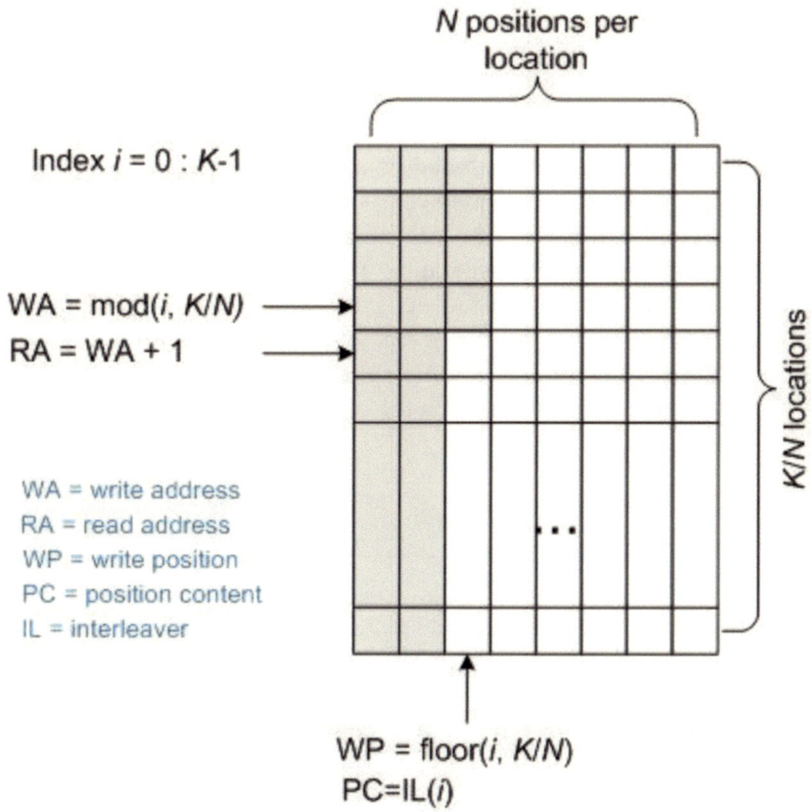

Figure 12. ILM memory writing procedure.

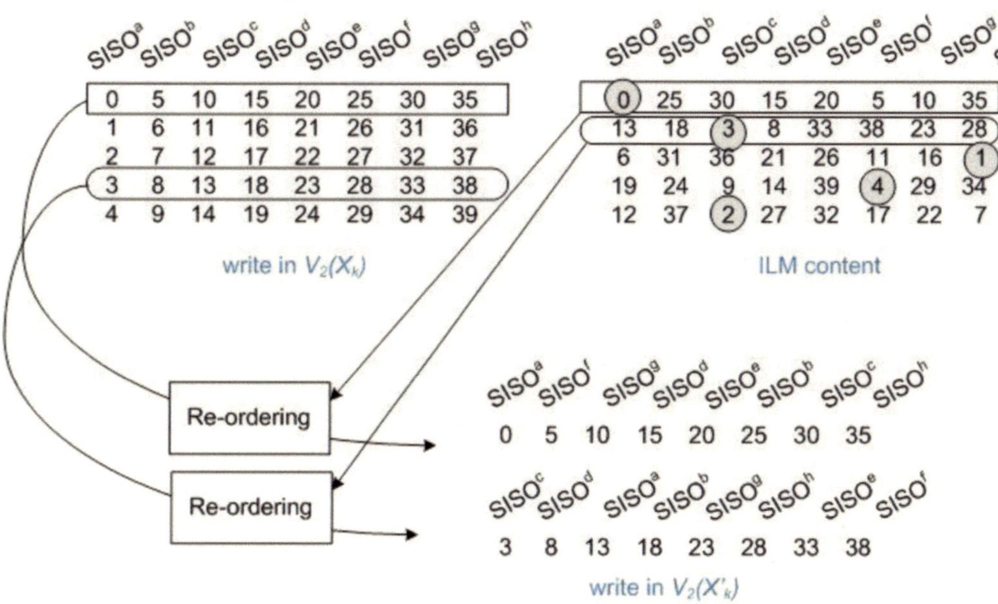

Figure 13. Virtual parallel interleaver.

and the adjacent even-indexed elements, respectively, the even-indexed elements and the adjacent odd-indexed values. The fourth example is the bitonic sorting. The two halves of the input data are sorted in opposite directions and then jointly processed to produce one complete sorted sequence.

The even-odd merge sorting method is based on a theorem saying that any list of $a = 4b$ (b natural) elements can be sorted if the following steps are applied: first, separate sorting is executed over the two halves of the list. After this step, the elements with odd index and the ones with even index are sorted separately. The last step consists in a comparing and switching procedure executed over all the elements $2n$ and $2n + 1$ ($n = 1,\ldots, a/2 - 1$). The demonstration of this theorem is available in Ref. [23]. An example for $N = 8$ is depicted in a graphical format shown in **Figure 14**. From a timing point of view, **Figure 15** depicts the case when $N = 2$ is used. Same comments as the ones for **Figure 11** apply.

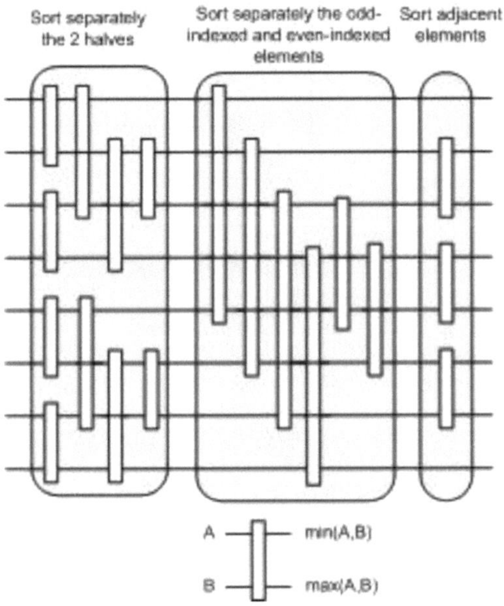

Figure 14. Even-odd merge sorting for $N = 8$.

In combination with the presented parallel decoding architecture, we also propose a simplified implementation for the interleaver block. As seen from Eq. (3), the arithmetic requirements for the computation of the memory addresses $\pi(i)$ consist of three multipliers, one adder and one divider (used for the extraction of the remainder associated with the *modulo* operation). For all possible K values associated with the division, the quotients range is very large, since the numerator and the denominator can have very big values (and often situated in different numerical ranges—up to billions). We propose an efficient method to reduce the arithmetic complexity associated with Eq. (3).

By introducing the notation

http://dx.doi.org/10.5772/67017

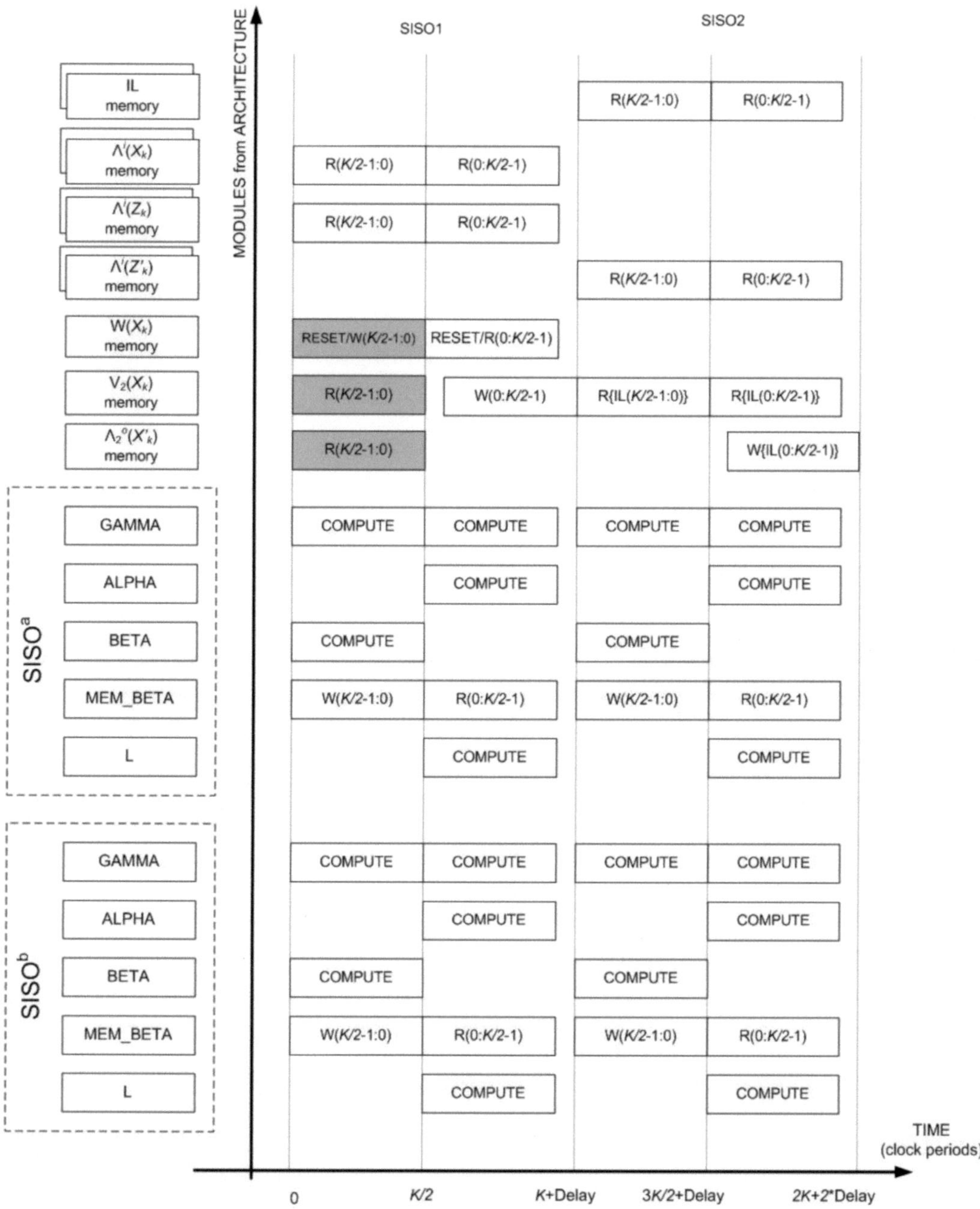

Figure 15. Time diagram for parallel turbo decoder ($N = 2$).

$$p(i) = f_1 i + f_2 i^2 \tag{23}$$

it can be observed that

$$\begin{aligned} &p(0) = 0, \\ &p(i) = p(i\text{–}1) \ + \ s_1 \ + \ s_2(i), \quad i > 0, \end{aligned} \tag{24}$$

where

$$\begin{aligned} &s_1 = f_1 \ \text{ and} \\ &s_2(i) = \begin{cases} 0, \ i = 0, \\ f_2, \ i = 1, \\ s_2(i\text{–}1) \ + \ 2f_2, \ i > 1 \end{cases} \end{aligned} \tag{25}$$

We can rewrite Eq. (3) using Eqs. (23) and (24)

$$\pi(i) = p(i) \ \bmod \ K = [p(i\text{–}1) \ + \ s_1 \ + \ s_2(i)] \ \bmod \ K \tag{26}$$

The multiplications are replaced by additions, which require less hardware resources. Nevertheless, the division is still necessary for the *modulo* operation. If we consider the *modulo* operator applied to a sum of elements expressed as

$$\left[\sum_k c_k\right] \bmod \ K = \left[\sum_k c_k \ \bmod \ K\right] \ \bmod \ K \tag{27}$$

we can decrease the arithmetic effort needed to obtain $\pi(i)$ in Eq. (26). The number of *modulo* operations becomes bigger, but the overall complexity of the corresponding divisions is reduced since smaller quotients are used. Consequently, using Eqs. (25)–(27), one can write:

$$s_3(i) = s_1 + s_2(i) = \begin{cases} 0, \ i = 0, \\ f_1 + f_2, \ i = 1, \\ s_3(i\text{–}1) \ + 2f_2, \ i > 1 \end{cases} \tag{28}$$

Using Eq. (29) in Eq. (26), the result is

$$\begin{aligned} \pi(i) = p(i) \ \bmod \ K &= [p(i\text{–}1) + s_3(i)] \ \bmod \ K \\ &= [p(i\text{–}1) + s_3(i\text{–}1) \ + 2f_2)] \ \bmod \ K \\ &= [\pi(i\text{–}1) + s_3(i\text{–}1) \bmod \ K + 2f_2 \bmod \ K] \ \bmod \ K \end{aligned} \tag{29}$$

All of the numerical values added in the last stage of Eq. (29) are lower than K and available recursively (during the processing of a distinct frame), such as $\pi(i\text{–}1)$ and $s_3(i\text{–}1) \bmod K$ or they can be predetermined and stored, like the case of $2f_2 \bmod K$.. The overall arithmetic complexity is reduced to $2K$ additions and $2K$ simplified modulo operations (i.e., each is resolvable using a comparison and a subtraction) for the address generation module. The method improves the solutions presented in [24, 25], by eliminating any multiplications or divisions. Additionally, the lower numerical range of the operators (with values lower than $2K$; i.e., values in the range of thousands) allows the usage of minimal resources for the representation of binary values.

http://dx.doi.org/10.5772/67017

6. Implementation results

6.1. WiMAX systems

The estimated system frequency when implementing the decoding structure on a Xilinx XC4VLX80-11FF1148 chip using the Xilinx ISE 11.1 tool is 125 MHz. The reserved chip area is around 3000 (8.37%) slices from a total of 35,840. The results are comparable with the assessments presented in [26].

The decoding latency and decoding rate corresponding to the above-mentioned clock frequency (see **Table 1**) are

$$Latency = 2L(2K + 10) \tag{30}$$

$$R_b = \frac{2K}{2L(2K + 10)T_{clk}} \tag{31}$$

		Latency [μs]			Rb [Mbps]		
Fclk [MHz]	K [di-bits]	L = 3	L = 4	L = 5	L = 3	L = 4	L = 5
125	24	2.78	3.71	4.64	17.24	12.93	10.34
125	240	23.52	31.36	39.2	20.41	15.31	12.24
125	2400	230.9	307.8	384.8	20.79	15.59	12.47

Table 1. Latency and throughput.

The implementation delay is represented by 10 clock periods per iteration and is added to the theoretical latency of the MAP algorithm (which is 4*KN* clock periods).

In **Figure 16**, the decoding performances are presented for a quadrature phase shift keying (QPSK) modulation, ½ rate, 1–4 iterations, a block size of 6 bytes (the smallest possible) and a transmission simulated through an additive white Gaussian noise (AWGN) channel. The results are depicted for the worst case scenarios, considering that the test was performed for the smallest block size.

6.2. LTE systems

Figures 11 and **15** show that the decoding latency is reduced in the case of parallel decoding with a factor almost equal to *N*. The presented implementation has an 11 clock period *Delay*, which is added for each forward trellis run (when the LLRs are computed). As a consequence, two such values must be considered during each iteration.

For serial decoding, the native latency is computed as follows: at the first semi-iterations, *K* clock periods required for the backward trellis run and another (*K* + *Delay*) clock periods for the forward trellis run and LLR computation. The value is then considered twice in order to take into account the second semi-iteration. By denoting *L* the number of executed iterations,

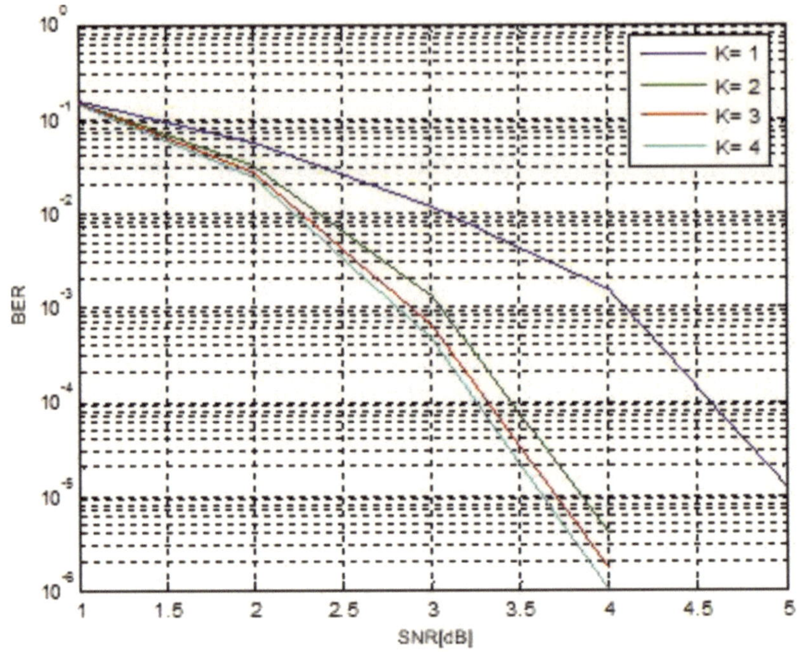

Figure 16. The impact of the number of iterations on decoding performances.

the numbers of clock periods required for a serial, respectively, a parallel block decoding operation result as:

$$Latency\_s = (4K + 2Delay)L \tag{32}$$

$$Latency\_p = (4K/N + 2Delay)L \tag{33}$$

When performing tests for the parallel decoding performances, a certain level of degradation was observed, since the forward and backward metrics are altered at the data block boundaries. In order to have similar performance as in the serial decoding case, a small overhead is accepted. By introducing an overlap at each parallel block boarder, the metrics computation gains a training phase. The minimum overlap window length is selected to cover the minimum standard defined data block (in this case $K_{min} = 40$ bits).

Figure 17 shows this situation, for the $N = 2$ setup. If we consider $N > 2$, which leads to blocks with K_{min} at both the left and right sides, the corresponding latency can be expressed as:

$$Latency\_po = \left(4(K/N + 2K_{min}) + 2Delay\right)L \tag{34}$$

For even-odd merge sorting network implementation, we can study the configuration $K = 40$ bits and $N = 8$. The input of the ILM content is represented by the 40 interleaved addresses organized in five memory locations and eight addresses for each location. The minimum-detected value for each ILM location (i.e., the natural-order memory location that will be

http://dx.doi.org/10.5772/67017

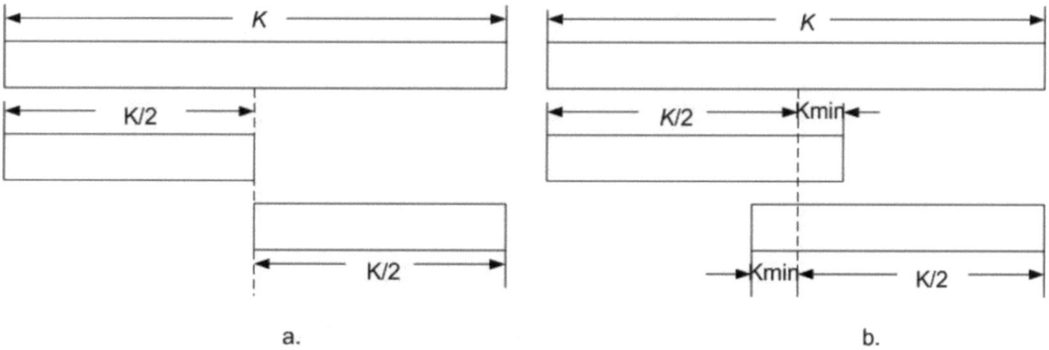

Figure 17. (a) Non overlapping split; (b) overlapping split.

accessed) is contained in the output of the sorting unit. Also, the module provides the order which will be used to send data read from natural-order memory location to the N decoding units. In this example, at the third clock period, the second ILM location is read, i.e., the addresses 6, 31, 36, 21, 26, 11, 16 and 1. The sorting module labels these addresses with an index, obtaining the pairs: (6, 0), (31, 1), (36, 2), (21, 3), (26, 4), (11, 5), (16, 6) and (1, 7). Then the addresses are arranges in an increasing order: (1, 7), (6, 0), (11, 5), (16, 6), (21, 3), (26, 4), (31, 1) and (36, 2). At the same time, the minimum address found at this location is sent at the output, 1 in this example. In conclusion, location number 1 is read from the natural-order data memory. The eight samples from the location 1 are distributed to the eight decoding units as indicated by the output index. The first sample from this location is sent to decoder unit 7, the second sample to decoder unit 0, the third one to decoder unit 5 and so on. As **Figure 18** shows that at the register transfer level (RTL), besides flip flops, the sorting unit includes only basic selection elements.

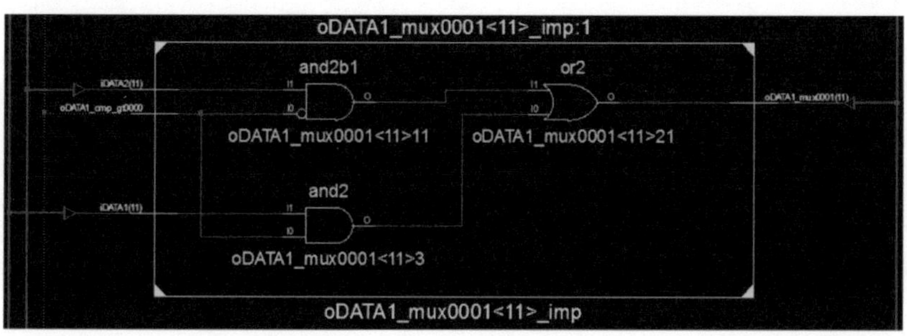

Figure 18. Basic selection element for binary inputs.

It can be seen in **Figure 19** that the sorting unit allows a pipeline data processing. Consequently, with a certain implementation delay (7 clock periods in the proposed scheme), the module provides a value belonging to the set of sorted indexes at each clock cycle.

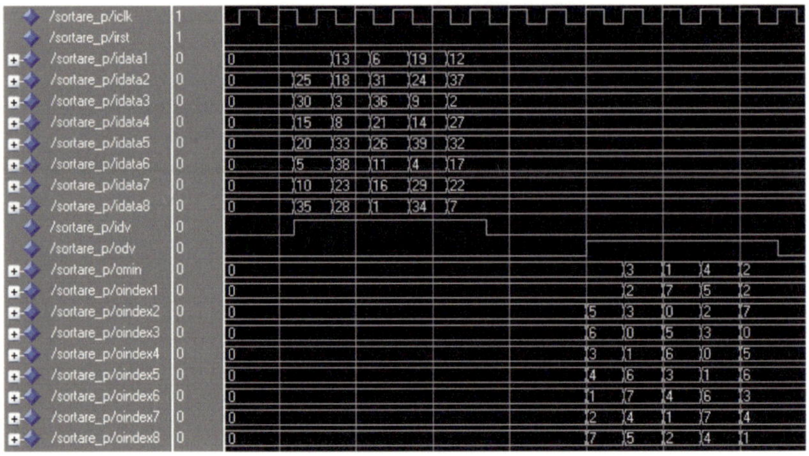

Figure 19. Even-odd merge sort – ModelSim simulation.

It is important to mention that the even-odd merge sorting was selected because it allows a pipeline functioning, consuming also lower resources than the other listed methods. Some comparative results were provided in [11, 27] in terms of used resources for the application-specific integrated circuit (ASIC).

In order to evaluate the performances, we used the very high speed hardware description language (VHDL), programming language. The code was tested using ModelSIM 6.5. For the generation of RAM/ROM memory blocks, Xilinx Core Generator 14.7 was employed and the synthesis process was accomplished using Xilinx XST from Xilinx ISE 14.7. Using the above-mentioned tools, the resulted values for the decoding structure when implemented on a Xilinx XC5VFX70T-FFG1136 are the following [28]: frequency of 310 MHz and 664 flip flops and 568 LUTs for the sorting unit, respectively, a frequency of 300 MHz, 1578 flip flop registers and 1708 LUTs for the interleaver.

The values listed in **Table 2** are obtained using Eqs. (32)–(34), when $N = 8$ is considered. One can observe that the overhead introduced by the overlapping split method is less important for bigger values of K, this being the scenario when a parallel approach is usually applied. The achieved overall system frequency is 210 MHz, with the longest signal propagation time required for the SISO unit.

Table 3 provides the corresponding throughput rate when the values from **Table 2** are used.

	Latency_s [μs]		*Latency_p* [μs]		*Latency_po* [μs]	
K	*L*					
	3	4	3	4	3	4
1536	88.08	117.4	11.28	15.04	15.85	21.14
4096	234.3	312.5	29.57	39.42	34.14	45.52
6144	351.4	468.5	44.2	58.9	48.7	56.02

Table 2. Latency values for $N = 8$, $L = 3$ or 4 and $K = 1536$, 4096 or 6144.

	Tput_s [Mbps]		*Tput_p* [Mbps]		*Tput_po* [Mbps]	
K	L					
	3	4	3	4	3	4
1536	17.43	13.07	136.1	102.0	96.86	72.64
4096	17.47	13.10	138.5	103.8	119.9	89.9
6144	17.48	13.11	139	104.2	125.9	94.4

Table 3. Throughput values for $N = 8$, $L = 3$ or 4 and $K = 1536$, 4096 or 6144.

As one can observe from **Table 3**, the serial decoding performance is similar to the theoretical one. Let us consider, for example, the case $L = 3$ and $K = 6144$. Considering the theoretical latency of $4KL$ clock periods, the theoretical throughput is 17.5 Mbps. After implementation, the obtained result for the proposed serial architecture is 17.48 Mbps.

The following performance graphs were obtained using a finite precision Matlab simulator. This approach was selected because the same outputs as the ModelSIM simulator are obtained in Matlab, while the testing time is considerably smaller.

All the simulation results were generated for the Max Log MAP algorithm. The illustrations present the bit error rate (BER) versus signal-to-noise ratio (SNR) expressed as the ratio between the energy per bit and the noise power spectral density.

Figure 20 presents the attained performances for the case of $K = 512$, $N = 2$, $L = 3$ and QPSK modulation, using the three discussed decoding methods, i.e., the serial one, the parallel without overlapped split one and the parallel with overlapped split one. **Figure 21** depicts the same performance comparison, this time for $K = 1024$ and $N = 4$.

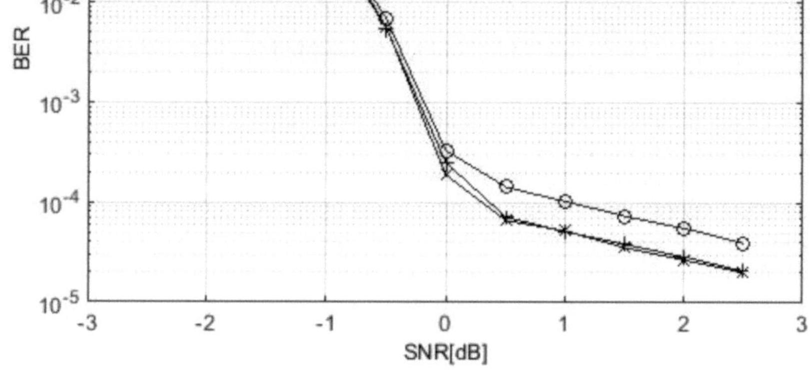

Figure 20. Comparative decoding results for QPSK, $L = 3$, $K = 512$, $N = 2$.

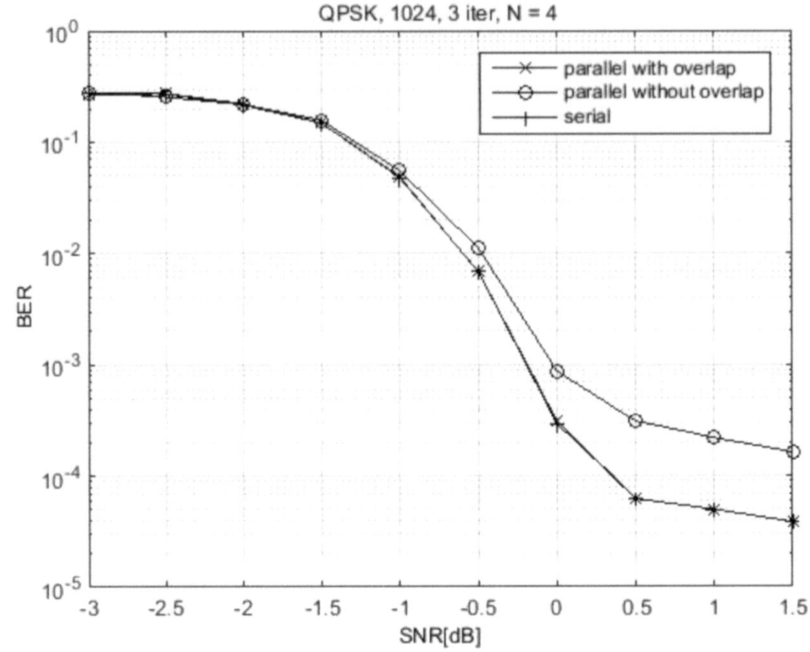

Figure 21. Comparative decoding results for QPSK, $L = 3$, $K = 1024$, $N = 4$.

Analyzing the results presented in **Figures 20** and **21**, one can conclude that the decoding performance obtained, when parallel decoding with the overlapped split method is used, is almost similar to the one for serial decoding. In contrast, the parallel decoding without the overlapped split method generates some loss in performance when compared to the serial decoding. This degradation is dependent on the parallelization factor N.

7. Conclusions

This chapter presented the most important aspects related to the FPGA implementation of a turbo decoder for WiMAX and LTE systems. The serial turbo decoder architectures for the two systems have been developed and efficiently implemented, important results being obtained especially for the proposed architectures of the interleaver/deinterleaver. For LTE systems, the interleaver memory ILM has been introduced. In this manner, the interleaver process effectively works only outside the decoding process itself.

The ILM has been written together with the input data, while the previous block was still under decoding. It should be outlined that this solution allows the transition from the serial to the parallel decoder in an efficient manner, involving only values that are concatenated at same memory locations. The parallel approach requires the same storing capacity (the number of BRAMs) and a single interleaver, thus adding only an even-odd merge sorting network. This unique interleaver has been implemented in an efficient configuration that uses only comparators and subtractors and no multipliers and dividers

http://dx.doi.org/10.5772/67017

The parallel decoding performances have been compared with the serial ones. In this context, certain degradation has been observed. In order to eliminate this degradation, a small overhead is accepted by the overlapping split that is applied to the parallel data blocks.

Acknowledgements

This work was supported by the UEFISCDI under Grant PN-II-RU-TE-2014-4-1880.

Author details

Cristian Anghel*, Cristian Stanciu and Constantin Paleologu

*Address all correspondence to: canghel@comm.pub.ro

Politehnica University of Bucharest, Romania

References

[1] C. Berrou, A. Glavieux and P. Thitimajshima, Near Shannon limit error-correcting coding and decoding: Turbo codes, *IEEE Proceedings of the International Conference on Communications*, Geneva, Switzerland, May 1993, pp. 1064–1070.

[2] C. Berrou and A. Glavieux, Near optimum error correcting coding and decoding: Turbo-Codes, *IEEE Transactions on Communications*, vol. 44, no. 10, pp. 1261–1271, Oct. 1996.

[3] C. Berrou and M. Jézéquel, Non binary convolutional codes for turbo coding, *Electronics Letters*, vol. 35, no. 1, pp. 9–40, Jan. 1999.

[4] M. C. Valenti and J. Sun, The UMTS turbo code and an efficient decoder implementation suitable for software-defined radios, *International Journal of Wireless Information Networks*, vol. 8, no. 4, pp. 203–215, Oct. 2001.

[5] C. Anghel, A. A. Enescu, C. Paleologu and S. Ciochina, CTC Turbo decoding architecture for H-ARQ capable WiMAX systems implemented on FPGA, *Ninth International Conference on Networks* ICN 2010, Menuires, France, April 2010.

[6] C. Anghel, A. A. Enescu, et al., FPGA implementation of a CTC Decoder for H-ARQ compliant WiMAX systems, *Proceedings of International Conference on Design & Technology of Integrated Systems*, DTIS 2007, Morocco, pp. 82–86.

[7] C. Anghel, V. Stanciu, C. Stanciu and C. Paleologu, CTC Turbo decoding architecture for LTE systems implemented on FPGA, *IARIA ICN 2012*, Reunion, France, 2012.

[8] S. Chae, A low complexity parallel architecture of turbo decoder based on QPP interleaver for 3GPP-LTE/LTE-A, http://www.design-reuse.com /articles/31907/turbo-decoder-architecture-qpp-interleaver-3gpp-lte-lte-a.html

[9] Y. Sun and J. R. Cavallaro, Efficient hardware implementation of a highly-parallel 3GPP LTE/ LTE-advance turbo decoder, *Integration, the VLSI Journal*, vol. 44, no. 4, pp. 305–315, Sept. 2011.

[10] D. Wu, R. Asghar, Y. Huang and D. Liu, Implementation of a high-speed parallel turbo decoder for 3GPP LTE terminals, ASICON '09, *IEEE 8th International Conference on ASIC*, pp. 481–484, 2009.

[11] C. Studer, C. Benkeser, S. Belfanti and Q. Huang, Design and implementation of a parallel turbo-decoder ASIC for 3GPP-LTE, *IEEE Journal of Solid-State Circuits*, vol. 46, no. 1, pp. 8–17, Jan. 2011.

[12] C. Anghel and C. Paleologu, Simplified parallel architecture for LTE-A turbo decoder implemented on FPGA, *Proceedings of the 9th International conference on Circuit, Systems, Signal and Telecommunications CCST 2015*, Dubai, pp. 102–111.

[13] C. Stanciu, C. Anghel and C. Paleologu, Efficient recursive implementation of a quadratic permutation polynomial interleaver for LTE systems, *Revue Roumaine, des Sciences Techniques - Serie Électrotechnique et Énergétique*, ISSN: 0035-4066, vol. 61, pp. 53–57.

[14] K. E. Batcher, Sorting networks and their applications," in *Proceeding of AFIPS Spring Joint Computer Conference*, vol. 32, 1968.

[15] C. Anghel, C. Stanciu and C. Paleologu, Sorting methods used in parallel turbo decoding for LTE systems, *2015 International Symposium on Signals, Circuits and Systems (ISSCS)*, 9–10 July, 4 p.

[16] Xilinx Virtex 5 family user guide, https://www.xilinx.com/support/documentation/user_guides/ug190.pdf

[17] Xilinx ML507 evaluation platform user guide, https://www.xilinx.com/products/boards/ml507/docs.htm

[18] https://standards.ieee.org/about/get/802/802.16.html

[19] 3GPP TS 36.212 V8.7.0 (2009-05) Technical Specification, "3rd Generation Partnership Project; Technical Specification Group Radio Access Network; Evolved Universal Terrestrial Radio Access (E-UTRA); Multiplexing and channel coding (Release 8)."

[20] P. Robertson, E. Villebrun and P. Hoeher, A comparison of optimal and sub-optimal MAP decoding algorithms operating in the log domain, *Proceeding of IEEE International Conference on Communications* (ICC'95), Seattle, pp. 1009–1013, June 1995.

[21] S. Papaharalabos, P. Sweeney and B. G. Evans, Constant log-MAP decoding algorithm for duo-binary turbo codes, *Electronics Letters*, vol. 42, no. 12, pp. 709–710, June 2006.

[22] J.-F. Cheng and T. Ottosson, Linearly approximated log-MAP algorithms for turbo decoding, *Vehicular Technology Conference Proceedings*, 2000. VTC 2000-Spring Tokyo. 2000 IEEE 51st vol. 3, pp. 2252–2256, 2000.

[23] Massachusetts Institute of Technology, Mathematics, last access date: November 2014, *math.mit.edu/~shor/18.310/batcher.pdf*

[24] R. Asghar, D. Wu, J. Eilert and D. Liu, Memory conflict analysis and a re-configurable interleaver architecture supporting unified parallel turbo decoding, *Journal of Signal Processing Systems*, vol. 60, no. 1, pp. 15–19, July 2010.

[25] S. Wang, L. Liu and Z. Wen, High speed QPP generator with optimized parallel architecture for 4G LTE-A system, *International Journal of Advancements in Computing Technology*, vol. 4, no. 23, pp. 355–364, July 2010.

[26] Xilinx, IEEE 802.16e CTC decoder core, DS137 (v2.3), July 11, 2006.

[27] E. Mumolo, G. Capello and M. Nolich, VHDL design of a scalable VLSI sorting device based on pipelined computation," *Journal of Computing and Information Technology - CIT 12*, vol. 12, no. 1, pp. 1–14, 2004.

[28] C. Anghel, C. Stanciu and C. Paleologu, LTE turbo decoding parallel architecture with single interleaver implemented on FPGA, Springer Verlag *Circuits, Systems and Signal Processing*, ISSN: 0278-081X, DOI 10.1007/s00034-016-0362-z, 2016.

Chapter 3

Motion Control with FPGA

Miguel Angel Martínez Prado,
Juvenal Rodríguez Reséndiz,
Diana Carolina Toledo Pérez,
Carlos Miguel Torres Hernández and
Gilberto Herrera Ruiz

Additional information is available at the end of the chapter

http://dx.doi.org/10.5772/67200

Abstract

The aim of this chapter is to provide an introduction to the field programmable gate array (FPGA)-based digital control system design for motion control. It is intended as a reference for the undergraduate students in science and engineering, professionals, and enthusiastic people who have a basic knowledge in discrete control theory and digital systems using reconfigurable logic. The scope of this chapter includes the analysis, simulation, and implementation of classic control algorithms. The presented topics serve as a foundation for the implementation of more complex systems. An experimental section is provided, which validates the proposed digital design.

Keywords: FPGA, motion-control, PID-control, VHDL, robotics

1. Introduction

The reconfigurable logic in industries opened countless opportunities especially in the field of control and automation. This technology facilitates the implementation of complex control algorithms with fast response.

Nowadays, the control system engineers require new tools for creating better electronic design for automation systems. Some modern tools that are available on the market allow the designer to create, simulate, and verify the desired hardware design. This can help in the evaluation of the complex system designs with fewer resources.

Among modern tools used by the controllers, the field programmable gate array (FPGA) provides a shorter processing time than the conventional methods like microprocessor- or microcontroller-based designs. Furthermore, it has benefits such as improved accuracy and efficiency of the algorithms.

In the industry, the FPGA technology began to be used by the designers in areas like telecommunications, signal processing, image processing, and control systems such as robotic arms and assembly lines. Later, this technology began to be utilized in applications where the fast processing of information is desired, such as medical equipment, robotics, aeronautics, etc. [1].

Proportional integral derivative (PID) controller is one of the most commonly used design, due to its simple design and its robustness with respect to the parameter uncertainty [2–4]. They are usually used in the speed controlling applications of direct current or permanent magnet motors, through pulse-width modulation (PWM) pulses [5], output current, voltage, or frequency. It is possible to find out in the scientific literature PID implementations on hardware [6–8] whose authors have demonstrate the effectiveness of their designs; however, most of these implementations are not easy to develop and for some cases they are destined to be implemented only in some FPGA families.

In order to enhance the designer experience, the FPGA card manufacturers incorporate multicore processors equipped with flash memory into their designs for enhancing the computing capacity and data parallel processing. In this way, the controllers can implement functions that require fast processing in hardware and computationally intensive algorithms into the processor.

Implementing functions into the FPGA chip, the platforms that are available on the market works with HDL codes, which decreases hardware resource use and therefore, at the same time, reduces the cost and energy consumption of the system. Moreover, these platforms manage simulators for assessment of the design before its implementation.

Section 2 of this chapter is related to the digital controllers, which describes the PID and other controllers in discrete form. Section 3 provides the hardware description of the PID controller in VHDL language, and finally, the fourth section provides the simulation and experimental validation, which demonstrates how to perform numerical simulations using Simulink and Modelsim. Furthermore, an experimental validation on a DC motor system is also provided.

2. Discretization of classical controllers

The proportional-integral derivative (PID) controller is widely used in industry due to its high performance with most of the plants even if they are nonlinear [2]. Besides, its parameters can be tuned empirically and still achieve a good performance. Due to the complexity of the algorithm, its implementation has been limited to microcontrollers or digital signal processors [3], and furthermore, most of the researchers who are experts in control theory do not have a deep knowledge on reconfigurable logic [4, 5].

PID controller has the following form:

http://dx.doi.org/10.5772/67200

$$u(t) = K_p\left[e(t) + \frac{1}{T_i}\int_0^t e(\tau)d\tau + T_d\frac{de(t)}{dt}\right] \tag{1}$$

where $e(t)$ is the difference between the desired ($w(t)$) and measured ($y(t)$) response of the system, i.e.:

$$e(t) = w(t) - y(t) \tag{2}$$

and $u(t)$ is the control signal, used to control the actuator's operation to obtain a desired closed-loop performance. Finally, the parameters K_p, T_i, and T_d are the proportional gain, integral, and derivative time constants, respectively. A more popular form of Eq. (1) can be obtained by using the Laplace transform as follows:

$$U(s) = \left(K_p + \frac{K_i}{s} + K_d s\right)E(s), \tag{3}$$

where $K_i = \frac{K_p}{T_i}$ and $K_d = K_p T_d$.

Eqs. (2) and (3) are time-dependent functions; therefore, they cannot be implemented directly into a digital system. In that case, it is necessary to find out the discrete form of Eq. (1) by applying numerical methods.

The proportional part of the equation does not require any additional transformation because it involves a simple multiplication, but the integral and derivative require a numerical approximation. First, the integral of the error function can be considered as the sum of the area of small rectangles of base longitude of T_s (which is commonly termed as the sampling period), and height $e(k)$ at a given time instant, $t = kT_s$, i.e.:

$$\int_0^t e(\tau)d\tau \approx T_s\sum_{i=1}^{k} e(i). \tag{4}$$

Similarly, the derivative term can be approximated as:

$$\frac{de(t)}{dt} \approx \frac{e(k) - e(k-1)}{T_s} \tag{5}$$

for a given time $t = kT_s$. Now substituting Eqs. (4) and (5) into Eq. (1), it is possible to rewrite the PID controller in its discrete form as:

$$u(k) = K_p\left\{e(k) + \frac{T_s}{T_i}\sum_{i=1}^{k} e(i) + \frac{T_d}{T_s}[e(k) - e(k-1)]\right\} \tag{6}$$

Eq. (6) provides the storage of error samples from $t = 0$ until $t = kT_s$, which can be easily implemented by software on a microprocessor or DSP target. It is common to have kilobytes of memory in microprocessor platforms and such storage does not carry any problem; however, when we deal with reconfigurable logic, it is of vital importance to save logic resources; therefore, a more suitable form of Eq. (6) is needed. Above is achieved by computing the differential term $\Delta u(k)$ instead of computing directly $u(k)$. Let us define the differential term $\Delta u(k)$ as:

$$\Delta u(k) = u(k) - u(k-1), \tag{7}$$

and

$$u(k-1) = K_p\left\{e(k-1) + \frac{T_s}{T_i}\sum_{i=1}^{k-1} e(i) + \frac{T_d}{T_s}[e(k-1) - e(k-2)]\right\} \tag{8}$$

Subtracting Eq. (8) from Eq. (6) yields:

$$\Delta u(k) = K_p\left\{e(k) - e(k-1) + \frac{T_s}{T_i}e(k) + \frac{T_d}{T_s}[e(k) - 2e(k-1) + e(k-2)]\right\} \tag{9}$$

From Eq. (7), it is possible to rewrite the control output $u(k)$ in terms of $u(k-1)$ and $\Delta u(k)$ as:

$$u(k) = \Delta u(k) + u(k-1). \tag{10}$$

It is worth to note that during the first iteration, i.e., for $t = kT_s = 0$, the term $u(k-1)$ becomes zero, while for subsequent iterations, this term holds the previously computed value of $u(k)$. Finally, substituting Eq. (9) in Eq. (10), the PID control law becomes

$$u(k) = K_p\left\{e(k) - e(k-1) + \frac{T_s}{T_i}e(k) + \frac{T_d}{T_s}[e(k) - 2e(k-1) + e(k-2)]\right\} + u(k-1) \tag{11}$$

The common terms in Eq. (11) can be grouped so that the control law takes the form of a digital filter, i.e.:

$$u(k) = q_0e(k) + q_1e(k-1) + q_2e(k-2) + u(k-1) \tag{12}$$

where

$$q_0 = K_p\left(1 + \frac{T_s}{T_i} + \frac{T_d}{T_s}\right)$$

$$q_1 = -K_p\left(1 + 2\frac{T_d}{T_s}\right)$$

$$q_2 = K_p\frac{T_d}{T_s}$$

Proceeding with the same analysis, the reader could easily derive the formulas for a proportional-integral (PI) digital controller, which has the form:

$$u(t) = q_0e(k) + q_1e(k-1) + u(k-1) \tag{13}$$

where $q_0 = K_p(1 + {}^{T_s}/_{T_i})$ and $q_1 = -K_p$. Similarly, the proportional-derivative (PD) controller may be written as:

$$u(t) = q_0 e(k) + q_1 e(k-1) + q_2 e(k-2) + u(k-1) \tag{14}$$

where

$$q_0 = K_p \left(1 + \frac{T_d}{T_s}\right)$$

$$q_1 = -K_p \left(1 + 2\frac{T_d}{T_s}\right)$$

$$q_2 = K_p \frac{T_d}{T_s}$$

Other controllers represented in the Laplace domain can be discretized by using approximations, e.g., the Tustin formulae:

$$s = \frac{2(z-1)}{T_s(z+1)} \tag{15}$$

For example, let us consider the following lead compensator:

$$\frac{U(s)}{E(s)} = k\frac{s+\omega_1}{s+\omega_2} \tag{16}$$

Substituting Eq. (15) in Eq. (16), we obtain:

$$\begin{aligned}
\frac{U(z)}{E(z)} &= k\frac{\dfrac{2(z-1)}{T_s(z+1)} + \omega_1}{\dfrac{2(z-1)}{T_s(z+1)} + \omega_2} \\
&= k\frac{2(z-1) + \omega_1 T_s(z+1)}{2(z-1) + \omega_2 T_s(z+1)} \\
&= k\frac{(\omega_1 T_s + 2)z + \omega_1 T_s - 2}{(\omega_2 T_s + 2)z + \omega_2 T_s - 2} \\
&= k\frac{(\omega_1 T_s + 2)z + \omega_1 T_s - 2}{z + \dfrac{\omega_2 T_s - 2}{\omega_2 T_s + 2}} \\
&= k\left(\frac{\omega_1 T_s + 2}{\omega_2 T_s + 2}\right)\frac{z + \dfrac{\omega_1 T_s - 2}{\omega_1 T_s + 2}}{z + \dfrac{\omega_2 T_s - 2}{\omega_2 T_s + 2}}
\end{aligned}$$

The above equation can be rewritten as:

$$\frac{U(z)}{E(z)} = K\frac{z + A}{z + B} \tag{17}$$

where

$$K = k\frac{\omega_1 T_s + 2}{\omega_2 T_s + 2}$$

$$A = \frac{\omega_1 T_s - 2}{\omega_1 T_s + 2}$$

$$B = \frac{\omega_2 T_s - 2}{\omega_2 T_s + 2}$$

From a digital point of view, Eq. (17) is still inconvenient. In order to obtain a suitable digital representation, it is necessary to represent this equation as a difference equation. This can be performed by multiplying the numerator and the denominator of the right-hand side of Eq. (17) by z^{-1}. This is equivalent to the shifting operation in the time domain, where the signal is delayed by one sample. Thus, the lead compensator takes the following form:

$$\frac{U(z)}{E(z)} = K\frac{1 + Az^{-1}}{1 + Bz^{-1}}$$

Further simplification yields

$$U(z)(1 + Bz^{-1}) = KE(z)(1 + Az^{-1})$$

Expanding terms:

$$U(z) + BU(z)z^{-1} = KE(z) + KAE(z)z^{-1}$$

Solving the above equation for $U(z)$ we have:

$$U(z) = KE(z) + KAE(z)z^{-1} - BU(z)z^{-1}$$

It is well known that:

$$E(z) = e(k) \tag{18}$$

and

$$E(z)z^{-1} = e(k - 1) \tag{19}$$

therefore, the discrete lead compensator filter can be expressed as:

$$u(k) = Ke(k) + KAe(k - 1) - Bu(k - 1) \tag{20}$$

which is quite similar to Eq. (14).

http://dx.doi.org/10.5772/67200

3. Hardware description

There are important features that the reader must consider before starting the description process. First, the nature of the feedback signal should be considered. If the sensor which measures the variable to be controlled has an analogue nature, it is necessary to use an analogue to digital converter (ADC) which has an output with a fixed bit width. In order to avoid performing arithmetic operations between signals of different bit width, it is strongly suggested that the setpoint or reference has the same bit width as the measured variable. Additionally, if the error signal has a wide bus width, let us say wider than 16 bits, this signal can be saturated in order to avoid wider bus widths in preceding computations.

The second aspect to consider is the number and characteristics of the embedded multipliers or DSP slices that the target device possesses. Most of the FPGAs available in the market have multipliers with a fixed bus width, 18 × 18 bits for instance. For any other bus-width, the synthesis tool shall use logic resources to build a customized multiplier instead of using those available in the hardware. This bad practice leads to major resource utilization.

In summary, the error signal and the controller gains shall have a bus width, which matches the bus width of the available embedded multipliers. For example, the available multipliers have an 18 × 18 bus width, the error signal has 16 bits width, and the controller gains are required to have a fixed point format 16.16. Then a possible solution that does not imply the usage of a customized multiplier is to expand the bus width of error and gains to 18 and 36 bits, respectively, being in the latter signal, 18 bits for the integer part and the remaining 18 bits for the fractional part. Thus, the synthesis tool would use two 18 × 18 embedded multipliers. Above is possible whenever the bus width of error signal and gains be the multiple of the bus width of the embedded multipliers and while the extension of the signals preserves their signs.

Finally, the controller output must be congruent with the nature of the actuators. If we deal with an analogue actuator, it is necessary to include a digital to analogue converter (DAC) which, as the case of the ADC, has a fixed bus width; therefore, the controller output must agree such width. However, depending on the polynomial degree of the filter and the bus width of the used multipliers, the controller output eventually could have a wider bus width; so, it would be necessary to saturate and truncate decimal data from the controller output signal.

In this section, we shall describe the design of a PID digital controller; however, the reader could easily modify the proposed design for most of the control above laws. The resultant design is implemented in VHDL; it is validated in a cosimulation environment, and finally, it is tested in a real-life application to control the position of a brushed DC servo motor.

The PID digital filter seen as a black-box module is depicted in **Figure 1**. The ERR signal represents the error signal, which is the difference between the setpoint and the feedback data. Similarly, the signals Q0, Q1, and Q2 are the controller gains, and their value depends on K_p, T_i, T_d, and T_s as it is explained in previous sections. On the other side, the UOUT signal is the filter output, which is feed forwardly to the actuator. CLK and RST are the master clock and master reset signals, respectively.

Table 1 summarizes the signals properties of the PID_Digital_Controller module. It is important to clear that the error signal and the controller gains have 16 and 32 bits of width,

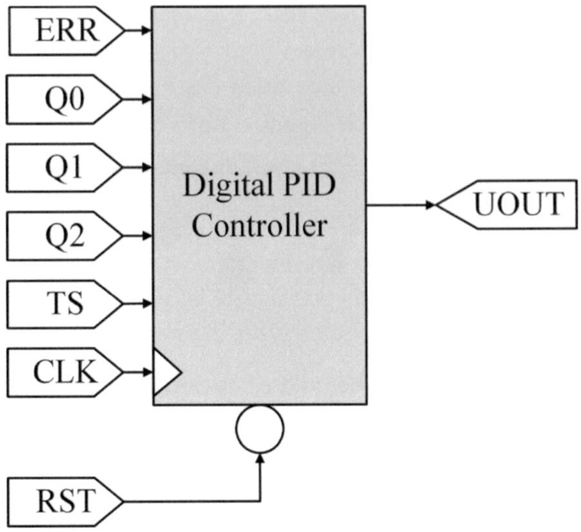

Figure 1. Black-box module of PID digital filter.

Signal name	Direction	Bus width	Description
RST	Input	1	Active low master reset
CLK	Input	1	Master clock
TS	Input	1	Sampling signal
ERR	Input	16	Two's complement error signal
Q0	Input	32	Filter coefficient q0
Q1	Input	32	Filter coefficient q1
Q2	Input	32	Filter coefficient q2
UOUT	Output	16	Controller output

Table 1. List of signals properties of PID digital controller.

respectively; above to match the standard data size of most of the common programming languages; however, these signals are internally expanded to 18 and 36 bits in order to use two 18 × 18 embedded multipliers as previously mentioned. Additionally, the controller gains are given in a fixed point format 16.16, i.e., the 16 most significant bits represent the integer part while the 16 less significant represent the fractional part.

Figure 2 illustrates a block diagram of the digital PID controller. This module comprises five standard load registers, two multiplexors, a multiplier, an adder, a saturator, and a finite state machine (FSM). White blocks represent pure combinational processes, whereas gray ones represent sequential and synchronous processes.

The data path starts at the input registers. At this point, the multiplexors bypass the corresponding signal error and controller gains selected by the SEL signal, which is driven by the FSM, the multiplier and adder accumulate this product with the previous result and so on

http://dx.doi.org/10.5772/67200

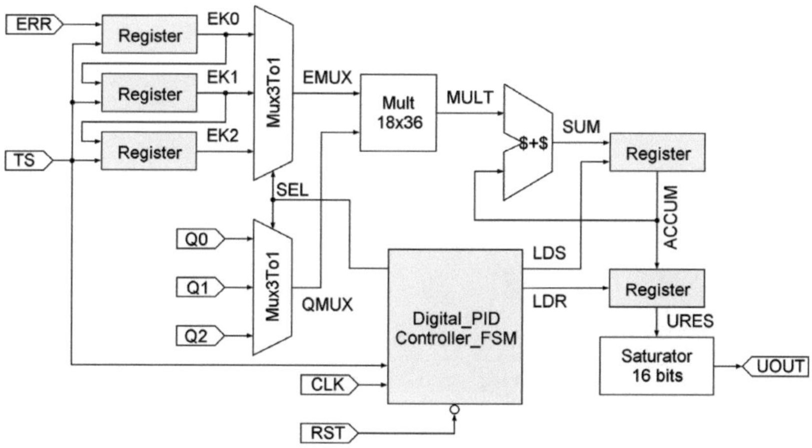

Figure 2. Block diagram of digital PID controller.

the next terms. At the final stage, a saturator trims the bus width of the controller output and saturates its value to 16 bits.

Signals EK0, EK1, and EK2 have 16 bits of width; however, at the multiplexor output, their sign is extended two bits, i.e., the EMUX signal has 18 bits of width. Similarly, the signals Q0, Q1, and Q2 have 32 bits, and at the multiplexor output, QMUX, these signals are extended to 36 bits. The product of error signal as per its corresponding coefficient has 54 bits; nevertheless, this signal is extended again in order to avoid a possible overflow because of the recurrent addition with previous results. Thus, given that there are three sums involved in the solution of the control algorithm, the signal MULT is extended three bits more to obtain a bus width of 57 bits finally. Signals ACCUM and URES also have a 57 bits bus width.

The pipelined structure of registers at the top-left corner, depicted in **Figure 2**, is planned to latch the error signals $e(k)$, $e(k-1)$, and $e(k-2)$ when signal TS is asserted. On the other hand, the register located at the top-right corner together with the adder perform the accumulation process through the assertion of signal LDS. And finally, the last register in the data path serves only as a holder for the final result of the algorithm. This latter register loads data when LDR is asserted.

Figure 3 illustrates the operation of the Digital_PID_Controller_FSM; it includes five states. The first state is an idle state, which waits for the assertion of the sampling signal TS. Second, third, and fourth states perform the multiplication and accumulation of the filter terms, and such partial results are added to the previous final result. The fifth state only asserts the signal LDR to latch the final result and jumps directly to the first state in order to repeat the whole process.

The following source code corresponds to the top-level entity of the design. Its architecture is structural since it is composed of many other components as mentioned above, in total there are 12 instances and 3 concurrent assignations.

Code 1. Digital_PID_Controller.vhd.

```
Library IEEE;

use IEEE.std_logic_1164.all;
```

```
Entity Digital_PID_Filter is
   port(
   RST : in std_logic;
   CLK : in std_logic;
   TS : in std_logic;
   ERR: in std_logic_vector(15 downto 0);
   Q0 : in std_logic_vector(31 downto 0);
   Q1 : in std_logic_vector(31 downto 0);
   Q2 : in std_logic_vector(31 downto 0);
   UOUT : out std_logic_vector(15 downto 0)
   );
end Digital_PID_Controller;
Architecture Structural of Digital_PID_Controller is
--Components declaration--------------------------------------------------------
Component Digital_PID_Controller_FSM is port(
   RST : in std_logic;
   CLK : in std_logic;
   TS : in std_logic;
   LDS : out std_logic;
   LDR : out std_logic;
   SEL : out std_logic_vector(1 downto 0));
end Component;
Component LoadRegister is generic(n : integer := 8);
port(
RST : in std_logic;
CLK : in std_logic;
LDR : in std_logic;
DIN : in std_logic_vector(n – 1 downto 0);
DOUT : out std_logic_vector(n – 1 downto 0));
end Component;
```

http://dx.doi.org/10.5772/67200

```
Component Multiplexor3To1 is generic(n : integer := 8);
   port(
   DIN0 : in std_logic_vector(n - 1 downto 0);
   DIN1 : in std_logic_vector(n - 1 downto 0);
   DIN2 : in std_logic_vector(n - 1 downto 0);
   SEL : in std_logic_vector(1 downto 0);
   DOUT : out std_logic_vector(n - 1 downto 0));
end Component;
Component Multiplier is generic(m, n : integer := 9);
   port(
   OPA : in std_logic_vector(m - 1 downto 0);
   OPB : in std_logic_vector(n - 1 downto 0);
   RES : out std_logic_vector((m + n - 1) downto 0));
   end Component;
Component Adder is generic(n : integer := 8);
   port(
   OPA : in std_logic_vector(n - 1 downto 0);
   OPB : in std_logic_vector(n - 1 downto 0);
   RES : out std_logic_vector(n - 1 downto 0));
end Component;
Component Saturator57To16 is port(
   DIN : in std_logic_vector(56 downto 0);
   DOUT : out std_logic_vector(15 downto 0));
end Component;
--Signals declaration---------------------------------------------------------
signal LDS : std_logic;
signal LDR : std_logic;
signal SEL : std_logic_vector(1 downto 0);
signal EK0 : std_logic_vector(15 downto 0);
signal EK1 : std_logic_vector(15 downto 0);
```

```
signal EK2 : std_logic_vector(15 downto 0);
signal EAUX : std_logic_vector(15 downto 0);
signal EMUX : std_logic_vector(17 downto 0);
signal QAUX : std_logic_vector(31 downto 0);
signal QMUX : std_logic_vector(35 downto 0);
signal MULT : std_logic_vector(53 downto 0);
signal MULE : std_logic_vector(56 downto 0);
signal USUM : std_logic_vector(56 downto 0);
signal ACCU : std_logic_vector(56 downto 0);
signal URES : std_logic_vector(56 downto 0);
begin
--Concurrent assignations-----------------------------------------------
  EMUX <= EAUX(15) & EAUX(15) & EAUX;
  QMUX <= QAUX(31) & QAUX(31) & QAUX(31) & QAUX(31) & QAUX;
  MULE <= MULT(53) & MULT(53) & MULT(53) & MULT;
--Component instances---------------------------------------------------
  U01 : Digital_PID_Controller_FSM port map(RST, CLK, TS, LDS, LDR, SEL);
  U02 : LoadRegister generic map(16) port map(RST, CLK, TS, ERR, EK0);
  U03 : LoadRegister generic map(16) port map(RST, CLK, TS, EK0, EK1);
  U04 : LoadRegister generic map(16) port map(RST, CLK, TS, EK1, EK2);
  U05 : Multiplexor3To1 generic map(16) port map(EK0, EK1, EK2, SEL, EAUX);
  U06 : Multiplexor3To1 generic map(32) port map(Q0, Q1, Q2, SEL, QAUX);
  U07 : Multiplier generic map(18, 36) port map(EMUX, QMUX, MULT);
  U08 : Adder generic map(57) port map(MULE, ACCU, USUM);
  U09 : LoadRegister generic map(57) port map(RST, CLK, LDS, USUM, ACCU);
  U10 : LoadRegister generic map(57) port map(RST, CLK, LDR, ACCU, URES);
  U11 : Saturator57To16 port map(URES, UOUT);
-----------------------------------------------------------------------
end Structural;
```

http://dx.doi.org/10.5772/67200

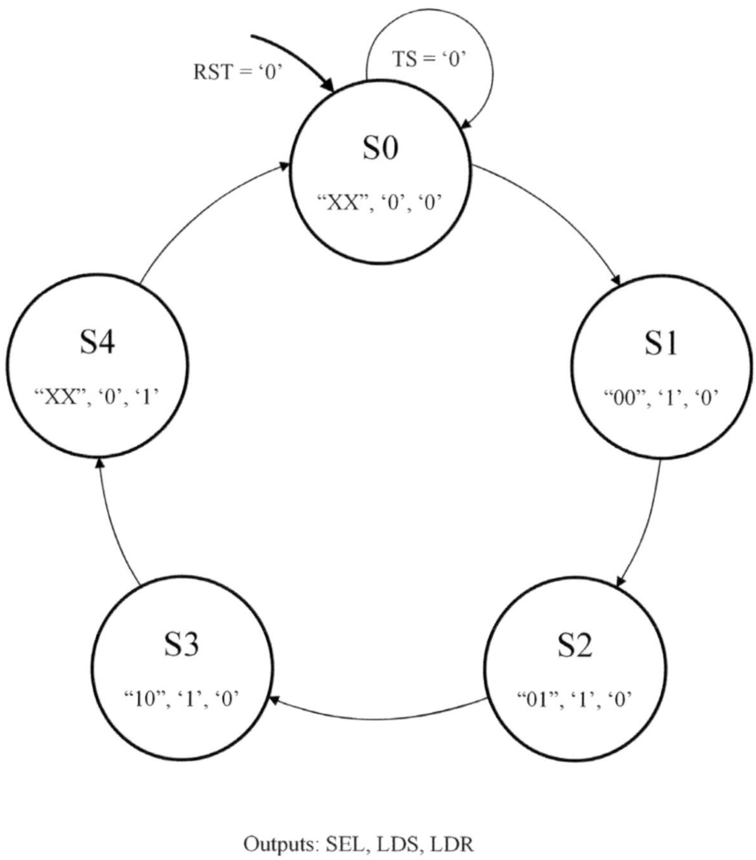

Figure 3. Finite state machine of the digital PID filter.

The following code corresponds to the implementation of the FSM depicted in **Figure 3**. It has a behavioral architecture since it is composed by a couple of processes; the first one includes the combinational logic, which performs the state transitions and sets the output logic. The second process emulates the behavior of a D-type flip-flop, which updates the data with each rising-edge of the master clock signal.

Code 2. Digital_PID_Controller_FSM.vhd.

```
Library IEEE;

use IEEE.std_logic_1164.all;

Entity Digital_PID_Controller_FSM is

  port(

  RST : in std_logic;

  CLK : in std_logic;
```

```
    TS : in std_logic;
    LDS : out std_logic;
    LDR : out std_logic;
    SEL : out std_logic_vector(1 downto 0)
    );
end Digital_PID_Controller_FSM;
Architecture Behavioral of Digital_PID_Controller_FSM is
signal Sp, Sn : std_logic_vector(2 downto 0);
begin
    combinational : process(Sp, TS)
    begin
case Sp is
when “000” =>
LDS <= '0';
LDR <= '0';
SEL <= “XX”;
if TS = '1' then
Sn <= “001”;
else
Sn <= Sp;
end if;
when “001” =>
LDS <= '1';
LDR <= '0';
SEL <= “00”;
Sn <= “010”;
when “010” =>
LDS <= '1';
LDR <= '0';
```

http://dx.doi.org/10.5772/67200

```
SEL <= "01";
Sn <= "011";
when "011" =>
LDS <= '1';
LDR <= '0';
SEL <= "10";
Sn <= "100";
when others =>
LDS <= '0';
LDR <= '1';
SEL <= "XX";
Sn <= "000";
end case;
   end process Combinational;
   Sequential : process(RST, CLK)
   begin
if RST = '0' then
Sp <= "000";
elsif CLK'event and CLK = '1' then
Sp <= Sn;
end if;
   end process Sequential;
end Behavioral;
```

The LoadRegister module is labeled as Register in **Figure 2**. The objective of this module is to store some particular value. While LDR is asserted, this module stores the value present at the input DIN, and the result is reflected in the output until the next clock event. When LDR is low, the register preserves the last latched value.

Code 3. LoadRegister.vhd.

```
Library IEEE;
use IEEE.std_logic_1164.all;
```

```
Entity LoadRegister is
generic(n : integer := 8);
port(
RST : in std_logic;
CLK : in std_logic;
LDR : in std_logic;
DIN : in std_logic_vector(n – 1 downto 0);
DOUT : out std_logic_vector(n – 1 downto 0)
);
end LoadRegister;
Architecture Behavioral of LoadRegister is
signal Qp, Qn : std_logic_vector(n – 1 downto 0);
begin
Combinational : process(Qp, LDR, DIN)
begin
if LDR = ‘1’ then
Qn <= DIN;
else
Qn <= Qp;
end if;
DOUT <= Qp;
end process Combinational;
Sequential : process(RST, CLK)
begin
if RST = ‘0’ then
Qp <= (others => '0');
elsif CLK'event and CLK = '1' then
Qp <= Qn;
end if;
```

http://dx.doi.org/10.5772/67200

```
end process Sequential;

end Behavioral;
```

The multiplexor in the following code allows directing each sample of the error signal with their corresponding coefficient.

Code 4. Multiplexor3To1.vhd.

```
Library IEEE;

use IEEE.std_logic_1164.all;

Entity Multiplexor3To1 is

generic(n : integer := 8);

port(

DIN0 : in std_logic_vector(n - 1 downto 0);

DIN1 : in std_logic_vector(n - 1 downto 0);

DIN2 : in std_logic_vector(n - 1 downto 0);

SEL : in std_logic_vector(1 downto 0);

DOUT : out std_logic_vector(n - 1 downto 0)

);

end Multiplexor3To1;

Architecture DataFlow of Multiplexor3To1 is

begin

With SEL Select DOUT <=

DIN0 when "00", DIN1 when "01", DIN2 when "10", (others =>'0') when others;

end DataFlow;
```

The following module performs an arithmetic sum between two vectors. It is worth to note that this module does not depend on the clock.

Code 5. Adder.vhd.

```
Library IEEE;

use IEEE.std_logic_1164.all;

use IEEE.std_logic_unsigned.all;

Entity Adder is

generic(n : integer := 8);
```

```
port(
OPA : in std_logic_vector(n – 1 downto 0);
OPB : in std_logic_vector(n – 1 downto 0);
RES : out std_logic_vector(n – 1 downto 0)
);
end Adder;
Architecture DataFlow of Adder is
begin
RES <= OPA + OPB;
end DataFlow;
```

Similarly, the multiplier performs an arithmetic product between two vectors; however, it is important to preserve the sign of the result; therefore, it is included in the IEEE.std_logic_arith library.

Code 6. Multiplier.vhd.

```
Library IEEE;
use IEEE.std_logic_1164.all;
use IEEE.std_logic_arith.all;
Entity Multiplier is
generic(m, n : integer := 9);
port(
OPA : in std_logic_vector(m – 1 downto 0);
OPB : in std_logic_vector(n – 1 downto 0);
RES : out std_logic_vector((m + n – 1) downto 0)
);
end Multiplier;
Architecture DataFlow of Multiplier is
begin
RES <= signed(OPA) * signed(OPB);
end DataFlow;
```

http://dx.doi.org/10.5772/67200

The last module is provided in the Code 7, which is used to limit the output. For this particular case, the controller output has been adjusted to 16 bits in order to convert this value to an analogue signal using a DAC. Since the controller gains are given in a fixed point format 16.16, the less significant bits of the controller output are trimmed, i.e., only the integer part of the control output is considered during the digital to analogue conversion.

Code 7. Saturator57To16.vhd.

```
Library IEEE;
use IEEE.std_logic_1164.all;
use IEEE.std_logic_arith.all;
Entity Saturator57To16 is
port(
DIN : in std_logic_vector(56 downto 0);
DOUT : out std_logic_vector(15 downto 0)
);
end Saturator57To16;
Architecture Behavioral of Saturator57To16 is
constant UMAX : std_logic_vector(56 downto 0) := ‘0’ & X”0000007FFF0000”;
constant UMIN : std_logic_vector(56 downto 0) := ‘1’ & X”FFFFFF80010000”;
begin
process(DIN)
begin
if signed(DIN) > signed(UMAX) then
DOUT <= UMAX(31 downto 16);
elsif signed(DIN) < signed(UMIN) then
DOUT <= UMIN(31 downto 16);
else
DOUT <= DIN(31 downto 16);
end if;
end process;
end Behavioral;
```

4. Simulation and experimental results

In this section, the designed PID controller is tested using the software and then experimental studies are carried out for a motion control application of a DC brushed servo motor. The software simulation was performed using Matlab Simulink and ModelSim. Both software applications can run with shared memory in order to perform the cosimulation process.

The first study consists of two control loops as illustrated in **Figure 4**. Both control loops have the same input and plant to be controlled, but the first one is implemented using a Simulink PID block (software implementation), whereas the second one is obtained using the VHDL implementation as described in the previous section (hardware implementation). The aim of this study is to compare the performance between the software- and hardware-based implementations.

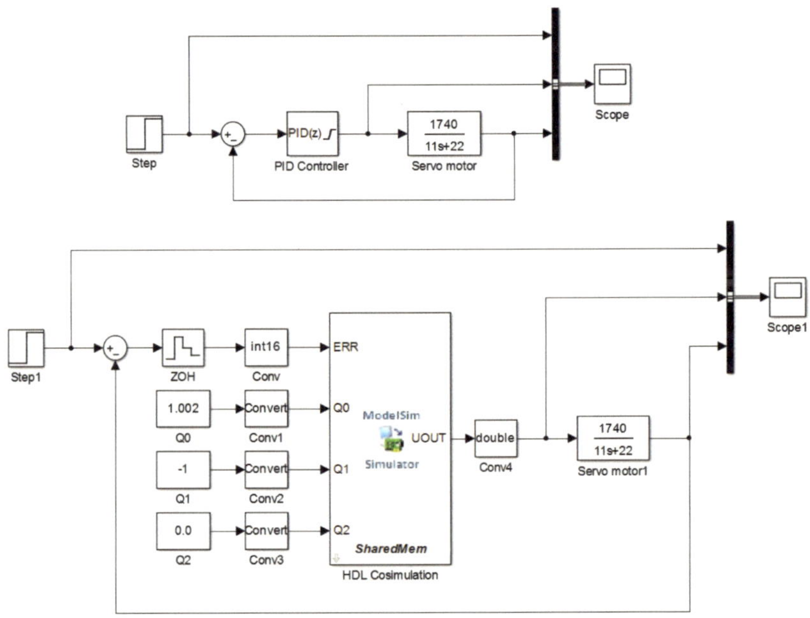

Figure 4. Simulink model for the test of the PID controller.

Before proceeding with the cosimulation, it is important to remark some differences between both PID implementations. The software implementation utilizes a floating point data type with double precision; also, it includes filtering algorithms to compute the derivative and integral term. On the other hand, the hardware implementation described above utilizes the backward rectangular method (BRM) to compute the integral and a simple two-point differentiation for the derivative term; furthermore, the data type utilized has a fixed point format.

The response of the tested control loops is depicted in **Figure 5**. There are three aspects to be considered from the output response. The first one is that the starting angle of the response for the case of the software implementation is higher than the hardware implementation, that is, the response of the software implementation control loop is faster. Second, the first plant reaches the setpoint, relatively faster than the second one, which proves the first assumption.

Finally, the control output of the second loop (hardware implementation) is noisy when compared with the software implementation.

Despite of the aforementioned difference, the performance of the hardware implementation could be acceptable for the control applications in industrial environments. Such kind of an application is described below.

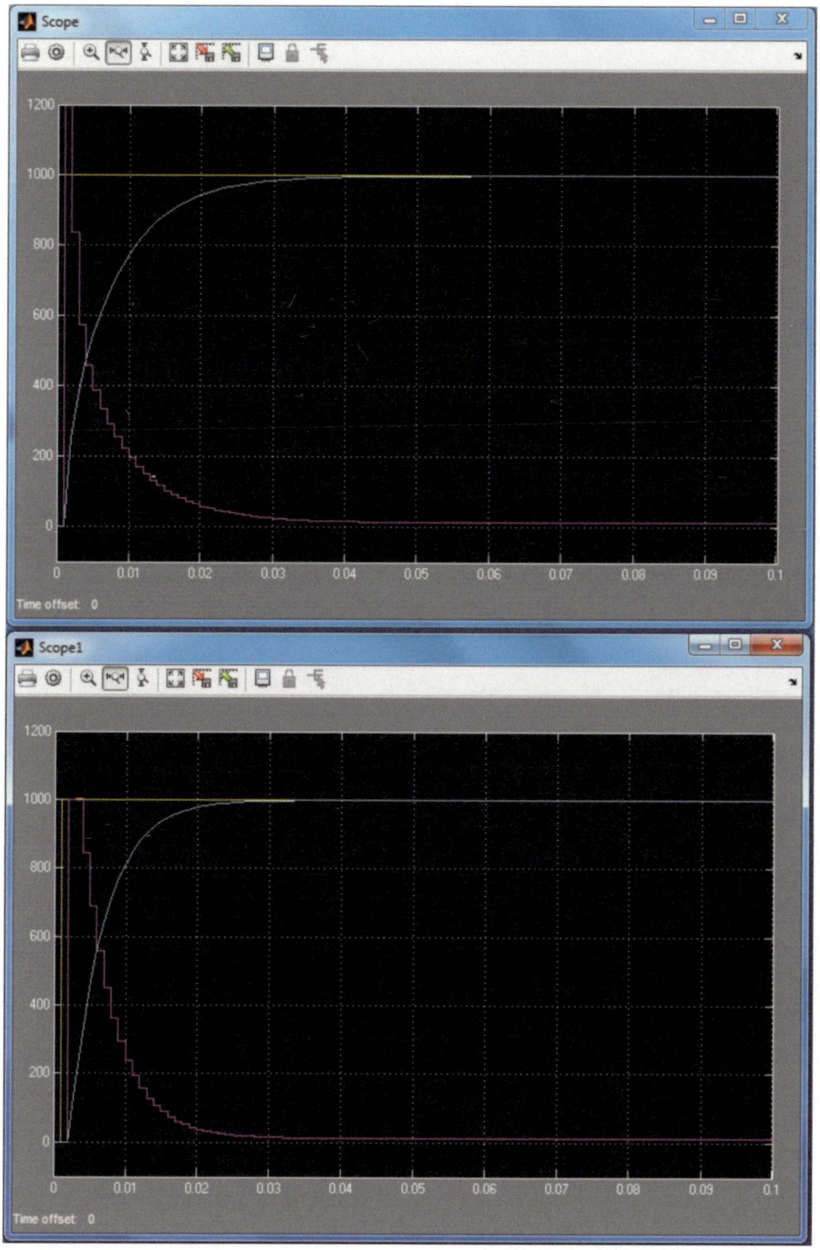

Figure 5. Simulation result of both PID implementations: Simulink (top) and VHDL (bottom).

The application was designed to perform the motion control of a three-degree of freedom robotic arm as illustrated in **Figure 6**. This robot is actuated by 12 V brushed DC motors combined with a 171.79:1 metal spur-gearbox and it has an integrated 48 counts per revolution (CPR) quadrature encoder on the motor shaft, which provides 8245.92 counts per revolution.

A power amplifier stage has been included, which consists of three Texas Instruments LMD18245 power amplifiers required to drive and control the current for the servo motors. These amplifiers can operate with an analogue current reference. For this reason, a digital to analogue converter (DAC) is required. For this application, the Analog Devices AD5668 is utilized, which has eight analogue outputs with a resolution of 16 bits.

The Servo_Controller module is mainly composed of a point of sum, a Digital_PID_Controller, a DAC-Driver, and an Encoder_Quadrature_Interface as shown in **Figure 7**. The reference position (REF) signal is treated as a 32-bit register whose value can be written directly from the PS. On the other hand, the motor position is measured using the Encoder_Quadrature_Interface module. This module uses the signals CHA and CHB from the encoder as the input, and it generates an output signal POS which has a width of 32 bits. Both signals REF and POS

Figure 6. Experimental setup.

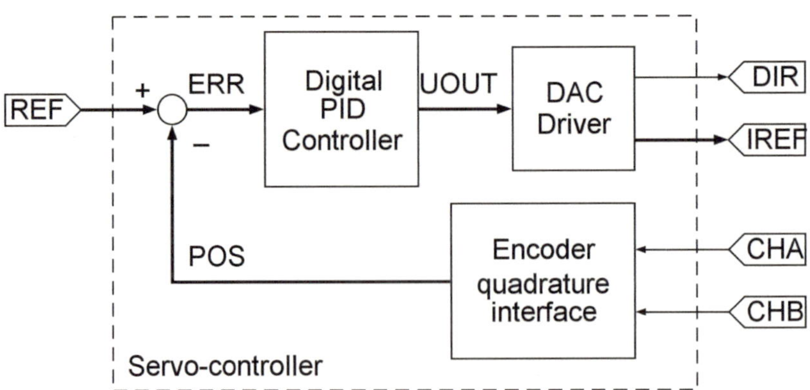

Figure 7. Servo controller block diagram.

http://dx.doi.org/10.5772/67200

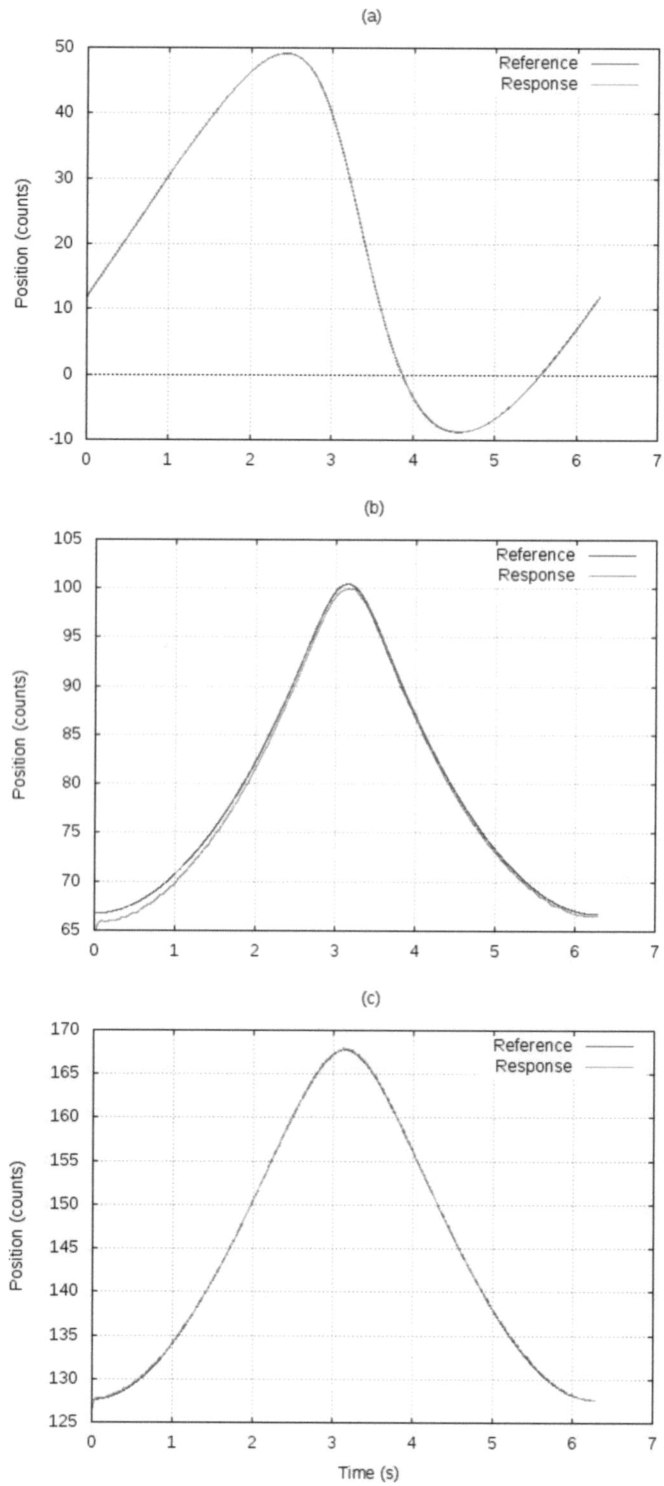

Figure 8. Response of the system to the trajectory commanded: (a) joint 1, (b) joint 2, and (c) joint 3.

are subtracted and trimmed to 16 bits to avoid any saturation in further computations. In **Figure 7**, this signal is labeled as ERR, which serves as the input to the Digital_PID_Controller.

A circular trajectory is consideredin order to evaluate the controller performance. The center of the circumference with a radius of 50 mm is located at (110.0, 0.0, 70.0) being the Z-coordinate constant through the whole movement, i.e., the entire movement is carried out only in the *X*-*Y* plane. First, the interpolation process generates each point along the circumference. This point is used to solve the inverse kinematics. Thus, the resulting angle set is converted to encoder counts and written to the setpoint registers of each servo controller. The sampling time for the generation of each point is chosen as $T_s = 0.001$ s.

Figure 8 shows the response of each servo controller to the generated path. It can be clearly seen that the first and the third joints closely follow the reference trajectory, whereas the second joint shows a larger variation from the commanded trajectory. This could be due to the influence of nonlinear dynamics of the servo motor or due to the gravity force. However, the Cartesian position of the robot remains very close to the commanded trajectory as can be seen in **Figure 9**.

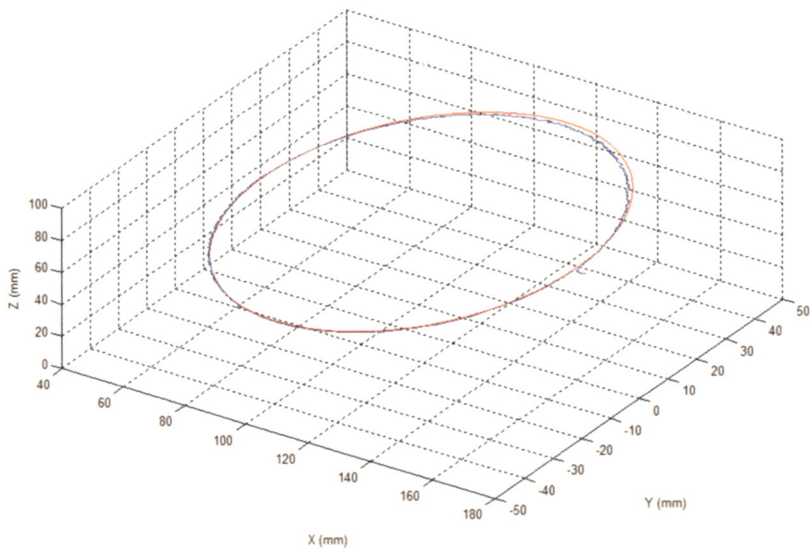

Figure 9. Cartesian response of the system.

5. Conclusions

The endeavor of this chapter is how to deal with theoretical and practical issues regarding the control systems. The readers who are not familiar with motion control systems can get a basic knowledge into reconfigurable logic circuit-based digital design. Generally, it is necessary to combine classroom and practical concepts; for that reasons, these sections are not only aimed for the students or professors but also for professionals who want to obtain a basic understanding about the closed-loop control design.

http://dx.doi.org/10.5772/67200

For educators, many concepts can be applied in courses as servo systems, programming, classical control, and digital control, to mention a few. Since FPGA technology is almost available in all engineering schools, there is no restriction to apply the code shown in this manuscript. In addition, sequential devices might be used too. It is due to the facility to translate HDL code to C code. In this sense, microcontrollers, digital signal processors, and digital signal controllers get a good approach to make a motion control task.

It is recommended to use standard compilers and hardware tools that do not demand high computational resources. It is because of the synthesis stage, which is often the hard part of the developing work.

Author details

Miguel Angel Martínez Prado, Juvenal Rodríguez Reséndiz*, Diana Carolina Toledo Pérez, Carlos Miguel Torres Hernández and Gilberto Herrera Ruiz

*Address all correspondence to: juvenal@uaq.edu.mx

Facultad de Ingeniería, Universidad Autónoma de Querétaro, Cerro de las Campanas SN, Col. Las Campanas, Querétaro, Qro, México

References

[1] Monmasson E, Cirstea M N. FPGA design methodology for industrial control systems - A review. IEEE Transactions on Industrial Electronics. 2007;**54**:1824–1842. DOI: 10.1109/TIE.2007.898281

[2] Zhao W, Kim B H, Larson A C, Voyles R M. FPGA implementation of closed-loop control system for small-scale robot. In: ICAR '05. Proceedings, 12th International Conference on Advanced Robotics, 2005; IEEE Xplore; July 2005;. pp. 70–77. DOI: 10.1109/ICAR.2005.1507393

[3] Ghosh S, Barai R K, Bhattarcharya S, Bhattarcharya P, Rudra S, Dutta A, Pyne R. An FPGA based implementation of a flexible digital PID controller for a motion control system. In: 2013 International Conference on Computer Communication and Informatics (ICCCI); 04 Jan–06 Jan 2013; Coimbatore, Tamil Nadu, India. IEEE Xplore; 2013; pp. 1–6. DOI: 10.1109/ICCCI.2013.6466277

[4] Xu Y, Zhao J, Huang J. Multiple linear motor control system based on FPGA. In: 2014 17th International Conference on Electrical Machines and Systems (ICEMS); 22 Oct–25 Oct 2014; Hangzhou, China. IEEE Xplore; 2014. pp. 2327–2331. DOI: 10.1109/ICEMS.2014.7013875

[5] Nandayapa M, Mitsantisuk C, Ohishi K. Improving bilateral control feedback by using novel velocity and acceleration estimation methods in FPGA. In: 2012 12th IEEE

International Workshop on Advanced Motion Control (AMC); 25 Mar–27 Mar 2012; Sarajevo, Bosnia and Herzegovina: IEEE Xplore; 2012. pp. 1–6. DOI: 10.1109/AMC.2012.6197024

[6] Bagni D, Mackay D. Floating-point PID controller design with Vivado HLS and system generator for DSP [Internet]. 2013. Available from: http://www.xilinx.com/support/documentation/application_notes/xapp1163.pdf [Accessed: 2016-02-25]

[7] Aboelaze M, Shehata MG. Implementation of multiple PID controllers on FPGA. In: 2015 IEEE International Conference on Electronics, Circuits, and Systems (ICECS); 7–9 Dec 2015; Egypt. IEEE Xplore; 2015. pp. 446–449. DOI: 10.1109/ICECS.2015.7440344

[8] Uzunović T, Žunić E, Badnjević A, Mioković I, Konjicija S. Implementation of digital PID controller. In: 2010 Proceedings of the 33rd International Convention MIPRO; 24–28 May 2010; Opatija, Croatia. IEEE Xplore; 2010. pp. 1357–1361.

Chapter 4

FPGA-Based Software-Defined Radio and Its Real-Time Implementation Using NI-USRP

Nikhil Marriwala , Om. Prakash. Sahu and
Anil Vohra

Additional information is available at the end of the chapter

http://dx.doi.org/10.5772/66272

Abstract

In this chapter, we propose a novel design of scalable and real-time data acquisition software architecture for software-defined radio (SDR) using universal software radio peripheral (USRP). The software has been designed and tested in multi-thread model, using LabVIEW, which guarantees real-time performance and efficiency. With the help of this design, we have been able to improve the stability of the system besides providing a reconfigurable and flexible architecture. Wireless transfer of sensitive data using communication is not a very safe option. In this chapter, we aim to provide a safe and private wireless transmission between two terminals using the SDR approach and verifying the results in real-world environment with the use of USRP. The novel design being presented here can be used to transfer (random data, text or an image) encoded with different forward error correction (FEC) codes, which is then verified at the receiving terminal and then decoded accordingly to produce the desired result.

Keywords: software-defined radio, universal software radio peripheral, forward error correction codes, biterrorrate (BER), signal-to-noise ratio (SNR), LabVIEW

1. Introduction to SDR

Software-defined radio (SDR) systems are those which can adapt to the future-proof solution and they cover both existing and emerging standards. An SDR has to possess elements of reconfigurability, intelligence and software programmable hardware [1]. As the functionality is defined in software, a new technology can be easily implemented in a software radio by means of a software upgrade. Channel equalization is an important subsystem in the software defined radio's receiver. For many years, modulation techniques have been extensively used for various wireless applications, but the modern communication system requires data transmitted at a higher rate and larger bandwidth [2].

This chapter discusses an SDR system built using LabVIEW for a generic transceiver. SDR provides an alternative to systems such as the third-generation (3G) and the fourth-generation (4G) systems. There are two frequency bands where the software-defined radio might operate in the near future, that is, 54–862 MHz [very high-frequency (VHF) and ultra-high frequency (UHF) TV bands] and 3–10 GHz [ultra-wideband (UWB) radios]. A software-defined radio comprises a programmable communication system where functional changes can be made by merely updating the software. SDRs can be reconfigured and can talk and listen to multiple channels at the same time. Normally, high-performance digital signal processors (DSPs) are used to serve as the baseband processor. SDR systems can be used in ubiquitous network environments because of its flexibility and programmability [3]. The use of digital signals reduces hardware, noise and interference problems as compared to the analogue signal in transmission, which is one of the main advantages of digital transmission.

In this chapter, the software simulator of the SDR transceiver has been designed using LabVIEW and has been tested in real time using the universal software radio peripheral (USRP). Digital modulation schemes namely the frequency shift keying (FSK), phase shift keying (PSK) and quadrature amplitude modulation (QAM) are chosen to be the modulation schemes of the designed software-defined radio system due to its easy implementation and widespread usage in wireless communication equipment. Digital modulation techniques are considered to be very common technology for transmission and reception in current and future wireless communications, especially in the VHF and UHF frequency bands giving excellent bit error rate (BER) versus signal-to-noise ratio (SNR) ratio with high data rates. The role of modulation techniques in an SDR is very vital given that the modulation techniques describe the central part of any wireless technology. They can be reconfigured and can talk and listen to multiple channels at the same time. The role of modulation techniques in an SDR is very crucial since modulation techniques define the core part of any wireless technology. SDR's inherent flexibility must, however, be planned for in advance via hardware and software considerations, ultimately resulting in increased code portability, improved communications system life cycles and reduced costs. The main interest in any communication group is the sure sending of signals of info from a transmitter to a receiver [4]. The signals are transmitted via a guide who corrupts the signal. It is needful that the distorting effects of the channel and noise are minimized and that the information transmitted through the channel at any given time is maximized [5]. The channel is subject to various types of dissonance, twisting and interference. Also, some communication systems have limitations on transmitter power. All of this may lead to various types of errors. Consequently, we may need some form of error control encoding in order to recover the information reliably.

1.1. Related technologies of SDR

In view of the fact that the field-programmable gate array (FPGA)'s prime function is to offer signal filtering and rate conversion from the analogue-to-digital converter (ADC) and to the system's digital-to analogue converter (DAC), its firmware can be customized to meet the particular needs of the user or just downloaded without modification, as a preconfigured file. Some of the USRP family support multi-radio cooperation using multiple-input, multiple output (MIMO) techniques. This is enabled by installing a MIMO interconnecting cable between two USRP devices [6].

To ensure reliable communication, forward error-correcting (FEC) codes are the main part of a communication system. FEC is a technique in which we add redundant bits to the transmitted data to help the receiver correct errors. There are two types of FEC codes: the convolutional codes and block codes. When we use block codes, they are defined by n and k, where n describes the total number of coded bits and k gives the number of input bits. In convolutional codes, the coding is applied to the entire data stream as one code word. In the year 1948, Shannon showed that arbitrarily reliable communication is only possible till the signal transmission rate does not exceed a certain limit which was termed as channel capacity [7]. After this different algebraic codes such as Golay codes, Bose-Chaud huri-Hocquenghem (BCH) codes and Reed-Solomon (RS) codes were created and used for error correction. The next series of codes originally referred as recurrent codes or convolutional codes were given which helped further to improve the error control coding [8]. The convolutional codes have efficient encoding and decoding algorithms and high performance over additive white Gaussian noise (AWGN) channels. Later on, concatenated-coding schemes were also given. Also, some weak points were there of convolutional codes during bursty transmissions which were later on reduced using Reed-Solomon codes by serially concatenating a convolutional code with an RS code. Development of turbo codes is the most recent discovery in the coding theory. Turbo codes show performance of near to Shannon limit with iterative decoding algorithms. Many iterative decoding algorithms came into existence such as Gallagher's low-parity density check (LDPC) codes [9]. Though these turbo codes exhibit excellent bit error performance but there are some problems associated with them such as these codes generate a certain number of low-weight code words which results in exhibition of an error 'floor' in the BER curve at high SNR. Also, the complexity of the soft-input, soft-output (SISO) decoder is such that low-cost decoders are unavailable for many commercial applications [10]. For these reasons, many applications still deploy RS codes because of its efficient decoder implementation and excellent error correction capabilities.

1.2. The benefits of multi-standard terminals

A multi-standard terminal (MST) is a subscriber unit that is capable of operation with a variety of different mobile radio standards. Although it is not strictly necessary for such a terminal to be implemented using software-defined radio techniques, it is likely that this approach is the most economic in many cases [11].

1.2.1. Economies of scale

Even if terminal adaption over the air or via third-party software was not possible or was not permitted by, for example, regulatory bodies, the production benefits of a software-defined radio approach could well justify its existence. The wide range of new and existing standards in the cellular and mobile marketplace has resulted in the adoption of a diverse range of subscriber terminal (and base-station) architectures for the different systems deployed around the globe [12].

1.2.2. Global roaming

The present proliferation of mobile standards and the gradual migration to third-generation systems means that a large number of different network technologies will exist globally for some time to come. Indeed, even in the case of 3G systems, where a concerted effort was made by international standards bodies to ensure that a single global standard was produced, there are still significant regional differences, in particular between the US and European offerings (and also, potentially, China). With this background, it is clearly desirable to produce a terminal which is capable of operation on both legacy systems and the various competing 3G standards. Indeed, it could be argued that this is the only way in which 3G systems will be accepted by users, since the huge cost of a full-coverage network roll-out will discourage many operators from providing the same levels of coverage (at least initially) as their existing 2G systems enjoy. A user is unlikely to trade in his or her 2G terminal for one with perhaps better services, but significantly poorer basic voice coverage [13]. This is indeed what is happening in virtually all current 3G deployments. Software-defined radio architecture represents a very attractive solution to this problem.

1.2.3. Upgrading the service

A powerful benefit of a software-defined radio terminal, from the perspective of the network operator, is the ability to download new services to the terminal after it has been purchased and is operational on the network. At present, significant service upgrades require the purchase of a new terminal, with the required software built-in and this clearly discourages the adoption of these new services for a period. The launch of general packet radio service (GPRS) data services on the GSM network is a good example of this. With SDR handset architecture, services could be downloaded overnight, when the network is quiet, or from a website in the same manner as personal computer (PC) software upgrades are distributed [12]. There are clearly a number of logistical issues with this benefit (e.g., what to do about phones which are turned off at the time of the upgrade or what happens if a particular phone crashes with the new software, perhaps just prior to requiring the phone for an emergency call—software which the phone user may not have wanted and so forth). Many of these problems have been solved by the PC industry and hence it is likely that this benefit will be realized in some manner with software-defined radios.

1.2.4. Adaptive modulation and coding

The ability to adapt key transmission parameters to the prevailing channel or traffic conditions is a further key benefit of a software-defined radio. It is possible, for example, to reduce the complexity of the modulation format, such as from 16-QAM (quadrature amplitude modulation) to quadrature phase-shift keying (QPSK) when channel conditions become poor, thereby improving noise immunity and decoding margin. It is also possible to adapt the channel-coding scheme to better cope with particular types of interference, rather than Gaussian noise,

http://dx.doi.org/10.5772/66272

when moving from, say, a rural cell to an urban one [14]. Many parameters may be adapted dynamically, for example, burst structure, modulation type, data rate, channel and source coding, multiple-access schemes and so forth.

1.3. Operational requirements: various operational requirements for SDR are as stated below:

1.3.1. Key requirements

The operational characteristics of an ideal multi-standard terminal include the following operations.

1.3.1.1. Software-definable operation

As outlined earlier, the key to many of the advantages of a multi-standard terminal lies in its ability to be reconfigured either during manufacture, prior to purchase, following purchase (e.g., after-market software), in operation (e.g., adaptation of coding or modulation), or preferably all four. This impacts primarily upon the digital and baseband sections of the terminal and will require the use of reprogrammable hardware as well as programmable digital signal processors in a power and cost-effective implementation.

1.3.1.2. Multi-band operation

The ability to process signals corresponding to a wide range of frequency bands and channel bandwidths is a critical feature of an MST. This will impact heavily on the radio frequency (RF) segments of the terminal and it is this area which is arguably the main technology limitation on software-defined radio implementation at present [15] (although processor power consumption and cost are still both major issues for SDR).

1.3.1.3. Multi-mode operation

Many multi-mode software-defined radios already exist, although they are often not promoted as such (since the other features/benefits of software-defined radio techniques are not exploited). The ability to change mode and, consequently, modulation, coding, burst structure, compression algorithms and signalling protocols is clearly an essential feature of an MST.

1.4. Digital aspects of a software-defined radio

1.4.1. Digital hardware

There exists a range of solutions to the digital-processing problem for a software-defined radio, each with its own characteristics and application areas. The digital-processing area is, in many respects, as challenging as the analogue processing described in detail in this book and the intention of this section is merely to highlight the options and their main characteristics. The two biggest issues at present are the power consumption and cost of the various options. In a base-station application, these are less of an issue (but are still a significant challenge); however, they are perhaps the main inhibitor to the widespread use of software-defined radio in handsets and other portable devices:

1. The use of reconfigurability as a method of providing upgrading, improvement or backwards compatibility (i.e., a smooth transition from a legacy system) is, however, a strong argument for flexible processing and SDR concepts. It is in this context that the processing options outlined in the following will be discussed.

Cost is also a multi-faceted issue. Most designs judge cost based almost exclusively on the cost of the target device used for the code (be it a processor or an FPGA). In the case of a very high-volume application (e.g., a handset), this might be a reasonable approach, although even here it could be somewhat short-sighted. In the case of a base-station design, however, there are many other considerations that will determine the overall cost of a design (particularly if lifetime cost is considered and not just purchase cost). As a summary, the factors that influence the cost of the digital elements of an SDR BTS include

a. Direct cost of the processing device itself.

b. Costs involved in the associated ancillary and interfacing devices (e.g., memory, clock circuitry and so forth).

c. Non-recurring expense (NRE). This is most obviously associated with application-specific integrated circuit (ASIC) or application-specific signal processor (ASSP) designs and includes mask-set costs, fabrication and so forth. These costs are rising dramatically as feature sizes reduce and are therefore making the break-even volume (compared to, say, FPGAs) much higher as time progresses.

d. Tools/training investment. Changing from one digital technology to another (e.g., from DSPs to FPGAs) may well involve a significant change of design personnel, or at the very least a degree of retraining. This will have an associated cost and also an opportunity cost as the time to market will be increased (see the following). Even changing from one manufacturer's processors to another may involve a loss of productivity while the development team familiarizes itself with a new feature set and the new tricks required to get the best out of a particular device.

e. Cooling. The cost of cooling can undergo step changes as the form of cooling required changes. The most obvious example is in going from convection cooling to forced-air cooling, with the cost of the fans now needing to be added to the bill of materials. Additional power consumption will also add to the cost of the power supply, although with modern switched-mode designs, this is usually small. It is, however, a much bigger issue in handset designs due to the increased requirements it places upon the battery and the user-acceptance issues of large batteries or reduced talk times.

f. Development time/resource. This is becoming an increasingly important aspect of cost, as product life cycles, even of base-station designs, reduce as each new design appears. The volume of units sold of a particular design is then lower and the cost of producing that design becomes an ever-larger proportion of its selling price. Techniques or architectures, which allow these designs to be generated quickly, or significant portions of designs to be reused between evolutionary models in a range (as well as across models in a given range), are clearly attractive, even if the devices upon which they are based are not the lowest-cost components available.

g. Flexibility. This is a benefit in terms of time to market for new products and hence a benefit in terms of opportunity cost. If full flexibility could be provided for the same cost as a fixed solution (e.g., a single-application ASIC), then it would be a simple decision to adopt a flexible approach. This is almost never true and hence a full business case must be developed for flexibility, in a given marketplace and each opportunity judged on its merits.

1.4.1.1. Digital signal processors

DSPs were arguably the original enabling technology for software-defined radio (other than perhaps in military circles where cost is less important). They have the advantage of complete flexibility, wide applicability and a wide availability of skilled practitioners in their software. They are also high-volume devices and hence the benefits of economies of scale may be realized across a large number of applications in a wide range of industries (not just wireless communications). This, in general, makes up for their lack of optimization for a given specific project or niche application area and allows them to be a realistic option for early prototyping and initial production volumes of a new design, as well as for the final volume product, in some cases.

They are best suited to the less computationally intensive forms of signal processing, rather than very high-speed front-end applications. They are often utilized for involved, off-line processing of data which has been acquired and undergone initial processing/storage by a different type of device (e.g., an FPGA or an ASIC) [16]. They are, however, well supported and also tend to come in backwards-compatible families, which allow development to take place on a state-of-the-art (SOTA) device, with the final application device being lower cost. This generally occurs for two reasons:

1. The SOTA device being used in development will not be SOTA by the end of the development cycle and hence will generally have reduced in cost. The volume of usage of the device will also have increased, which will also help to reduce its cost.

2. Developers tend to pick a device for their development systems which is definitely large enough to meet the requirement in question. It is often the case that once development is nearing completion, the design will have been optimized such that it may be executed in a lesser member of the same device family. This will have an associated cost benefit. In larger systems, it may be the number of devices that can be reduced; however, this will still result in a lower overall cost.

1.4.1.2. Field-programmable gate arrays

FPGAs have undergone a revolution in recent years, both in performance and in cost. From humble beginnings as simple, flexible glue logic in complex digital designs, they are now a credible processing platform in their own right and able to rival ASIC solutions in many areas (and act as a low-cost prototyping mechanism for ASIC designs). They have also undergone a revolution in volume pricing, which means that they are no longer consigned to the prototype and initial volume parts of the product life cycle, but can now be used throughout volume production, in some applications [17]. It is also possible to convert from an FPGA to a quasi-ASIC, with a highdegree of confidence of success and a relatively low NRE (and hence break-even volume). FPGAs are therefore challenging and displacing ASICs in traditional ASIC application areas.

Furthermore, they provide much more flexibility than can be cost-effectively built into an ASIC, thereby fitting with the requirements of SDR very well. In common with DSPs, they also tend to come in families, thereby, again, allowing an initial design to take place on a large (potentially over specified) device with the final device being chosen to just fit the processing requirement. It is also possible (but not necessarily economic) to add IP processor cores into an FPGA (or an FPGA-derived ASIC). This makes possible a single-chip solution in some applications and this may be important for size or reliability reasons (with the improved reliability coming from the reduction in devices and soldered joints).

1.4.1.3. ASICs

The main issue with utilizing ASICs (or, more correctly, application-specific signal processors) within an SDR system lies in their lack of flexibility (or conversely the cost of adding flexibility). There are many methods by which flexibility may be introduced within an ASSP and these include the following:

a. Provision of multiple toolbox functions with flexible input parameters. An example would be a QAM modulator that had an input variable to configure it from 16 to 256 QAM, for example [18].

b. Provision of hardware for all current modulation formats, coding schemes and so forth in a single (large!) ASSP, with the ability to select between the different paths. This is not strictly flexible in the generic sense; however, it is flexible in its range of functionality—the user will not care how he is provided with service over a range of standards, just that he obtains service at a low cost. The major disadvantage with this option is that it is not really future-proof, unless the system designer has an extraordinary insight into the future trend in mobile communications (and can convince his or her management that he or she is right).

c. A combination of one or both of the above with some programmable DSP functionality (e.g., using an embedded DSP core). The key here is in providing enough DSP power to be useful and provide a degree of future-proofing, without designing essentially a DSP device—it would almost certainly be lower cost to buy an off-the-shelf DSP device from a volume vendor.

1.5. About NI-USRP

The NI USRP-2922 can be used for various communication applications, including

http://dx.doi.org/10.5772/66272

a. WiFi

b. WiMax

c. S-band transceivers

d. 2.4-GHz ISM band transceivers.

The NI USRP-2922 as shown in **Figure 1** has the following basic characteristics:

i. Frequency range of 400 MHz to 4.4 GHz

ii. Tunable RF transceiver

iii. Transmits and receives bandwidth up to 40 MHz

iv. High-speed ADC and DAC for streaming baseband I and Q signals to a host PC over gigabit Ethernet.

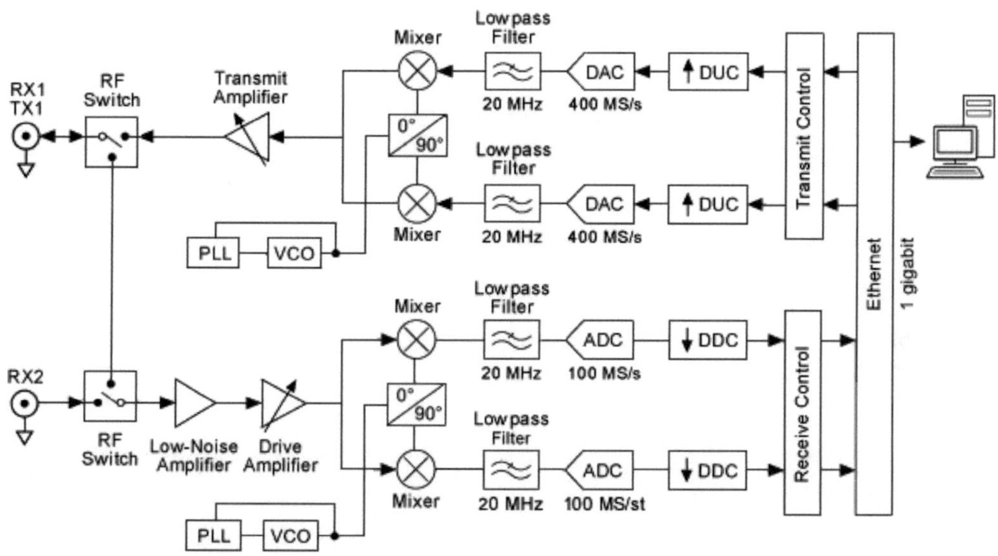

Figure 1. Block diagram of NI USRP-2922.

The RF switch allows transmit and receive operations to occur on the same shared antenna. On the NI USRP-2922, one antenna is designated receive-only.The NI USRP is also capable of receiving the signal, where the received signal is mixed down from RF using a direct-conversion receiver to baseband I/Q components. The digitized I/Q data follows parallel paths through a digital down-conversion (DDC) method that mixes, filters and decimates the input signal to a user-specified rate. Gigabit Ethernet connection is used to pass the down-converted samples to the host computer.

1.5.1. Receive trail

- **i.** The received signal is amplified using the low-noise amplifier and drive amplifier amplify.
- **ii.** The voltage-controlled oscillator (VCO) is controlled by the phase-locked loop (PLL) so that the device clocks and local oscillator (LO) can be frequency-locked to a reference signal.
- **iii.** The mixer down-converts the signals to the baseband in-phase (I) and quadrature-phase (Q) components.
- **iv.** To reduce the noise and high-frequency components in the signal, low-pass filter is used.
- **v.** The analogue-to-digital converter digitizes the I and Q data.
- **vi.** The digital down-converter (DDC) mixes, filters and decimates the signal to a user-specified rate.
- **vii.** A standard gigabit Ethernet connection is used to pass the down-converted samples.

1.5.2. Transmit trail

- **i.** The host computer synthesizes baseband I/Q signals and transmits the signals to the device over a standard gigabit Ethernet connection.
- **ii.** The digital up converter (DUC) mixes, filters and interpolates the signal to 400 MS/s.
- **iii.** The digital-to-analogue converter converts the signal to analogue.
- **iv.** The low-pass filter reduces noise and high-frequency components in the signal.
- **v.** The mixer is used to up-convert the received signals to a user-specified RF frequency.
- **vi.** The PLL controls the VCO so that the device clocks and LO can be frequency-locked to a reference signal.
- **vii.** The transmit amplifier amplifies the signal, which is then transmitted through the antenna.

1.6. Design of a generic transceiver for FPGA-based SDR

In this section, the building blocks of a generic transceiver for FPGA-based SDR modem system built in LabVIEW have been explained. The designed system consists of two parts: the transmitter section and the receiver section. The transmitter section consists of four modules which are a message source module, pulse-shaped filter module, QAM modulator module and Gaussian noise module. The receiver has been designed using an adaptive filter module, QAM demodulator module, sync and tracking module. A brief description of each block follows. The front panel of generic transceiver for SDR modem system is shown in **Figure 2**.

http://dx.doi.org/10.5772/66272

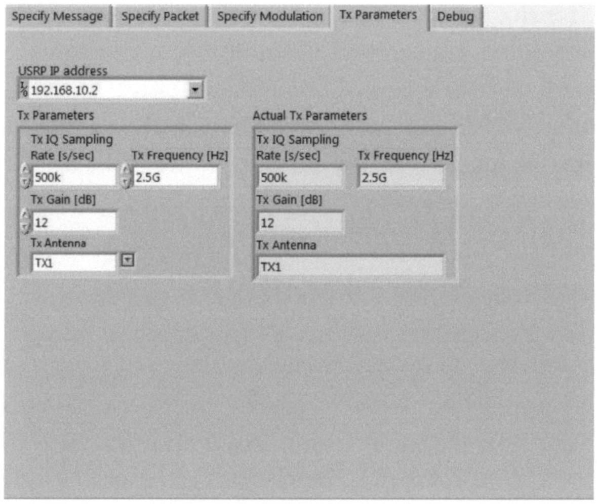

Figure 2. Front panel of transmitter, Tx parameter.

In **Figure 2**, under Tx parameter is used to select the parameters such as IQ sampling rate, Tx frequency, Tx gain and so on of the transmitter. In the **USRP, IP address one s**elects the IP address of the transmitter device such as 192.168.10.2. In the **Tx IQ sampling rate [s/s] field, we must s**et the sampling rate to 500 k for text, random data and 2M for image transmission. **Tx frequency [Hz] is used to s**et the frequency to 2.5 G (the user can use any frequency from 400 MHz to 4.4 GHz for USRP2922). In the **Tx gain [dB]** field, we enter Tx gain 12 and we can take upto 31. Now, one need to select the Tx antenna as Tx1. In the specify message window in the transmitter front panel, we can select the type of the file to be transmitted;it consists of three files: random data, text and image as shown in **Figure 3**.

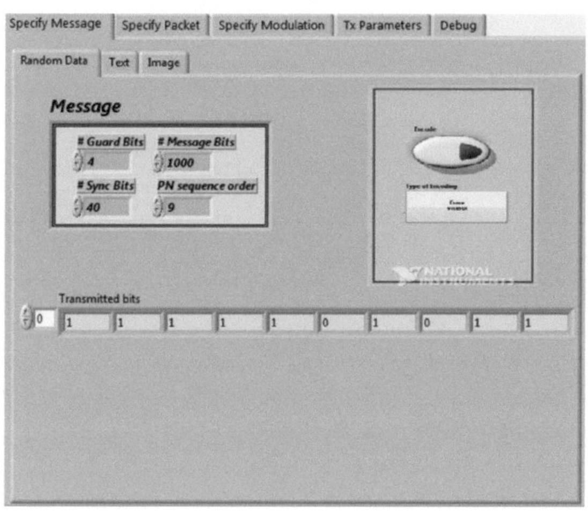

Figure 3. Specify message/random data window.

Random data can be transmitted by selecting the random data field. The message field is set to set the data to be transmitted. Transmitted bits indicate the bits which have to be transmitted. Encode button is meant to encode the transmitted data. Type of encoding is used to select the type of encoding required either convolution viterbi or turbo coding. Similar to the random data, one can select text or image window in order to transmit text or image, respectively. **Figures 3** and **4** show text and image window, respectively.

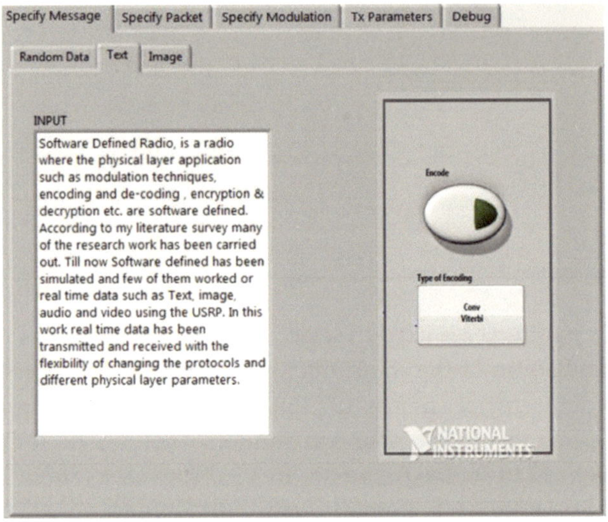

Figure 4. Specify message/text window.

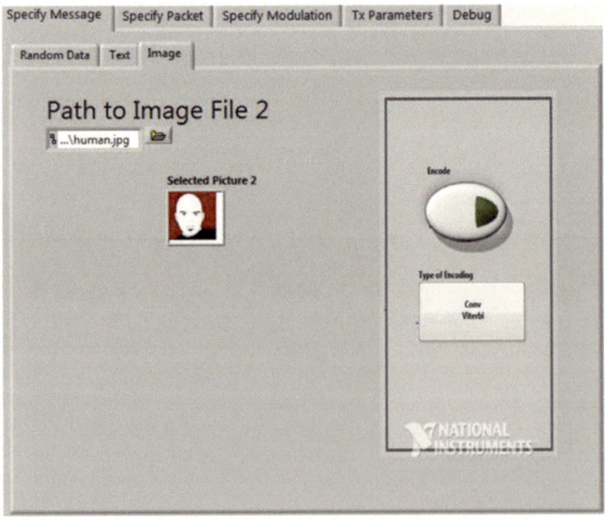

Figure 5. Specify message/image window.

http://dx.doi.org/10.5772/66272

We can write the text to be transmitted in the input window as shown in **Figure 4**. Encode button is used to provide encoding scheme. One needs to press the encode button to encode the transmitted data. The type of encoding field is used to select the type of encoding required either convolution viterbi or turbo coding as shown in **Figure 4**. Similarly, one can transmit the image as shown in **Figure 5**. We need to select the path or browse the path of the image and the selected picture is indicated in the front panel.

Pulse-shaping filter parameters field is used to set the system filter parameters as shown in **Figure 5**:

a. Tx filter: Used to select the type of filter required raised cosine filter or root raised cosine or none.

b. Alpha: Set the alpha to 0.50.

c. Filter length: Set filter length to 6.

Samples per symbol: Set samples per symbol to 8.

Symbol rate [symbols/s]: It indicates the symbol rate **M-FSK:** Set the FSK parameters as shown in **Figure 5**.

a. M-FSK: Set M-FSK to 4 (can be changed).

b. FSK deviation [Hz]: Set FSK deviation to 25.00 k.

c. Symbol phase continuity: Select symbol phase continuity to continuous.

M-PSK: Set M-PSK to 4. **M-QAM:** Set M-QAM to 8. Now, select the specific packets window from the transmitter front panel to set the message bits as shown in **Figure 6**.

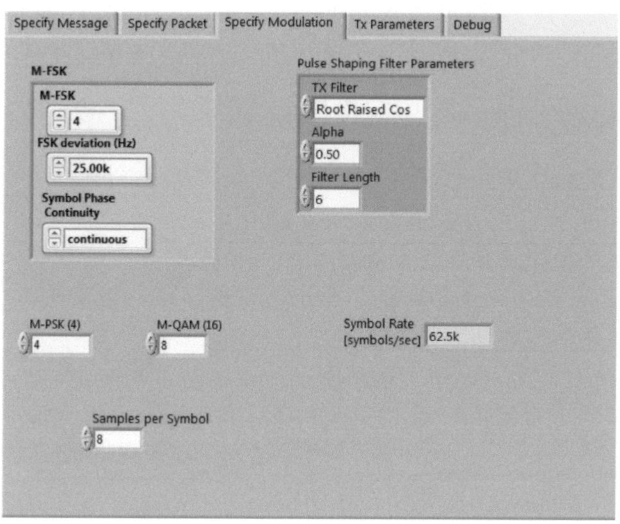

Figure 6. Specify modulation window.

Specify packet window is used to set the packet parameters such as guard bits, sync bits, message bits and packed pad (samples). Message bits need to be set 128 bits for random data, text and 4096 bits for image. The PN sequence order for sync bits is also to be specified in the window as shown in **Figure 7**.

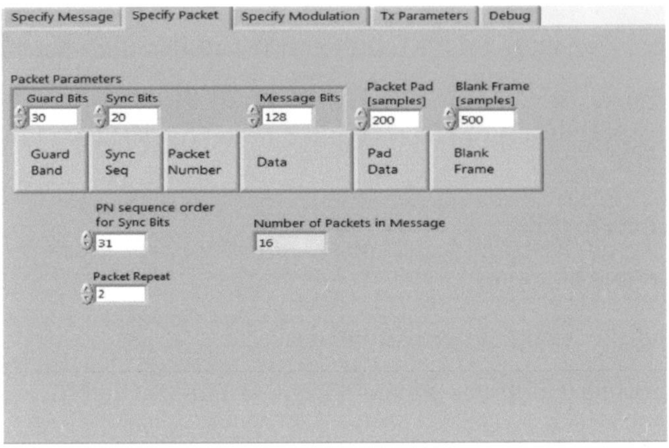

Figure 7. Specify packet window.

Now at the receiver end we need to receive the sent message for that the receiver must be set to certain parameters as shown in **Figure 8**. The front panel of the receiver looks as shown in **Figure 8**; select Rx parameter window, specify the IP address, Rx parameter sand acquisition duration [s].

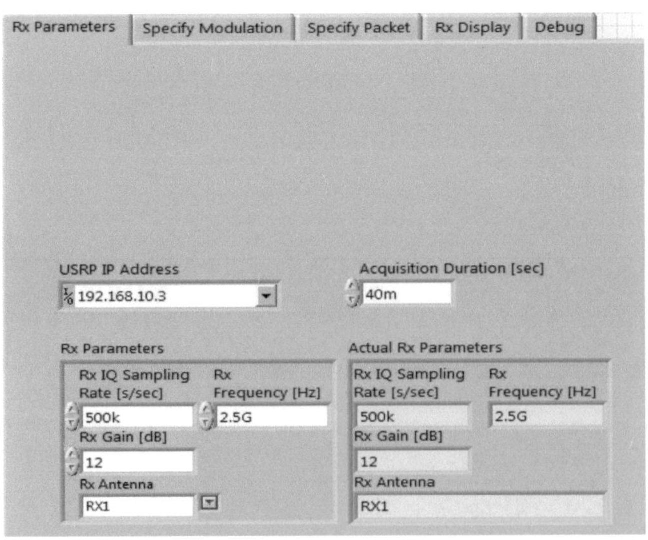

Figure 8. Front panel of USRP receiver/Rx parameter.

http://dx.doi.org/10.5772/66272

Rx parameter is used to select receiver Rx parameter. USRPIP address for the receiver is specified in the IP address of the receiver device. Rx IQ sampling rate [s/s] is set to 500k to provide the required sampling rate. Rx frequency [Hz] is used to set the frequency to2.5G. Rx gain [dB] is set to12, same as gain for the Tx. Rx antenna field is used to select the Rx antenna as Rx1. In the field acquisition duration [s], set duration 40 m for random data and text, 56 m for image. Now one must select Rx display as per selected specify message in the transmitter. If text is selected at transmitter, then text should be selected at receiver similar to random data and image as shown in **Figures 9–11**.

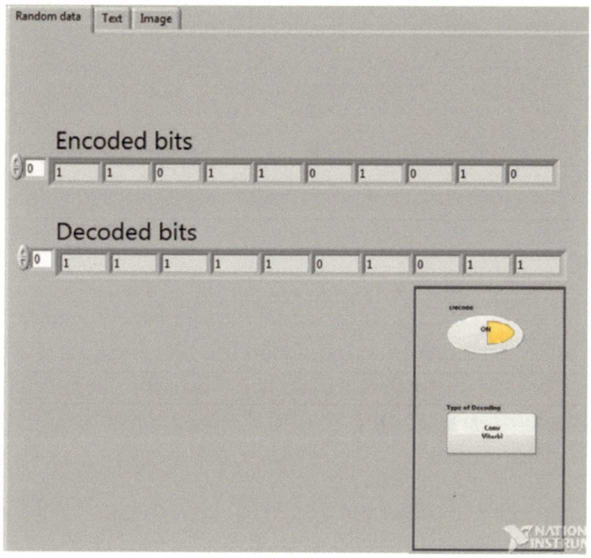

Figure 9. Rx display/random data Rx.

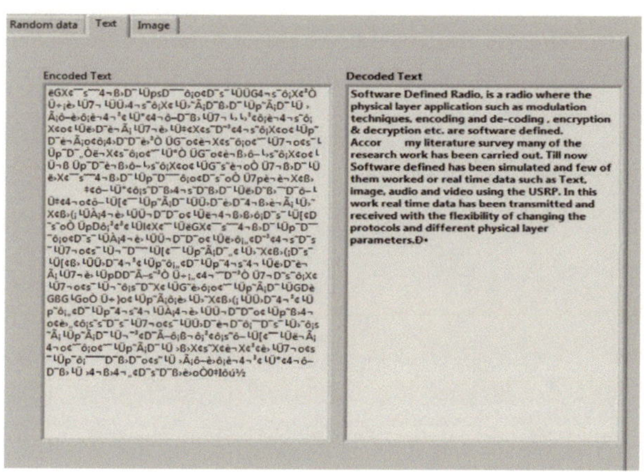

Figure 10. Rx display/text Rx.

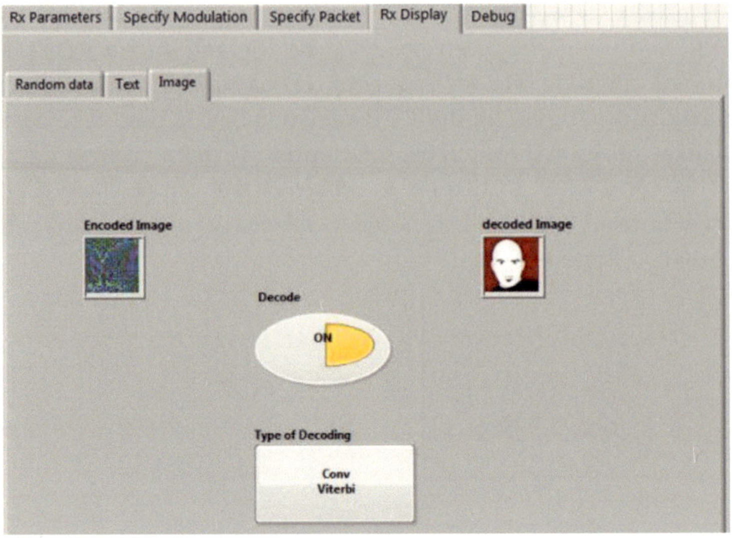

Figure 11. Rx display/image.

Random data: Click on random data in order to receive random data. Encoded bits in this window indicate the bits which are encoded. Decoded bits indicate the bits that are decoded from the transmitted data. Decode button is used to decode the received data. Type of decoding field is used to select the type of decoding required either convolution viterbi or turbo coding which is selected at the transmitter.

For receiving the text, we need to select the text field. Encoded text field indicates the encoded text. Decoded text field indicates the decoded text.

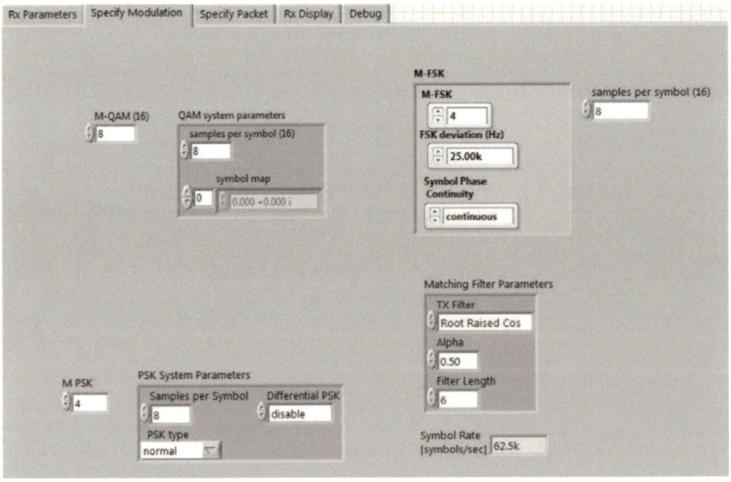

Figure 12. Specify demodulation window.

http://dx.doi.org/10.5772/66272

For receiving the image, we need to select the image field. Encoded image field indicates the encoded image. Decoded image field indicates the decoded image. Decode button is used to decode the received data. Type of decoding needs to be set to either convolution viterbi or turbo coding which is selected at the transmitter. Now, select the specify demodulation window, as shown in **Figure 12**, all the parameters to be set exactly the same as transmitter modulation.

Matching filter parameters: Set the matching filter parameters as shown in **Figure 12**:

a. Tx filter: Select the type of filter required raised cosine filter or root raised cosine or none.

b. Alpha: Set the alpha to 0.50.

c. Filter length: Set filter length to 6.

Samples per symbol: Set samples per symbol to 8. **Symbol rate [symbols/s]. M-FSK:** Set the FSK parameters as shown in **Figure 13**:

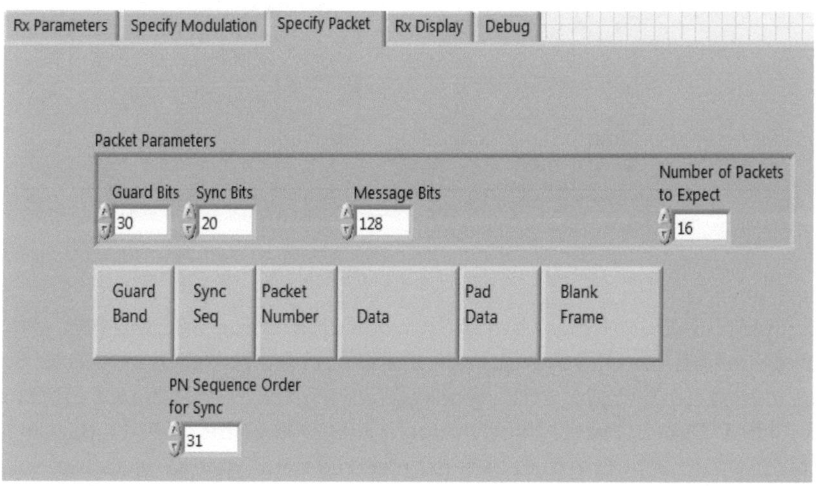

Figure 13. Receiver/specify packet window.

a. M-FSK: Set M-FSK to (any M-ary value)

b. M-PSK: Set M-PSK to (any M-ary value)

c. M-QAM: Set M-QAM to (any M-ary value)

d. FSK deviation [Hz]: Set FSK deviation to 25.00 k

e. Symbol phase continuity: Select symbol phase continuity to continuous.

Now, select the specify packet window from the receiver front panel, as shown in **Figure 13**.

Specify packet specifies the packet same as transmitter, that is, 128 for random data and text, for image 4096. Message bits field holds the number of bits, same as in the transmitter. Number of packets to be expected must be entered in this window which is the same as in the transmitter. Now, select output parameters window, which shows the constellation diagram, eye diagram and BER Vs E_b/N_o with respective received data as shown in **Figure 14**.

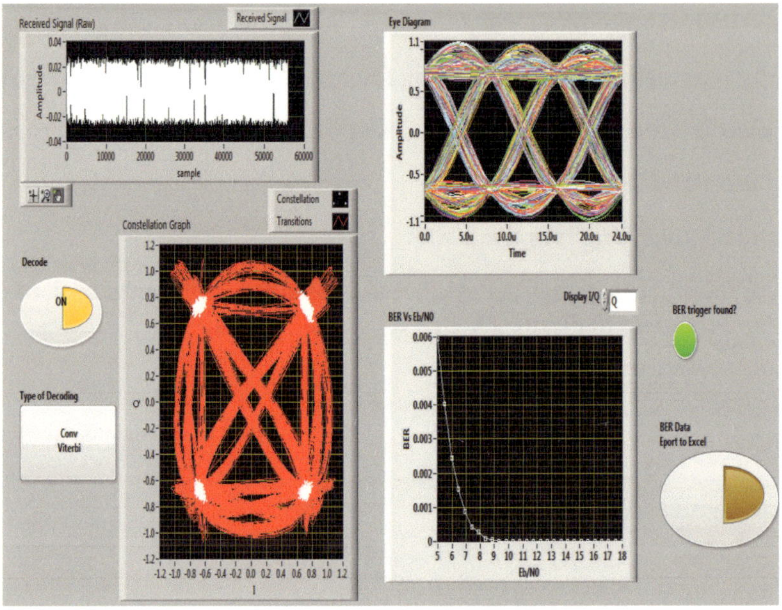

Figure 14. Output parameters.

Constellation graph is used to indicate the constellation of the received bits with respect to data. Eye diagram indicates the eye diagram of the received bits with respect to data. BER Vs E_b/N_o indicates the BER Vs E_b/N_o of the received bits with respect to data. BER Data Export to Excel: Click on BER Data Export to Excel Button it will export BER Vs E_b/N_o data to Excel sheet. Decode: Press the decode button to decode the received data. Type of decoding: Select the type of decoding required either convolution viterbi or turbo coding which is selected at the transmitter.

2. Conclusion

USRP can be effectively combined to create an SDR transceiver. The LabVIEW/USRP combination presents an opportunity to enhance communications education by enabling a low-cost, hands-on approach with live signals for realistic, real-world demonstrations, laboratory exercises, capstone design projects and cutting-edge research.

The universal software radio peripheral family of products is a popular platform for hardware-based research and test bed validations conducted by universities in the software-defined radio and cognitive radio (CR) fields. The USRP offers a simpler, scalable and easier to use combined

http://dx.doi.org/10.5772/66272

platform that will both broaden the accessibility of the technology and platform for hands-on applications and spur further adoption and use within university communication systems classrooms, teaching laboratories and their natural follow-on coursework (e.g., SDR, CR, digital communications, wireless communications and satellite communications).This chapter describes SDR system built using LabVIEW and testing the output using real-time data. Through this chapter, we have tried to cover emerging SDR standards and the technologies being used to specify and support them. We have proposed expanding the SDR definition and discussed the issues pertaining to the design of a multi-band flexible receiver and a linearized transmitter using broadband quadrature techniques. Here, we have described the case study of the SDR by designing an SDR generic transceiver in LabVIEW highlighting the multi-modulation approach for the SDR. The design uses forward error correction codes namely convolution codes and turbo codes for enhanced security for the data being transmitted. The proposed design is entirely reconfigurable in nature and supports multiple M-ary modulation schemes which can be changed accordingly by the user any time during the runtime. The biggest advantage of this design is that we have used phase tracking for identification of the constellation points. The analysis of the case study proves that turbo coding gives a much improved and better minimization of the data errors than the convolution coding.

Author details

Nikhil Marriwala[1*], Om. Prakash. Sahu[2] and Anil Vohra[3]

*Address all correspondence to: nikhilmarriwala@gmail.com

1 Electronics and Communication Engineering Department, University Institute of Engineering and Technology, Kurukshetra University, Kurukshetra, Haryana, India

2 Electronics and Communication Engineering Department, NIT, Kurukshetra, Haryana, India

3 Electronics Science Department, Kurukshetra University, Kurukshetra, Haryana, India

References

[1] J. Mitola, "Software and DSP in Radio," IEEE Communications Magazine, Volume: 38, Issue: 2, Feb 2000, ISSN 0163-6804.

[2] M. N. O. Sadiku and C. M. Akujuobi, "Software-defined radio: a brief overview," *IEEE Potentials*, vol. 23 no. 4, pp. 14–15, 2004.

[3] J. C. Lyke, C. G. Christodoulou, G. A. Vera and A. H. Edwards, "An introduction to reconfigurable systems," *Proc. IEEE*, vol. 103, no. 3, pp. 291–317, 2015.

[4] N. Marriwala, O. P. Sahu and A. Vohra, "Novel design of a low cost flexible transceiver based on multistate digitally modulated signals using Wi-Fi protocol for software defined radio," *Wirel. Pers. Commun.*, 2016, 87: 1265. doi:10.1007/s11277-015-3052-4

[5] N. Marriwala, O. P. Sahu and A. Vohra, "Design of a hybrid reconfigurable Software Defined Radio transceiver based on frequency shift keying using multiple encoding schemes," *Egypt. Informatics J.*, vol. 17, no. 1, pp. 89–98, 2015.

[6] N. Shahin, N. J. Lasorte, S. A. Rajab and H. H. Refai, "802.11g channel characterization utilizing labview and NI-USRP," *Conf. Rec. – IEEE Instrum. Meas. Technol. Conf., The Depot Minneapolis*, MN, USA, pp. 753–756, 2013.

[7] "A mathematical theory of communication," C. E. Shannon, System, vol. 27, no. July, 1928, pp. 379–423, 1948.

[8] N. Marriwala, "LabVIEW based design implementation of M-PSK transceiver using multiple forward error correction coding technique for software defined radio applications," *J. Electr. Electron. Eng.*, vol. 2, no. 4, p. 55, 2014.

[9] Y. Z. Y. Zhu and C. Chakrabarti, "Architecture-Aware LDPC code design for multiprocessor software defined radio systems," *IEEE Trans. Signal Process.*, vol. 57, no. 9, pp. 3679–3692, 2009.

[10] H. Shehab and W. Ismail, "The development & implementation of Reed Solomon codes for OFDM using software-defined radio platform," *Int. J. Comput. Sci. Commun.*, vol. 1, no. 1, pp. 129–136, 2010.

[11] P. F. Morlat, A. Luna, X. Gallon, G. Villemaud, J. M. Gorce and C. Insa-lyon, "Structure and implementation of a SIMO multi-standard multi-channel SDR receiver," HAL archives open, inria-00412054, version 1, pp. 283–286, 2008.

[12] P. B. Kenington, *TEAM LinG*. Boston, London: Artech House, 2005.

[13] M. Imran Anwar, S. Virtanen and J. Isoaho, "A software defined approach for common baseband processing," *J. Syst. Archit.*, vol. 54, no. 8, pp. 769–786, 2008.

[14] T. Javornik, M. Mohorčič, A. Švigelj, I. Ozimek and G. Kandus, "Adaptive coding and modulation for mobile wireless access via high altitude platforms," *Wirel. Pers. Commun.*, vol. 32, no. 3–4, pp. 301–317, 2005.

[15] C. B. Haskins and W. P. Millard, "Multi-band software defined radio for spaceborne communications, navigation, radio science and sensors," *IEEEAC Conf. IEEE*, vol. 5, pp. 1–9, 2010.

[16] D. Wu, J. Eilert and D. Liu, "Implementation of a high-speed MIMO soft-output symbol detector for software defined radio," *J. Signal Process. Syst.*, vol. 63, no. 1, pp. 27–37, 2009.

http://dx.doi.org/10.5772/66272

[17] I. Hatai and I. Chakrabarti, "FPGA implementation of a digital FM modem for SDR architecture," *Comput. Devices Commun. 2009. CODEC 2009. 4th Int. Conf.*, Kolkata, India, no. August, 2009.

[18] G. Baldini, T. Sturman, A. R. Biswas and M. Street, "Security aspects in software defined radio and cognitive radio networks: a survey and a way ahead," *IEEE Commun. Surv. TUTORIALS*, vol. 14, no. 2, pp. 355–379, 2012.

Chapter 5

Design Trade-Offs for FPGA Implementation of LDPC Decoders

Alexandru Amaricai and Oana Boncalo

Additional information is available at the end of the chapter

http://dx.doi.org/10.5772/66085

Abstract

Low density parity check (LDPC) decoders represent important throughput bottlenecks, as well as major cost and power-consuming components in today's digital circuits for wireless communication and storage. They present a wide range of architectural choices, with different throughput, cost, and error correction capability trade-offs. In this book chapter, we will present an overview of the main design options in the architecture and implementation of these circuits on field programmable gate array (FPGA) devices. We will present the mapping of the main units within the LDPC decoders on the specific embedded components of FPGA device. We will review architectural trade-offs for both flooded and layered scheduling strategies in their FPGA implementation.

Keywords: forward error correctionLDPC decoder, FPGA, digital circuits

1. Introduction

Low density parity check (LDPC) codes are a class of capacity approaching codes which provide increased error correction capability for both binary symmetric channel (BSC) and binary-input additive white Gaussian noise (BIAWGN) channel models [1]. Therefore, LDPC codes are used in a wide range of standards for both wireless communication [2]—WiFi, WIMAX, DVB-S2, etc—as well as for FLASH-based storage systems [3].

Decoding of LDPC codes is performed in an iterative manner, using message passing algorithms [4, 5]. These algorithms rely on simple computations—additions and comparisons on a small number of bits—which are performed on dedicated computational nodes. Although the node level computational complexity is low, LDPC codes implemented in communication and storage standards employ thousands or tens of thousands of such computational nodes, which leave a wide range of design options and trade-offs for the implementation of decoding architectures [2]. These trade-offs take into account the throughput, error correction capability, cost of the hardware implementation, and power consumption.

In this chapter, we will present the most important architectural options for both flooded and layered LDPC decoders implemented on field programmable gate array (FPGA) devices. The implementation of LDPC decoders on such devices is motivated by the increased flexibility of FPGAs, which make them suitable to implement highly versatile solutions in both wireless communications—such as software defined radios—and storage systems—such as software defined storage—as well high level of parallelism degree for fixed-point computations, which ensure the possibility of obtaining high throughputs for the decoders.

This book chapter is organized as follows: Section2 presents the algorithms for the LDPC decoding, as well as the strategies for it; Section3 summarizes the main features and building blocks of modern FPGA devices; Section4 presents the implementation and design trade-offs for FPGA-based flooded LDPC decoding architectures; layered architectures are detailed in Section5; last section is dedicated to the concluding remarks.

2. Theoretical background of LDPC decoding

LDPC codes are a class of linear algebraic codes, defined by a sparse parity check matrix H [1]. The LDPC code can also be represented by a bipartite graph, called the Tanner graph [6]. This graph contains two types of nodes: variable or bit nodes—corresponding to the columns in the H matrix and the codeword bits—and check nodes—corresponding to the rows in the H matrix and the parity check equations. A check node is connected to a variable node if the corresponding value in the parity check matrix is nonzero. **Figure 1** depicts a simple parity check matrix and its associated Tanner graph. LDPC decoding is performed in an iterative manner, consisting of the message exchange between the check and variable nodes along the edges of the Tanner graphs in several rounds or iterations. This type of decoding is called message passing (MP) decoding [4]. LDPC codes defined in communication or storage standards use parity check matrices consisting of thousands of columns, such as the 2304 columns for WiMAX, 64800 columns for DVB-S2, or 1944 columns for WiFi. The number of nonzero entries on each column represents the variable node degree—d_v —, while the number of the nonzero elements on each row in the H matrix represents the check node degree—d_c. An LDPC code is said to be regular if all the rows/columns in the parity check matrix contain an equal number of nonzero entries; otherwise, the LDPC code is irregular.

In order to enable efficient hardware implementations, quasi-cyclic LDPC (QC-LDPC) codes are used in most of the standards [7]. These subclasses of LDPC codes present highly structured parity check matrices, defined by blocks of circulant matrices. A QC-LDPC code is defined by a base matrix B, consisting of -1 elements and nonnegative elements. The parity check matrix H is obtained from the matrix B in the following way: -1 elements are expanded by $z * z$ all 0 matrix, while nonnegative elements within the matrix B are expanded by the $z * z$ identity matrix permutated with the nonnegative element value. The coefficient z is known as the expansion factor for the QC-LDPC code. **Figure 2** depicts the B matrix for the WiMAX

http://dx.doi.org/10.5772/66085

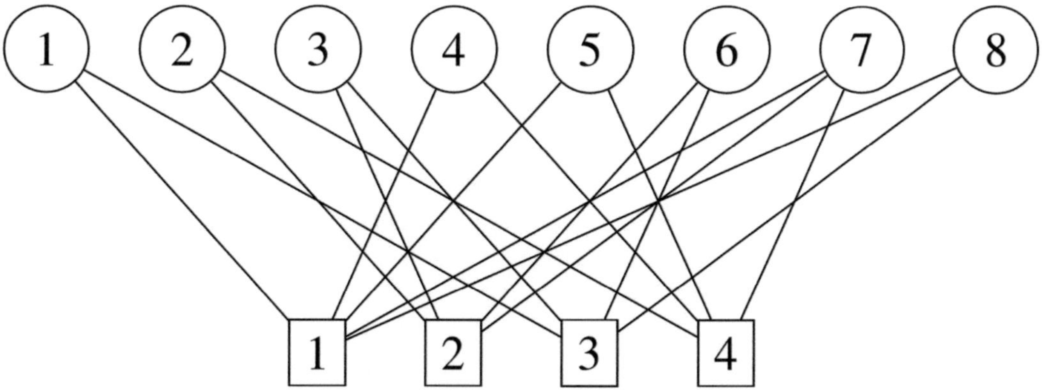

$$H = \begin{bmatrix} 1 & 0 & 0 & 1 & 1 & 0 & 1 & 1 \\ 0 & 1 & 1 & 0 & 0 & 1 & 1 & 0 \\ 1 & 0 & 1 & 0 & 0 & 1 & 0 & 1 \\ 0 & 1 & 0 & 1 & 1 & 0 & 1 & 0 \end{bmatrix}$$

Figure 1. Parity check matrix and its associated Tanner graph.

-1	94	73	-1	-1	-1	-1	-1	55	83	-1	-1	7	0	-1	-1	-1	-1	-1	-1	-1	-1	-1	-1
-1	27	-1	-1	-1	22	79	9	-1	-1	-1	12	-1	0	0	-1	-1	-1	-1	-1	-1	-1	-1	-1
-1	-1	-1	24	22	81	-1	33	-1	-1	-1	0	-1	-1	0	0	-1	-1	-1	-1	-1	-1	-1	-1
61	-1	47	-1	-1	-1	-1	-1	65	25	-1	-1	-1	-1	-1	0	0	-1	-1	-1	-1	-1	-1	-1
-1	-1	39	-1	-1	-1	84	-1	-1	41	72	-1	-1	-1	-1	-1	0	0	-1	-1	-1	-1	-1	-1
-1	-1	-1	-1	46	40	-1	82	-1	-1	-1	79	0	-1	-1	-1	-1	0	0	-1	-1	-1	-1	-1
-1	-1	95	53	-1	-1	-1	-1	-1	14	18	-1	-1	-1	-1	-1	-1	-1	0	0	-1	-1	-1	-1
-1	11	73	-1	-1	-1	2	-1	-1	47	-1	-1	-1	-1	-1	-1	-1	-1	-1	0	0	-1	-1	-1
12	-1	-1	-1	83	24	-1	43	-1	-1	-1	51	-1	-1	-1	-1	-1	-1	-1	-1	0	0	-1	-1
-1	-1	-1	-1	-1	94	-1	59	-1	-1	70	72	-1	-1	-1	-1	-1	-1	-1	-1	-1	0	0	-1
-1	-1	7	65	-1	-1	-1	-1	39	49	-1	-1	-1	-1	-1	-1	-1	-1	-1	-1	-1	-1	0	0
43	-1	-1	-1	-1	66	-1	41	-1	-1	-1	26	7	-1	-1	-1	-1	-1	-1	-1	-1	-1	-1	0

Figure 2. Base matrix for WiMAX rate ½ LDPC code [2].

LDPC code, rate ½, with 2304 columns and 1152 rows, and an expansion factor of 96. A horizontal layer of H matrix is defined as the set of z consecutive rows which correspond to one row within the base matrix. Composite layers, consisting of integer multiples of z rows within the parity check matrix, may be also used.

MP LDPC decoding may be performed using different scheduling strategies. These strategies indicate the order in which the check node and variable node computations are performed during the decoding iterations [13]. Two types of strategies may be employed: flooded and layered. Flooded decoding represents the conventional approach for decoding: each iteration consists of the update of messages at the check nodes, which subsequently pass their output messages to the variable nodes, which, in turn, update their corresponding messages [5]. Using this strategy, both the variable nodes and the check nodes are updated once per iteration. The layered scheduling consists of splitting the parity check matrix in horizontal layers; these layers are processed in a serial manner, while the check node updates within the same layer are processed in a similar manner with the flooded scheduling [8]. The variable node updates are performed after each layer processing. Therefore, in layered scheduling, the updates per iteration at variable node level are equal to the number of layers. Layered scheduling has two major advantages with respect to flooded: (i) faster convergence and (ii) reduced memory requirement [8]. The flooded approach has the advantage of increase resilience to faults in the hardware architectures [9], as well as the possibility for very high throughputs due to the high level of parallelism at the decoder level.

LDPC decoding can be performed by different types of algorithms, with different error correction capabilities. These can be split into two major classes [13]:

1. Hard-decision algorithms: These algorithms rely on 1-bit messages exchanged between the processing units. Such algorithms include bit-flipping, gradient descent and probabilistic gradient descent bit-flipping, Gallagher-A and Gallagher-B. The advantage of these algorithms is represented by the low requirements in terms of resource usage and power consumption. Their main drawback is represented by their low error correction capability with respect to soft-decision algorithms, for both BSC and BIAWGN channel models.

2. Soft-decision algorithms: These algorithms use messages quantized on several bits (usually between 3 and 7), which are exchanged between the variable nodes and check nodes. The hardware implementations for soft-decision algorithms are significantly more costly with respect to the hard-decision versions. However, using soft decoding, LDPC codes are able to have the capacity approaching error correction capabilities which make them suitable candidates for a wide range of communication standards.

In this chapter, we will discuss the implementation aspects related to the soft-decision-based LDPC decoders. The most important class of soft-decision LDPC decoding is represented by the min-sum (MS) algorithm [13] and its variants: offset MS (OMS) [10], normalized MS (NMS) [10], self-correcting MS (SCMS) [11], and finite alphabet iterative decoding (FAID) [12]. In these algorithms, the following messages are used [13]:

1. Input log-likelihood-ratio (LLR): These messages represent the input from the communication channel. For BSC channel model—used in storage systems—the input LLR is on 1 bit, while for BIAWGN channel model—used in wireless communication—the input LLR is quantized on several bits. The input LLR is denoted as γ and is quantized on $quant(\gamma)$ bits.

2. Variable node messages: These messages are the outputs of the check node units and serve as inputs for the variable node units. These messages are denoted as α and are quantized on $quant(\alpha)$ bits.

3. Check node messages: These messages represent the output of the variable nodes and are the inputs for the check nodes. These messages are denoted as β and are quantized on $quant(\beta)$ bits.

4. A posteriori LLR (AP-LLR): These messages represent the output of each decoding iteration/layer. The output of the decoder is given by the sign of the AP-LLR. It is denoted as $\tilde{\gamma}$.

Flooded MS decoding of LDPC codes consists of several iterations, where each variable node message—and check node message—is updated once. Each iteration consists of the following steps [5, 13]:

1. Variable node update

$$\alpha_{i,j} = \gamma_i + \sum_{k\epsilon\{C(i)\backslash j\}} \beta_{i,k}, \forall j \in C(i) \tag{1}$$

$$\tilde{\gamma}_i = \gamma_i + \sum_{k\epsilon C(j)} \beta_{i,k}, \forall j \in C(i) \tag{2}$$

2. Check node update

$$sign(\beta_{l,j}) = \prod_{k\epsilon\{V(l)\backslash j\}} sign(\alpha_{l,k}), \forall j \in V(l) \tag{3}$$

$$|\beta_{l,j}| = \min(\alpha_{l,k}), \forall j \in V(l), k\epsilon\{V(l)\backslash j\} \tag{4}$$

$C(i)$ denotes all the check node messages connected to the variable node i, while $V(l)$ denotes all the variable node messages connected to the check node l. The number of variable nodes is equal to number of columns in the parity check matrix, while the number of check nodes is equal to the number of rows in the H matrix.

Layered decoding is performed layer by layer, each layer consisting of the following steps [8, 13]:

1. Variable node update

$$\alpha_{i,j} = \tilde{\gamma}_i - \beta_{i,j}, \forall j \in V(i) \tag{5}$$

2. Check node update

$$sign(\beta_{i,j}) = \prod_{k\epsilon\{V(i)\backslash j\}} sign(\alpha_{i,k}), \forall j \in V(i) \tag{6}$$

$$|\beta_{i,j}| = \min(\alpha_{i,k}), \forall j \in V(i), k\epsilon\{V(i)\backslash j\} \tag{7}$$

3. AP-LLR update

$$\tilde{\gamma}_i = \alpha_{i,j} - \beta_{i,j}, \forall j \in V(i) \tag{8}$$

Both for flooded and layered scheduling, decoding is stopped either when a codeword is found—all the parity check equations are satisfied—or when the maximum number of iterations is reached.

The MS decoding, in both layered and flooded strategies, comprises of simple arithmetic operations, performed on small operands (3–8 bits). The variations of the MS algorithms target decoding performance improvement. OMS and NMS are based on the fact that the minimum computation at the check node level represents an overestimation of the check node message [10]. Therefore, both approaches try to reduce the value of the check node message computed by the check node unit.

The OMS approach uses a -1 addition from the absolute value of the $\beta_{i,j}$ in order to reduce its value. The check node computation in the OMS algorithm becomes:

$$sign(\beta_{i,j}) = \prod_{k\epsilon\{V(i)\backslash j\}} sign(\alpha_{i,k}), \forall j\in V(i) \tag{9}$$

$$|\check{\beta}_{i,j}| = \min(\alpha_{i,k}), \forall j\in V(i), k\epsilon\{V(i)\backslash j\} \tag{10}$$

$$|\beta_{i,j}| = |\check{\beta}_{i,j}| - 1 \tag{11}$$

The NMS approach uses scaling of the absolute value of the $\beta_{i,j}$ in order to reduce its value, by a normalization factor λ (usually with the values of 0.75 or 0.875) multiplication. The check node computation in the NMS algorithm becomes:

$$sign(\beta_{i,j}) = \prod_{k\epsilon\{V(i)\backslash j\}} sign(\alpha_{i,k}), \forall j\in V(i) \tag{12}$$

$$|\check{\beta}_{i,j}| = \min(\alpha_{i,k}), \forall j\in V(i), k\epsilon\{V(i)\backslash j\} \tag{13}$$

$$|\beta_{i,j}| = \lambda * |\check{\beta}_{i,j}| \tag{14}$$

SCMS represents an approach which aims at improving the error correction capability by erasing the variable node messages which change their sign after an iteration [11]. The erasure process cannot be performed in two consecutive iterations. The modification of the variable node update for a layered scheduling for the SCMS algorithm is:

$$\alpha_{i,j}^{new} = \tilde{\gamma}_i - \beta_{i,j}, \forall j\in V(i) \tag{15}$$

$$e_{i,j}^{new} = (!e_{i,j}^{old}) \& (sign(\alpha_{i,j}^{new}) \oplus sign(\alpha_{i,j}^{old})) \tag{16}$$

$$\alpha_{i,j} = \begin{cases} \alpha_{i,j}^{new}, e_{i,j}^{new} = 0 \\ 0, e_{i,j}^{new} = 1 \end{cases} \tag{17}$$

FAID decoding aims at improving the error floor region of the LDPC decoding. It changes the variable node operations, by implementing nonlinear dedicated function for the variable node message update, based on the channel information and the check node messages [12]. For a flooded scheduling, the variable node processing becomes:

http://dx.doi.org/10.5772/66085

$$\alpha_{i,j} = FAID(\gamma_i, \beta_{i,k}), \forall j \in C(i), \forall k \epsilon \{C(i) \backslash j\} \quad (18)$$

$$\tilde{\gamma}_i = \gamma_i + \sum_{k \epsilon C(j)} \beta_{i,k}, \forall j \in C(i) \quad (19)$$

The implementation of the *FAID* function is done using dedicated look-up tables (LUT). The complexity of these tables is dependent on the check node message quantization and the variable node degree d_v.

3. Architectural components of FPGA devices

FPGAs are digital devices with a programmable structure. This programmable structure provides FPGAs with very high flexibility, which makes them the ideal candidates for prototyping, as well as products with very low time-to-market constraints or applications which require high degree of flexibility. Furthermore, FPGAs have a built-in structure which allows a high degree of parallelization for applications that rely on fixed-point computations.

The main digital building blocks of modern FPGA devices are the configurable logic block (CLB), the embedded memory block RAM (BRAM), and the DSP block. DSP blocks implement 18 bit or wider multiplication, multiply-accumulate or multiply-add fused, and addition operations [14]. Because they are optimized for operand sized of 18 bit or more, and mainly for multiplication-based operations, they are of little use for the implementation of LDPC decoders.

CLBs are the main logic resource, which implement both sequential and combinational logic elements [15]. Usually, CLBs are composed of several slices, each of the slice being composed of a look-up table (LUT) and a D flip-flop, plus additional dedicated logic, such as logic and dedicated wire for ripple carry addition. The combinational logic is implemented using LUT, with modern FPGAs having six-input LUTs. Therefore, in a LUT and flip-flop pair, six-input combinational functions have the same cost as one or two input combinational functions. For specific families, the LUT can also be used as a memory circuit such as the distributed RAM in Xilinx FPGAs. The D flip-flop is used as the basic sequential logic. Because the combinational logic is paired with the D flip-flop in the same structural unit, pipelining can be easily and without significant resource consumption implemented in modern FPGA devices.

Another important feature of modern FPGAs is represented by the built-in memory blocks [16]. For large memories, FPGAs include the block RAM, which is block of 9 or 18 kbits. They have configurable width (9, 18, 36, or 72 bit), with the depth of the BRAM being determined by the width (for an 18 kbit BRAM and 72 bit word, the depth is 512 words). The number of BRAMs for a design is highly dependent on the width and the depth of required memory. For example, a memory which requires 96 bit words, and only 64 words, will consume 2 BRAM blocks, although the number of memory bits is significantly less with respect to the number of memory bits in a BRAM. Another important issue of the BRAM block is the number of read/

write ports: it is optimized for 1 read and 1 write port. The maximum number of memory ports for a BRAM is 2 read and 2 writes, but with limitations in the size of the word. For memories with few bits, and/or memories with a high number of ports, the distributed RAM implemented in CLBs is used.

From an LDPC decoder perspective, the FPGA implementation will make use of the CLBs for the implementation of the processing nodes and the routing network, and memories, either BRAM or distributed RAM.

4. Flooded LDPC decoders

The straightforward LDPC decoder architecture is represented by the hardware implementation of the corresponding Tanner graph. This type of architecture is known as the fully parallel decoder [17]. It consists of:

1. Processing nodes: A fully parallel decoder contains a number of variable node units equal to the number of columns in the parity check matrix and a number of check node units equal to the number of rows in the H matrix.

2. Routing network: The routing network is represented by wires which connect the variable node units with the check node units, according to the parity check matrix.

Although this kind of architecture is straightforward, the main problem arises due to the routing network. For LDPC codes that have thousands of rows and columns in the parity check matrix, the routing network involves tens of thousands of connections between the variable node units and check node units. Furthermore, the H matrix presents an irregular structure, which makes the interconnections component highly irregular. This will further contribute to the increase in cost, as well as reduction in the maximum operating frequency—due to the routing delay across the routing components of the FPGA. Another disadvantage of fully parallel LDPC decoder is the low flexibility: the decoder is specific to a LDPC code, and a slight modification in the code leads to the entire decoder redesign. Furthermore, these types of architecture cannot easily accommodate features such as multi-rate decoder, which is desired due to the fact that each communication and storage standard uses multiple LDPC codes with different rates. The main advantage of this architecture is represented by its high throughput, due to low number of clock cycles required for an iteration [17].

In order to reduce the complexity of these decoders, one approach relies on the reduction of the wires between the check node unit and variable node units. One such solution relies on the bit-serial decoder: the check node messages and the variable node messages are sent bit by bit to their corresponding processing unit [18]. Thus, the connection between a variable node unit and a check node unit consists of only two wires, instead of a $quant(\alpha)$ bit and a$quant(\beta)$ bit wires. This decoder trades throughput for reduced cost. Other solution relies on reduced quantization for the messages [19, 20]. The reduced quantization leads to a reduced number of wires between the processing units and thus to a reduction in the interconnection network. These solutions trade the error correction capability for reduced cost.

The other approach to reduce the complexity and the cost of the flooded LDPC decoder relies on the serialization of the check node and variable node operations at different levels. Thus, partially parallel flooded architectures are employed [21–28]. These partially parallel decoders exploit the regular structure of the QC-LDPC codes in order to obtain regular, low complexity architectures. Because serialization is employed at different levels, messages have to be stored in dedicated memory units. Stored messages have to be routed from the memory blocks to the processing units according to the LDPC matrix. In order to provide a flexible way for message routing, barrel shifters are employed. The read/write addresses for the memories, as well as the shift amounts employed in routing, are generated from a dedicated control unit. The main components for a partial parallel flooded decoder are as follows:

1. Processing nodes: The number of variable node units and check node units is dependent on the different parallelism degrees at different level. Furthermore, the number of inputs and outputs for such units can also vary, depending on how many messages can be processed each clock cycle.

2. Routing network: The routing is implemented using barrel shifters. The number and size of the barrel shifters may vary with message quantization, circulant size of the base matrix, different level parallelization degrees, etc. High-frequency pipelined barrel shifters may be implemented without additional cost in modern FPGA devices due to the LUT and D flip-flop pair which compose the basic component of the CLB.

3. Memory blocks: Memory blocks are used to store both the input LLRs and the check node and variable node messages. Usually, high degrees of parallelism—increased throughput —require wide memory words and multi-port memories. In many implementations, the multi-port memories are replaced by independent memory banks, which can be easily mapped on the FPGA BRAM blocks.

4. Control unit: The control unit is used to generate the shift amounts, the read/write memory addresses, as well as the control signals for the processing units. The shift amounts and the memory addresses are code dependent; this kind of information is usually stored in dedicated ROM memories.

For a quasi-cyclic LDPC decoder, two types of partial parallel flooded architectures have been proposed:

1. Parallel circulant, serial row/column processing: In this type of architecture, a number of z rows/columns are processed in parallel, while the rows and columns of the base matrix are processed sequentially [21–24]. This decoder is depicted in **Figure 3**. This kind of architecture requires z variable node units and z check node units. The memory words will consist of z messages. An important design parameter is represented by the parallelism degree at the processing node level—the number of processed messages per clock cycle. For the variable node unit, the maximum parallelism degree is d_v, while for the check node unit is d_c. Increasing parallelism at the processing node level will greatly influence the FPGA resource consumption of the decoder. This is due to the increased number of barrel shifters, which will lead to an increase in the conventional slice-based resource consumption, as well as for the increase in the number of memory ports, or the number of memory banks.

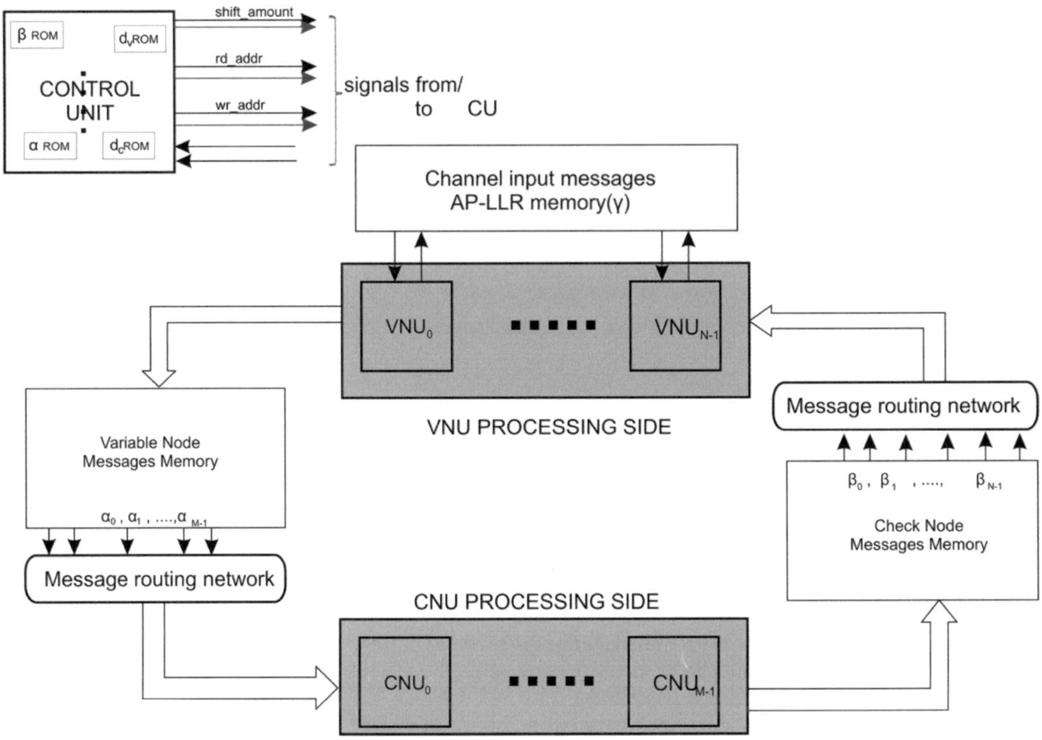

Figure 3. Parallel circulant, serial row/column processing flooded architecture.

Increasing the number of memory ports will lead to the implementation of the message memories with distributed RAM, while the increase in the memory banks will lead to an increase in the number of BRAM blocks.

2. Serial circulant, parallel row/column processing: In this kind of architecture, the rows/columns of the base matrix are processed in parallel, while the elements corresponding to a vertical/horizontal layer are processed sequentially [24–28]. This type of architecture is depicted in **Figure 4**. The number of check node units is equal to the number of rows in the B matrix, while the number of variable node units is equal to the number of columns in the base matrix. The number of columns in the base matrix gives also the number of input LLR message memories, while the variable and check node messages are stored in a $d_v nr\_col(B)$ memory blocks. Each memory has a depth equal to the circulant size and a width equal to the message quantization. This type of memory organization is suitable for FPGA devices, as each memory block maps to a BRAM block. This kind of decoder does not use dedicated routing circuits, as the routing of the messages between the memory blocks and the processing units is done via the offset address within each memory block. The processing units are fully parallel, as the read/write operations are done from d_v or d_c memory blocks. In order to increase the throughput, vectorization technique is proposed [25, 26]. This technique relies on packing multiple messages within a single memory word, which to be processed in parallel. Increasing the vectorization degree will lead to alignment problems,

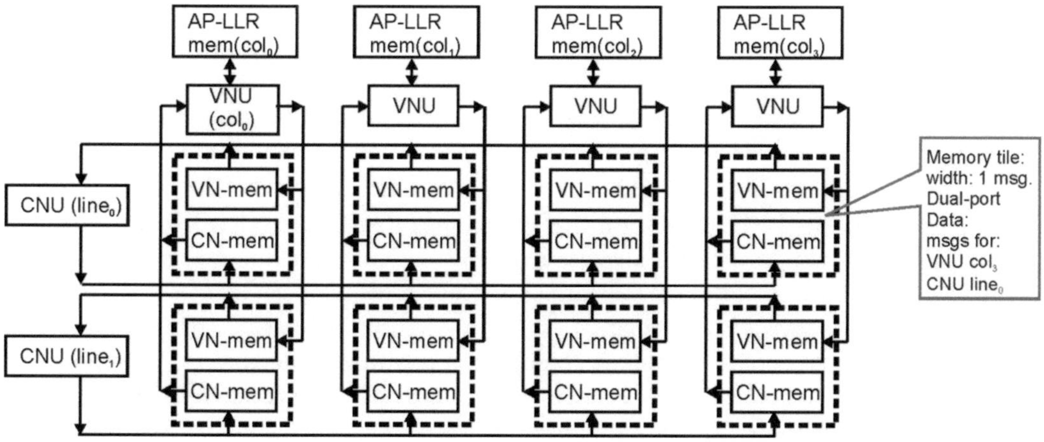

Figure 4. Serial circulant, parallel row/columns processing flooded architecture.

which lead to increased additional logic, as well as the number of stall clock cycles. Therefore, the maximum number of packed messages used with vectorization has been limited to four.

Partial parallel flooded FPGA architectures have two drawbacks:

1. Idle times for processing units: A major disadvantage of flooded decoder is represented by significant idle times for both variable node and check node units, during the variable node processing, the check node units, and vice-versa. Therefore, during one decoding iteration, only half of the decoder is utilized. Two strategies are employed:

   a. Processing two different codewords in parallel [22, 23]—while variable nodes compute the variable node messages for one codeword, the check nodes compute the check node messages for a second codeword; this solution implies small changes in the control unit, a double memory for the input LLR messages, and the hard-decision bits, with the advantage of a double throughput.

   b. Using waiting time minimization algorithms [25, 26]—using these algorithms, the order in which the rows/columns within the base matrix or within the parity check matrix are processed can be determined, without having data hazards and memory conflicts when performing the variable node and check node updates; therefore, almost simultaneous variable node and check node processing can be achieved; a second optimization obtained by employing these types of algorithms is represented by reduced memory usage; because data hazards and memory conflicts are avoided, the check node messages and variable node messages can be stored in the same memory locations.

2. Low usage of BRAM memories: In parallel circulant, serial row/column processing architectures, the memory word for the variable node messages is $zquant(\alpha)$, while the number of memory words is $d_v nr\_col(B)$. For LDPC code with circulant size of 96, 24 columns in the base matrix, $d_v = 3$, and message quantization of 4 bits, the word size is 384 bits, while the number of words in the memory is 72. For BRAM blocks consisting of 72 bits memory

words and 512 words, this kind of configuration results in the usage of 6 BRAM block, with only 72/512 utilization for each BRAM. For the second type of flooded architectures, for the same LDPC code, for each memory block required to store the variable node messages, the memory word is of 4 bits, while the number of words is 96. Also, in this case, it can be observed that the BRAM has poor usage. Several approaches have been proposed to address this issue. One is to use multiple codewords. The solution in [23] targets the increase in the memory words within the BRAM. The codewords are processed in serial. This solution achieves increase in the BRAM utilization for the same logic usage and throughput. The solution in [28] targets increase in the memory word size stored in the BRAM and addresses serial circulant, parallel row/column processing architectures. In the same memory word are stored messages from multiple codewords. The number of processing units is increased in order to process in parallel the codewords. This solution results in an increase of CLB logic usage, as well as throughput increase. Also for the serial circulant, parallel row/column processing architectures in [26] are presented folding, which aims at storing in the same BRAM messages associated with different columns/rows within the base matrix.

It can be observed that FPGA implementations of flooded architectures present a wide range of architectural variations, with different parallelism degrees at different levels, which aim at different throughput/cost/error correction capability trade-offs. The fully parallel solution presents increased throughput, but high cost due to routing, as well as low flexibility. Partial parallel solutions use memories for message storage. For these architectures, BRAM-based memory units are targeted in the FPGA implementation. However, employing BRAM blocks leads to several challenges related especially to the low usage of these.

5. Layered LDPC decoders

Layered architectures have been proposed first in [8], with the main goal of reducing the required memory bits. In the case of a layered decoder, two types of messages require memory storage: the AP-LLR messages and the check node messages. A typical layered LDPC decoder [29–33], depicted in **Figure 5**, contains the following components:

1. Processing units: The processing in the layered scheduling consists of the computation of the variable node messages, computation of the check node messages, and the AP-LLR update. The variable node message is computed from the AP-LLR and the check node message. The check node message is computed in the same way as in flooded scheduling, while the AP-LLR is updated from the new values of the variable node and check node messages. Because messages do not require routing between processing nodes—as in flooded—and just routing between memories and processing units, a combined unit—variable-check unit—is employed for processing. The number of processing units in the typical layered decoder is equal to the number of rows which constitute one layer, which is usually given by the circulant size. A combined unit contains an adder to perform the variable message computation, a FIFO buffer, used for routing the updated variable node

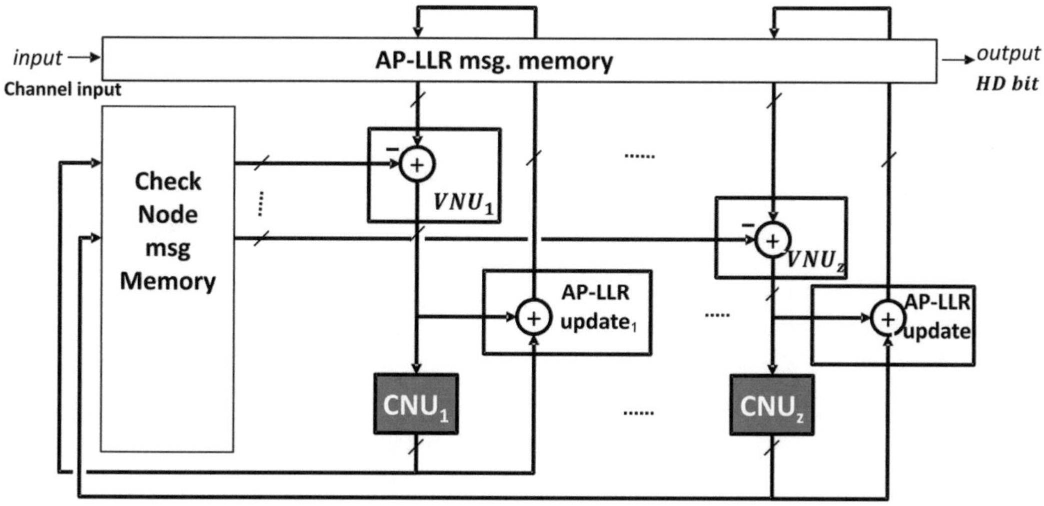

Figure 5. Layered decoding architecture.

message to the AP-LLR update, a comparator for updating the check node message, and the addition unit for the AP-LLR update [29–32]. Specific FPGA optimization can be implemented within the combined processing unit, which includes the use of the 6-input LUT within the CLB for comparator implementation—the comparator is implemented as ROM memories [30]—as well as the usage of the dedicated shift register chains for the implementation of the FIFOs. The processing unit has as inputs d_c AP-LLR messages and d_c check node messages, and outputs d_c updated AP-LLR messages and d_c updated check node messages. An important parameter for the entire decoding architecture is represented by the parallelism degree at the variable-check unit level, which represents the number of AP-LLR messages processed each clock cycle (maximum parallelism degree is equal to d_c). A higher degree of parallelism requires more simultaneous AP-LLRs read/write, as well as routing, which leads to increased number of memory ports or memory banks, and barrel shifters for routing [33].

2. Memory blocks: Layered decoders require the storage of two types of messages: AP-LLRs and check node messages. The AP-LLRs are messages which are routed between different processing units between layer processing. The check node messages are specific to each processing unit: these do not require routing from a processing unit to another between different layers. Therefore, the AP-LLR memory is a shared, global memory, while the check node message memories are local to each processing unit. Regarding the AP-LLR memory, the memory word for each bank is of $quant(\tilde{\gamma})$, while the maximum depth of this memory is equal to the number of columns in the base matrix. Regarding BRAM implementation of the AP-LLR memory, a drawback is represented by the low usage of the embedded block memory. Regarding the check node messages, two variants for their storage are used: (i) uncompressed form, when the β messages in their conventional two's complement format, and (ii) compressed form [34]. The compressed check node message is based on the fact that d_c−1β messages within a row corresponding to a row in the parity

check matrix have the same absolute value, equal to the minimum of the α messages connected to the corresponding check node unit, while the d_c-th check node message absolute value is equal to the second minimum. Therefore, a compressed check node message can be used, consisting of the signs, first minimum, second minimum, and the index of the first minimum. Regarding the FPGA implementation, the compressed form is suitable for shift register-based implementation in conventional CLB logic, while the uncompressed form is suitable for BRAM implementation. However, in BRAM-based implementation of the check node message, memory in compressed form is proposed for layered decoder with serial processing at processing node level. Routing from the BRAM blocks containing the check node messages to the processing units is achieved using large shift registers.

3. Routing network: Routing network is implemented using barrel shifters. The number of barrel shifters is dependent on the degree of parallelism in the processing unit. For each AP-LLR input of the processing unit, a pair of barrel shifters—one for routing read messages and one for routing the update message required for write—is required.

4. Control unit: The control unit is responsible for the generation of read/write addresses for the two memories, the shift amounts for the barrel shifter, as well as the control signals corresponding to the processing units. As in the case of the flooded decoders, ROM type of memories is used to embed the LDPC code information, from which are computed the memory addresses, as well as the shift amounts.

A major issue in the layer architecture is represented by the data hazards. Depending on the LDPC code, read-after-write (RAW) data hazards may affect the AP-LLR update: the updated value of the AP-LLR has not been written into the memory, before it is read for a new layer processing [35]. The problem of data hazards is aggravated by the usage of pipeline stages, both in the barrel shifters and in the processing units.

6. Conclusions

This book chapter presents an overview of the main design trade-offs in the implementation of LDPC decoders on FPGA devices. We detail how the main architectural choices for both flooded and layered scheduling strategies map on the built-in resources of modern FPGA devices. The main conclusions which can be drawn from this survey are as follows:

1. The degree of parallelism at processing node level has a major influence in the resource consumption of the LDPC decoder: it gives the number of barrel shifters used for routing, as well as the number of memory ports or memory banks used for message storage.

2. Routing represents an important factor in the cost/performance of the LDPC decoder; high-performance pipelined barrel shifter-based routing can be advantageously implemented in modern FPGA devices using conventional CLB resources.

3. Memories for message storage in partial parallel flooded LDPC decoder or layered decoders can be implemented using embedded BRAM blocks; the main problem is represented by the low usage of the memory bits within the BRAM.

The implementation of LDPC decoders on FPGA devices has a wide range of architectural and design parameters, which present different throughput/cost/error correction capability trade-offs. Furthermore, many FPGA-specific optimizations may be applied in the LDPC decoder design, such as the message memory mapping or optimization in the processing units.

Regarding the future use of the LDPC codes and decoder architectures, throughput and flexibility will represent highly important features. Regarding throughput, future wireless communication will require tens or hundreds of Gbps, which will impose new architectural challenges. Furthermore, the use of software-defined radios and software-defined flash will require highly flexible architectures, which can adapt code rate, quantization, as well as other features.

Acknowledgements

This work has been supported by bilateral UEFISCDI-ANR project DIAMOND.

Author details

Alexandru Amaricai* and Oana Boncalo

*Address all correspondence to: alexandru.amaricai@cs.upt.ro

Computer and Information Technology Department, University Politehnica Timisoara, Timisoara, Romania

References

[1] R. G. Gallagher. Low Density Parity Check Codes. MIT Press; 1963

[2] P. Hailes, L. Xu, R. Maunder, B. M. al-Hashimi, L. Hanzo. A survey of FPGA-based LDPC decoders. IEEE Communications Surveys and Tutorials. 2016;**18**(2):1098–1125. doi:10.1109/COMST.2015.2510381

[3] K. Zhao, W. Zhao, H. Sun, T. Zhang, X. Zhang, N. Zheng. LDPC-in-SSD: making advanced error correction codes work effectively in solid state drives. In: Proceedings of the 11th USENIX conference on File and Storage Technologies FAST'13; USENIX Association, Berkeley; 2013. pp. 243–256.

[4] T. J. Richardson, R. L. Urbanke. The capacity of low-density parity. IEEE Transactions on Information Theory. 2001;**47**(2):599–618. doi:10.1109/18.910577

[5] F. R. Kschischang, B. J. Frey. Iterative decoding of compound codes by probability propagation in graphical models. IEEE Journal on Selected Areas in Communications. 1998;**16**(2):219–230. doi:10.1109/49.661110

[6] R. Tanner. A recursive approach to low complexity codes. IEEE Transactions on Information Theory. 1981;**27**(5):533–547. doi:10.1109/TIT.1981.1056404

[7] M. P. C. Fossorier. Quasicyclic low-density parity-check codes from circulant permutation matrices. IEEE Transaction on Information Theory. 2004;**50**(4):1788–1793. doi:10.1109/TIT.2004.831841

[8] D. E. Hocevar. A reduced complexity decoder architecture via layered decoding of LDPC codes. In: IEEE Workshop on Signal Processing Systems; 13–15 October; IEEE; Austin, Texas, USA, 2004. pp. 107–112. doi:10.1109/SIPS.2004.1363033

[9] C. L. KameniNgassa, V. Savin, D. Declercq. Analysis of min-sum based decoders implemented on noisy hardware. In: Asilomar Conference on Signals, Systems and Computers; Pacific Groove, California, USA, 2013. pp. 866–870. doi:10.1109/ACSSC.2013.6810411

[10] J. Chen, M. P. C. Fossorier. Near optimum universal belief propagation based decoding of low-density parity check codes. IEEE Transaction on Communication. 2002;**50**(3):406–414. doi:10.1109/26.990903

[11] V. Savin. Self-corrected min-sum decoding of LDPC codes. In: IEEE International Symposium on Information Theory; IEEE; Toronto, Canada, 2008. pp. 146–150. doi:10.1109/ISIT.2008.4594965

[12] S. K. Planjery, D. Declercq, L. Danjean, B. Vasic. Finite alphabet iterative decoders—Part I: decoding beyond belief propagation on the binary symmetric channel. IEEE Transactions on Communications. 2013;**61**(10):4033–4045. doi:10.1109/TCOMM.2013.090513.120443

[13] V. Savin. LDPC decoders. In: D. Declerq, M.P.C. Fossorier, E. Biglie, editors. Channel Coding: Theory, Algorithms, and Applications; Elsevier; 2015. doi:10.1016/B978-0-12-396499-1.00004-2

[14] Xilinx. 7 Series DSP48E1 Slice User Guide—UG479. 2014.

[15] Xilinx. 7 Series FPGA Configurable Logic Block User Guide—UG474. 2014.

[16] Xilinx. 7 Series Memory Resources User Guide—UG473. 2014.

[17] V.Torres, A. Perez-Pascual, T. Sansaloni, J. Valls. Fully-parallel LUT-based (2048, 1723) LDPC code decoder for FPGA. In: 19th IEEE International Conference on Electronics, Circuits and Systems (ICECS); IEEE; 2012. p. 408–411. doi:10.1109/ICECS.2012.6463663

[18] A. Darabiha, A. C. Carusone, F. R. Kschischang. A bit-serial approximate min-sum LDPC decoder and FPGA implementation. In: 2006 IEEE International Symposium on Circuits and Systems; 21–24 May; IEEE; Island of Kos, Greece, 2006. doi:10.1109/ISCAS.2006.1692544

[19] V. A. Chandrasetty, S. M. Aziz. An area efficient LDPC decoder using a reduced complexity min-sum algorithm. Integration, The VLSI Journal. 2012;**45**(2):141–148. doi:10.1016/j.vlsi.2011.08.002

[20] A. Balatsoukas-Stimming, A. Dollas. FPGA-based design and implementation of a multi-GBPS LDPC decoder. In: 22nd International Conference on Field Programmable Logic and Applications (FPL); IEEE; Oslo, Norway, 2012. doi:10.1109/FPL.2012.6339191

[21] C. Beuschel, H.-J.Pfleiderer. FPGA implementation of a flexible decoder for long LDPC codes. In: 2008 International Conference on Field Programmable Logic and Applications; IEEE; Heidelberg, Germany, 2008. pp. 185–190. doi:10.1109/FPL.2008.4629929

[22] A. Blad, O. Gustafsson. FPGA implementation of rate-compatible QC-LDPC code decoder. In: 20th European Conference on Circuit Theory and Design (ECCTD); IEEE; Linkoping, Sweden, 2011. p. 777–780. doi:10.1109/ECCTD.2011.6043844

[23] A. Amaricai, O. Boncalo, I. Mot. Memory efficient FPGA implementation for flooded LDPC decoder. In: 23rd Telecommunications Forum Telfor (TELFOR); Belgrade, Serbia, 2015. pp. 500–503. doi:10.1109/TELFOR.2015.7377516

[24] Z. Wang, Z. Cui. A memory efficient partially parallel decoder architecture for quasi-cyclic LDPC codes. IEEE Transactions on Very Large Scale Integration (VLSI) Systems. 2007;**15**(4):483–488. doi:10.1109/TED.2007.895247

[25] Y. Chen, K. Parhi. Overlapped message passing for quasi-cyclic low-density parity check codes. IEEE Transactions on Circuits and systems—I: Regular Papers. 2004;**51**(6):1106–1113. doi:10.1109/TCSI.2004.826194

[26] X. Chen, J. Kang, S. Lin, V. Akella. Memory system optimization for FPGAbased implementation of quasi-cyclic LDPC codes decoders. IEEE Transactions on Circuits and Systems I: Regular Papers. 2011;**58**(1):98–111. doi:10.1109/TCSI.2010.2055250

[27] X. Chen, Q. Huang, S. Lin, V. Akella. FPGA-based low-complexity high-throughput tri-mode decoder for quasi-cyclic LDPC codes. In: 47th Annual Allerton Conference on Communication, Control, and Computing; IEEE; Monticello, Illinois, USA, 2009. pp. 600–606. doi:10.1109/ALLERTON.2009.5394917

[28] S. Nimara, O. Boncalo, A. Amaricai, M. Popa. FPGA architecture of multi-codeword LDPC decoder with efficient BRAM utilization. In: IEEE 19th International Symposium on Design and Diagnostics of Electronic Circuits and Systems (DDECS); IEEE; Kosice, Slovakia, 2016. doi:10.1109/DDECS.2016.7482452

[29] S. Mhaske, H. Kee, T. Ly, A. Aziz, P. Spasojevic. High-Throughput FPGA-based QC-LDPC Decoder Architecture. In: IEEE 82nd Vehicular Technology Conference (VTC Fall); IEEE; Boston, Massachusetts, USA, 2015. doi:10.1109/VTCFall.2015.7390967

[30] O. Boncalo, A. Amaricai, A. Hera, V. Savin. Cost-efficient FPGA layered LDPC decoder with serial AP-LLR processing. In: 24th International Conference on Field Programmable Logic and Applications (FPL); IEEE; Munich, Germany, 2014. doi:10.1109/FPL.2014.6927474

[31] S. Kim, G. E. Sobelman, H. Lee. A reduced-complexity architecture for LDPC layered decoding schemes. IEEE Transactions on Very Large Scale Integration (VLSI) Systems. 2011;**19**(6):1099–1103. doi:10.1109/TVLSI.2010.2043965

[32] K. Zhang, X. Huang, Z. Wang. High-throughput layered decoder implementation for quasi-cyclic LDPC codes. IEEE Journal on Selected Areas in Communications. 2009;**27** (6):985–994. doi:10.1109/JSAC.2009.090816

[33] O. Boncalo, P. Mihancea, A. Amaricai. Template-based QC-LDPC decoder architecture generation. In: 10th International Conference on Information, Communications and Signal Processing (ICICS); Singapore, 2015. doi:10.1109/ICICS.2015.7459838

[34] O. Boncalo, A. Amaricai, P. Mihancea, V. Savin. Memory trade-offs in layered self-corrected min-sum LDPC decoders. Analog Integrated Circuits and Signal Processing. 2016;**87**(2):169–180. doi:10.1007/s10470-015-0639-3

[35] Z. Wu, K. Su. Updating conflict solution for pipelined layered LDPC decoder. In: IEEE International Conference on Signal Processing, Communications and Computing (ICSPCC); IEEE; Xi'an, China, 2016. doi:10.1109/ICSPCC.2015.7338879

Chapter 6

Design of Digital Advanced Systems Based on Programmable System on Chip

Nordin Aranzabal, Adrián Suárez, José Torres, Raimundo García-Olcina, Julio Martos, Jesús Soret, Abraham Menéndez and Pedro A. Martínez

Additional information is available at the end of the chapter

http://dx.doi.org/10.5772/66579

Abstract

This chapter fills up an advanced analysis of the state-of-the-art design in programmable SoC systems, giving a critical overall vision for every designer to implement real time operating systems and concurrent processing. The content of the chapter is divided in the next four main sections.

- First the evolution timeline of FPGA based systems is covered from its beginning until the last AP SoC chips. They are complex devices and it is necessary to have a well-known understanding to utilise them in the more efficient form possible.
- The more important advance digital systems structures and architectures are described. The embedded AP SoCs are analysed and main design methodologies are covered, focusing in hardware and co-design strategies.
- In this section is described the development of a real open source application that covers the fundamental parts in the design of a SoC system, ranging from the hardware development until the software design involving the embedded operating system and the user interface application.
- Finally, the system described in the last section is tested in a real scientific experiment and the results are evaluated.

As conclusions the advantages of SoC systems when running an embedded Linux for interfacing FPGA based designs are highlighted.

Keywords: embedded systems, field programmable gate array (FPGA), system on chip (SoC), codesign hardware/software, embedded Linux, real-time instrument

1. Introduction

FPGAs are devices involved in a continuous evolution in order to offer more features and a better performance. The main characteristics of the FPGAs include the following:
High integration capacity that allows to implement complex digital systems in a single circuit.

- Flexible architecture that is easily adapted to each application.
- Have specific resources for performing arithmetic circuits, by reducing delays and making them more efficient.
- Include specific logical resources to generate internal memory units.
- Reconfigurability, it is possible to change the block function almost in real time.
- Generate adaptive circuits.
- Hardware description programming, ability to run multiple applications in parallel.

FPGAs have stopped being a simple architecture because they now represent powerful integrated systems with a lot of possibilities and different families to choose. In recent years, FPGAs have experimented a great evolution since the first important change that carried out when appeared Virtex II Family in 2001 based on look up tables (LUT) until today's new technologies [1].

It is necessary to know FPGA internal logic architecture in order to make the most of their advantages:

- Performance
- Time to market
- Prize
- Reliability
- Maintenance

Thereby, a review of block diagram and internal functionality will be presented in the next sections, beginning by reviewing the timeline since Virtex II family until new SoCs such as Zynq. **Figure 1** shows the more important Virtex II blocks and components.

One of the most important characteristics of Virtex-II device is that it featured a large number of 18 Kb block RAM memories. The block RAM memory is a true dual-port RAM, offering fast, discrete, and large blocks of memory in the device. The memory was organized in columns, and the total amount of block RAM memory depended on the size of the Virtex-II device. As shown in **Figure 2**, it was also formed by distributed RAM block and high-performance interfaces with external memories such as DDR SDRAM, ZBT SRAM, and QDR SRAM.

http://dx.doi.org/10.5772/66579

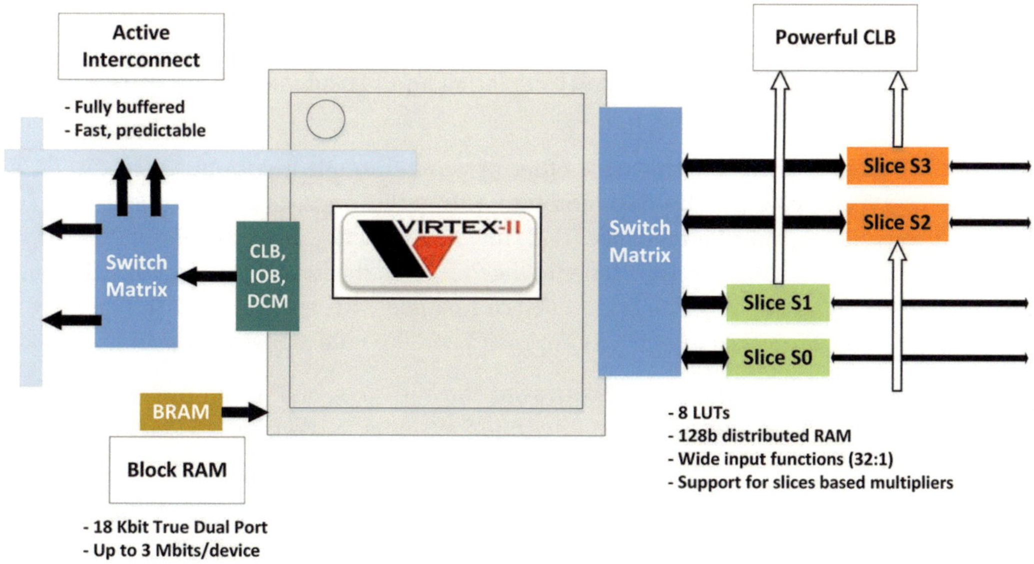

Figure 1. Virtex II internal block diagram.

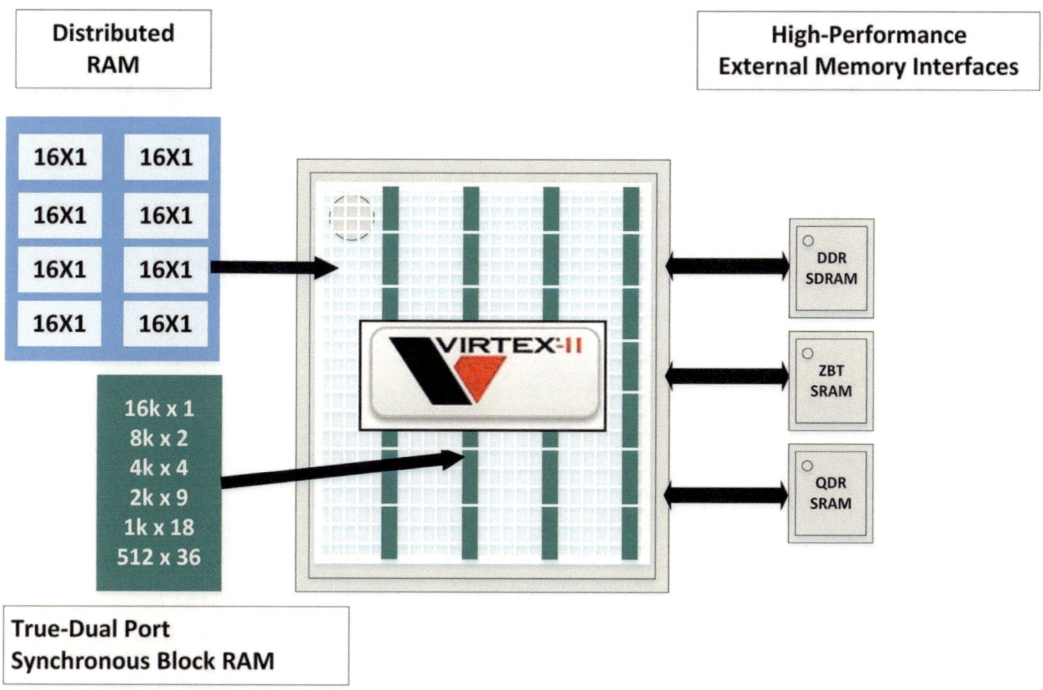

Figure 2. Virtex II memory block distribution and interface.

Another interesting feature included in Virtex II FPGAs was the dedicated 18 × 18 bits hardware multipliers that allow us to implement MAC functions. These modules make possible to carry out two's complement signed operations.

Regarding to Virtual II clock circuits, each FPGA has 16 clock global multiplexors, which manage the clock signal provided from an input port, digital clock manager (DCM), or interconnection local line. The DCMs are provided by external input terminals and allow the FPGA to delay and amplify the clock signal.

Those FPGA models communicate with other systems through input/output blocks (IOB) which are based on two input flip-flops and four output flip-flops, as shown in **Figure 3**. The IOBs include a digital control impedance (DCI) that allows FPGA to set a configurable output impedance in order to adapt the impedance to the PCB track that will be connected. Moreover, it is possible to configure the termination to be compatible with receivers and transmitters in own FPGA, so that, the signal integrity is improved and the reflections are suppressed.

Once described the more important features of the Virtex II, one of the high-end FPGAs that supposed a great evolution in the integrated systems market, next FPGA evolutions will be explained.

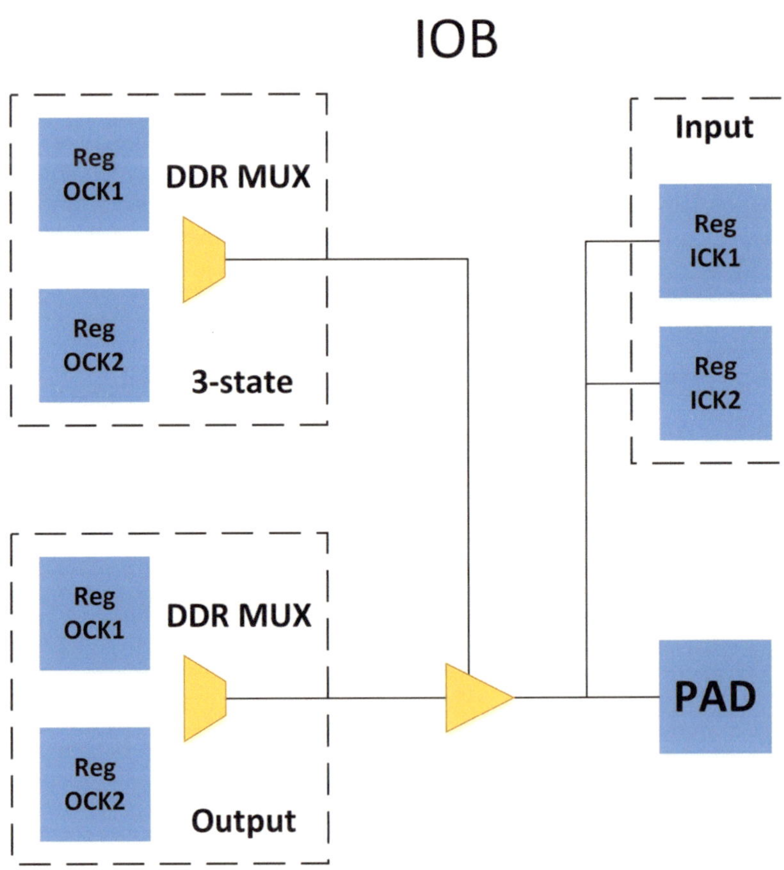

Figure 3. Input/output blocks architecture.

The next model was the Virtex II Pro which added to its predecessor the RocketIO multi-gigabit transceiver (MGT) blocks able to manage very-high data rate (2488–10,312 Gbps) through optical fiber or Gigabit Ethernet. Furthermore, this model could work with selectable 8, 16, and 32 bit buses, included up to 24 transceivers, integrated up to four PowerPC hardware processors that worked at 300 MHz and added more hardware multipliers, I/O terminals, and internal memory. These features are shown in **Figure 4**.

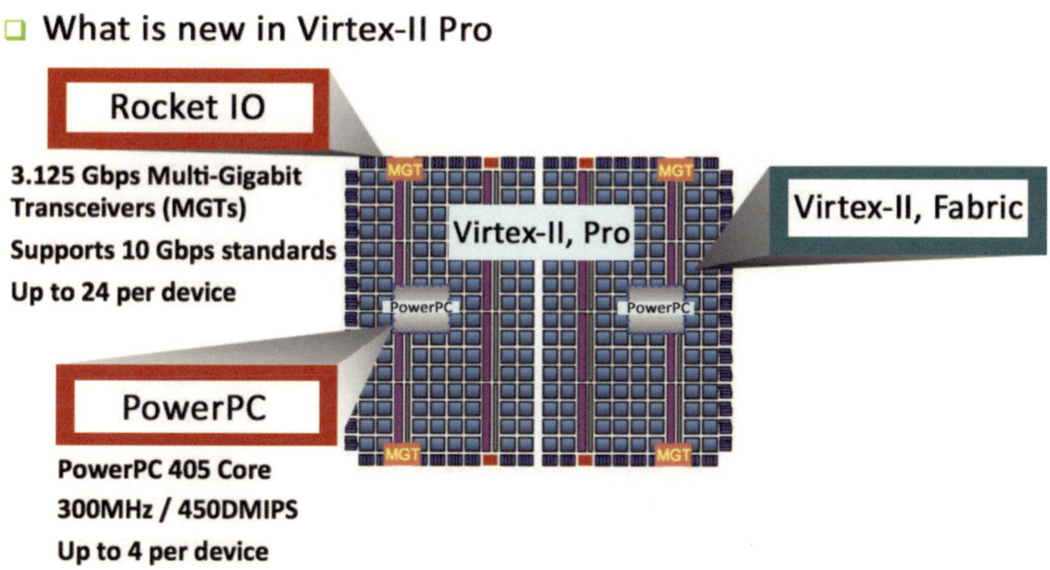

Figure 4. Improvements included in Virtex-II Pro model.

Next evolution came with the Virtex 4 and 5 models being that they added dedicated digital signal processors (DSPs) which made possible to carry out more complex computation with a better performance, as shown in **Figure 5**. Moreover, they included other improvements such as:

- Phase-matched clock dividers (PMCD), serial-to-parallel and parallel-to-serial interfaces integrated in the I/O terminals.
- Ethernet media access controller (TEMAC).
- BRAM memories set as FIFOs.
- Regional clock buffers, an advanced clock distribution network.
- Multi-gigabit transceivers up to 11.1 Gbps.
- Voltage and temperature monitoring (Virtex 5).
- Less power consumption.

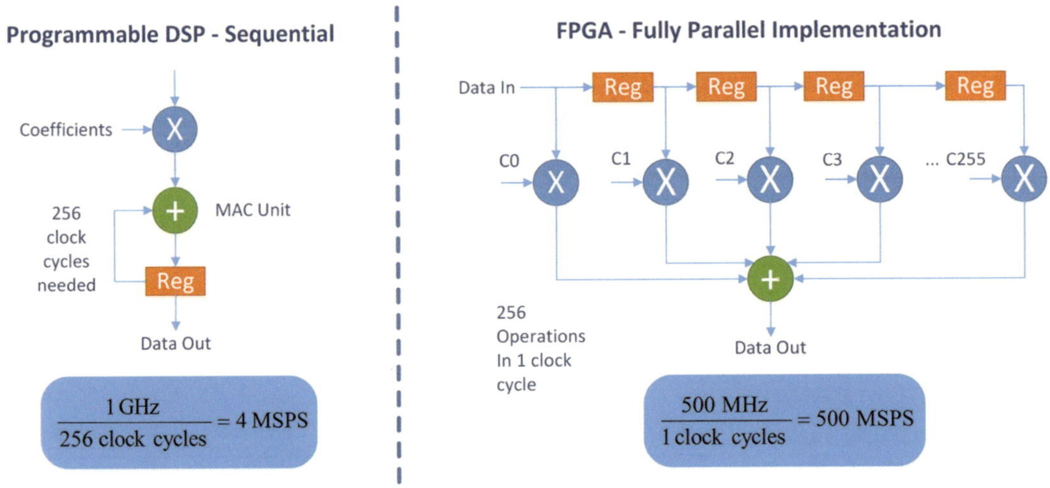

Figure 5. Performance comparison between programmable DSP and new-dedicated DSP.

Nevertheless, the change that stirred up the Xilinx FPGAs took place with the new families focused on low power consumption and high performance. Thereby, it disappears the concept high cost family (Virtex) and low cost family (Spartan), and Xilinx FPGA series are classified as shown in **Figure 6**.

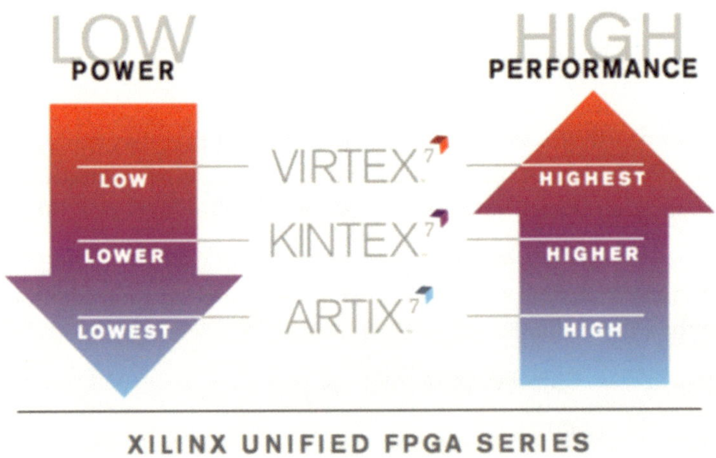

Figure 6. New Xilinx FPGA series range (picture courtesy of Xilinx).

This change was further started immediately after Xilinx promoted to combine a FPGA with external microprocessors by combining various components in a single chip. The new proposed technology was the Zynq-7000 line of 28 nm SoC devices that combine an ARM core with a FPGA [2], as shown in **Figure 7**. This was also joined by a change in the software tool because new models had to be programmed with Vivado Design Suite instead of the traditional ISE design platform given that it had not been developed to handle the capacity and complexity of designing with a FPGA with a hardware microprocessor core. The new software tool includes high-level synthesis functionality that allows engineers to compile the co-processors from a C-based description.

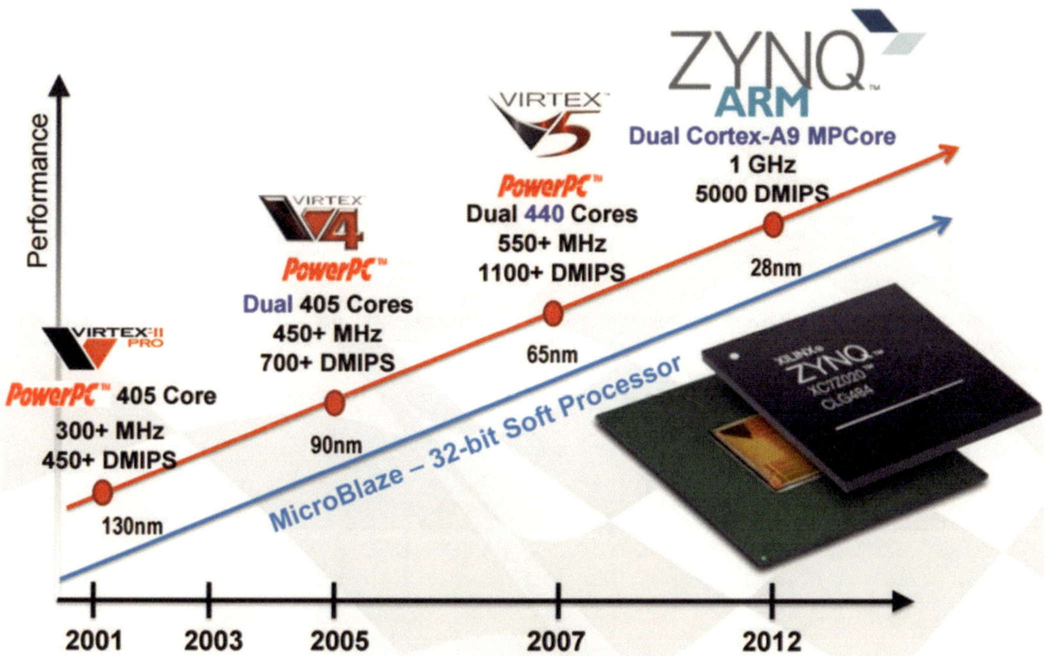

Figure 7. Processors evolution of Xilinx FPGAs (picture courtesy of Xilinx).

The Zynq-7000 family of system-on-chip (SoCs) represents a new concept because it integrates a complete Cortex-A9-processor-based 28 nm system. The Zynq architecture is different from other devices that combine both programmable logic and embedded processors because this kind of systems are focused on the processor instead of FPGA platform, as shown in **Figure 8**. For software developers, Zynq-7000 appears the same as a standard, fully featured ARM processor-based system-on-chip (SOC), booting immediately at power-up and capable of running a variety of operating systems independently of the programmable logic. Next generation of Xilinx SoCs was introduced by the Zynq-7100 model because it integrates digital signal processing (DSP) to meet emerging programmable systems integration requirements.

Therefore, the apparition of a large amount of specific new components, a lot of parameters such as input/output technology, clock signals, DSP blocks, flexibility, scalability, performance, integration, software or hardware processor, consumption or cost have to be taken into account to choose which is the better solution for each case as shown in **Figure 9**.

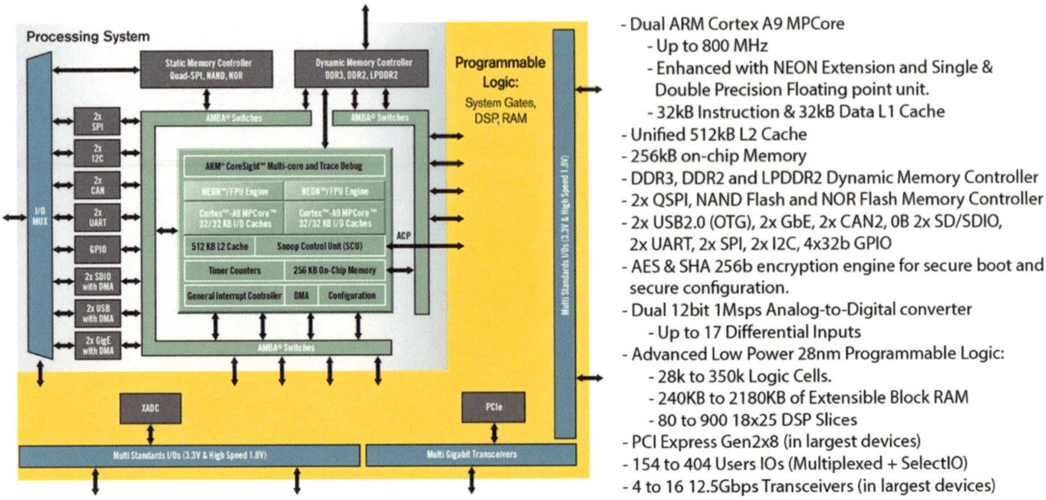

Figure 8. Architecture of SoCs based on programmable logic part (orange) and processing system such as ARM (blue) (picture courtesy of Xilinx).

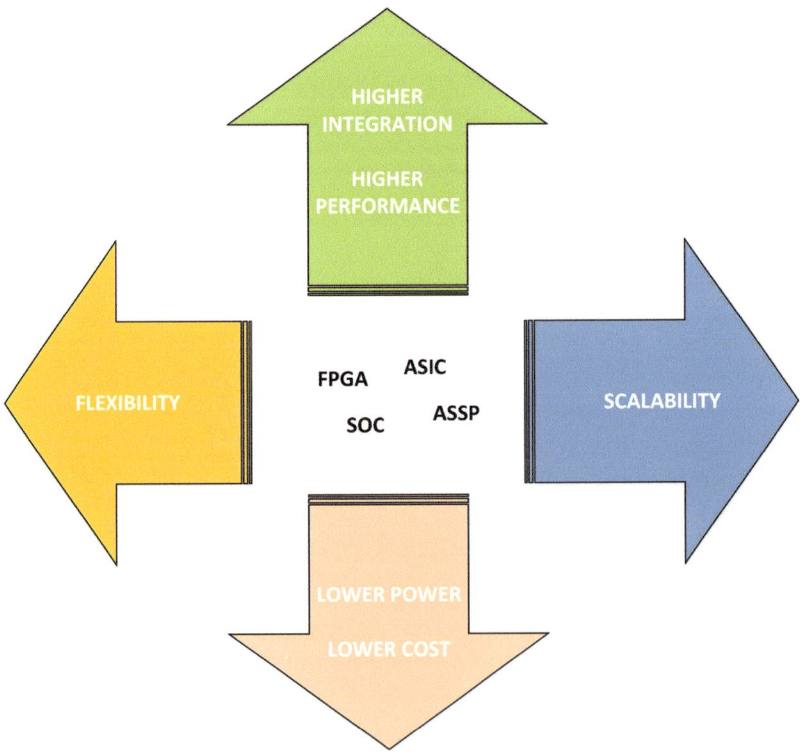

Figure 9. SoCs features.

2. Programmable SoC structures and architectures: hardware and software codesign methodology

2.1. Structures and architectures

Advance digital systems are those in which the core is a standard computational device or a combination of them. **Figure 10** shows two examples of this kind of devices integrated into an electronic design. Nowadays, the more important computational devices that take part in the front line of the embedded design market are the following:

- Field programmable gate array (FPGA): A digital integrated circuit with programmable logic to be configured by the designer.
- Application-specific integrated circuit (ASIC): A customized embedded circuit designed for a specific application, rather than a general-purpose.
- Microprocessor: An embedded circuit that incorporates a central processing unit (CPU) functions and operates on numbers and symbols represented in the binary numeral system. It is a clock driven, multipurpose, register-based, programmable electronic device that accepts binary data as input, processes it according to instructions stored in its memory, and gives results as output. The logic can be either combinational or sequential.
- Microcontroller: An integrated circuit embedding a processor core, programmable input/output peripherals and memory.
- Digital signal processor (DSP): A microprocessor with an optimized architecture in order to perform the computational operations in digital signal processing.
- Complex programmable logic device (CPLD): An integrated device with programmable logic that derives between programmable array logics (PALs) and FPGAs and the architectural characteristics of both.

Figure 10. Examples of advance digital systems.

To confront the development of those advance digital systems, two methods can be differentiated. The functionality of the desired application can be reached either by a software or hardware method. However, it is fundamental to have a remarkable understanding of both methodologies in order to obtain the best performance and effectiveness in the designs.

Software method:

- Driven by standard circuit processors (microprocessor, microcontrollers or DSPs).
- Involves programming languages associated with a fixed set of instructions.
- System flow runs sequentially.

Hardware method:

- Based on hardware circuits customizable by the user (CPLDs, FPGAs, and ASICs).
- Involves hardware description languages.
- Implements concurrent processing.

The software method is limited by the sequential flow of the instructions and the particular architecture. While the hardware method is limited by the difficulty that implies the design using the hardware programming languages, the time required for develop specific hardware systems and the high cost of those designs. Currently, in order to overcome the drawbacks of both methods, alternative methods must be determined.

This intermediate option is the most utilized nowadays and is known as integrated system, embedded system, or system on a chip (SoC). A SoC is an integrated circuit that includes a combination of the elements below in one single chip:

- One or several operating units (microprocessors/DSPs/FPAs…).
- Different standard interface circuits (UART, SPI, I2C, Ethernet…).
- One of the several memory units (RAM/FLASH…).
- An analog-to-digital converter (ADC).
- Different specific circuits for the application (audio/graphics/automotive…).

Currently, the options that enable to develop an application based on a programmable system on a chip such as the devices shown in **Figure 11**:

- Those who are already in the market and integrate a microcontroller, a programmable digital and analogic parts and communication ports: FIPSOC (Sidsa), Fusion (Actel), FPSC (Lattice), and PSoC (Cypress).
- Those that through a FPGA of high logic capacity permit to integrate a microcontroller and customizable communication ports: Xilinx and Altera.

The increase in the capacity of integration in the manufacturing of the chips has allowed the development and evolution of the embedded systems. **Figure 12** depicts an example of how the integration capacity has been increased over the time.

http://dx.doi.org/10.5772/66579

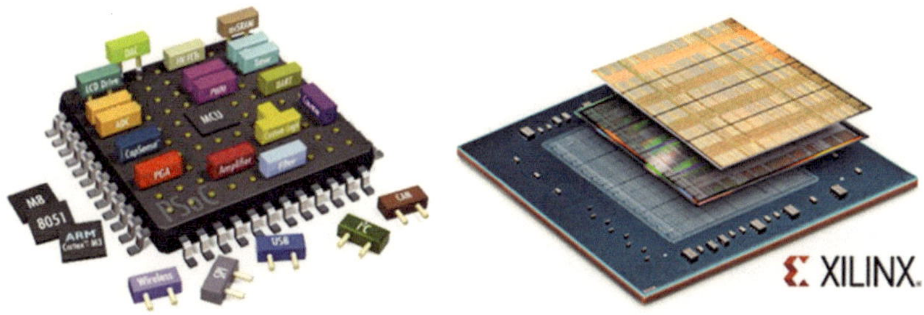

Figure 11. Examples of SoCs (picture courtesy of Cypress and Xilinx).

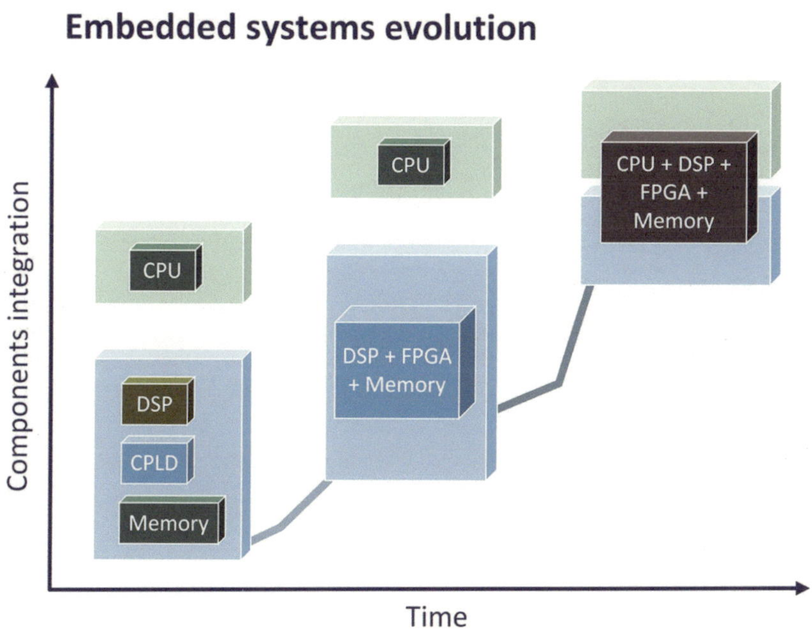

Figure 12. Embedded systems evolution in terms of components integration.

In a design of an application using an embedded system, the following factors need to be taken into consideration:

- The "hardware" is different for each application.
- Some applications may require an operating system (RTOS).
- A compact code implementation is desirable to reduce the size of the program memory.
- A combination between high-level (C) languages and low-level languages (VHDL) is required in order to optimize the processing speed.

In this section, among the presented embedded systems, we will be addressed in more detail the ones based on FPGAs of Xilinx, also known as systems on a programmable chip (SoPCs).

Xilinx, one of the main manufacturers of FPGAs, provides two alternatives to carry out a SoPC-based design:

- Hardware microprocessor block. PowerPC of 32 bits in FPGAs Virtex 2 pro, Virtex 4 FX, and Virtex 5 FXT ARM Dual-Core Cortex-A9 of 32 bits in Zynq.
- Software microprocessor block. Picoblaze of 8 bits in any FPGA. Microblaze of 32 bits in the families Spartan, Virtex, and the new Artix and Kintex.

For a better understanding of the different architectures of the hardware or software microprocessors involved in SoPC designs, various diagrams are presented in **Figures 13–16**. It is always important to take time studying and comprehending the architecture of each microprocessor for being able to choose the most effective solution for the desired application.

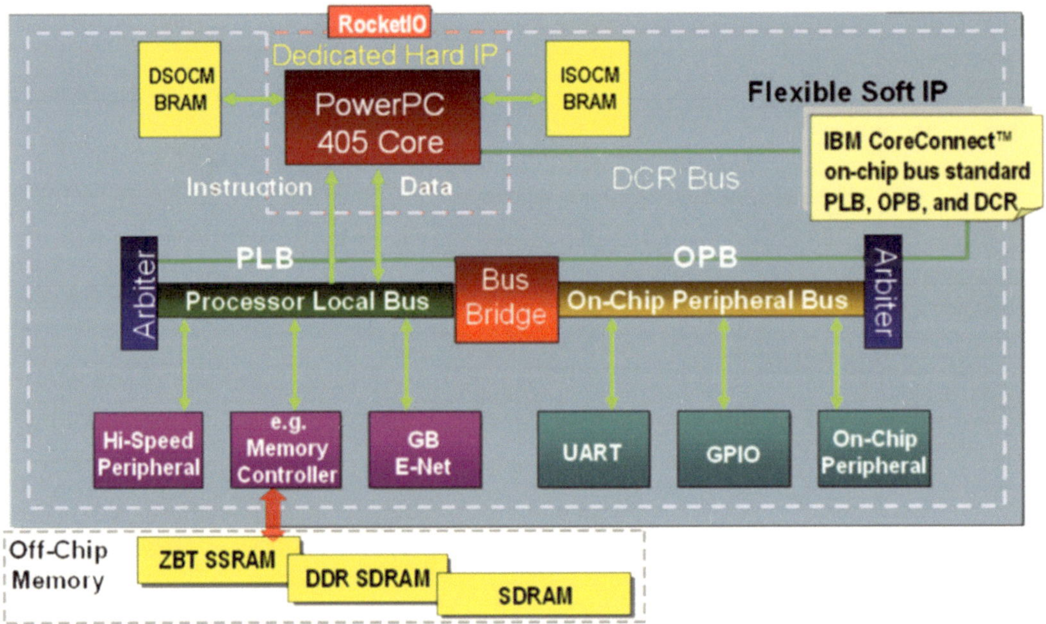

Figure 13. Architecture of an embedded system "hardware" based on PowerPC.

Concretely, from the previous architecture diagrams, it is relevant to recognize the possibilities, advantages, and disadvantages that offer hardware and software microprocessors.

There is only one possible option when choosing a given hardware microprocessor block:

- Commercial microprocessor core designed to be implemented in FPGAs. Advantages: the alternative of having a commercial microprocessor widely used before in the market (8051, PIC, ...). Lower development time.

  Disadvantages: higher cost and a fixed architecture of the microprocessor.

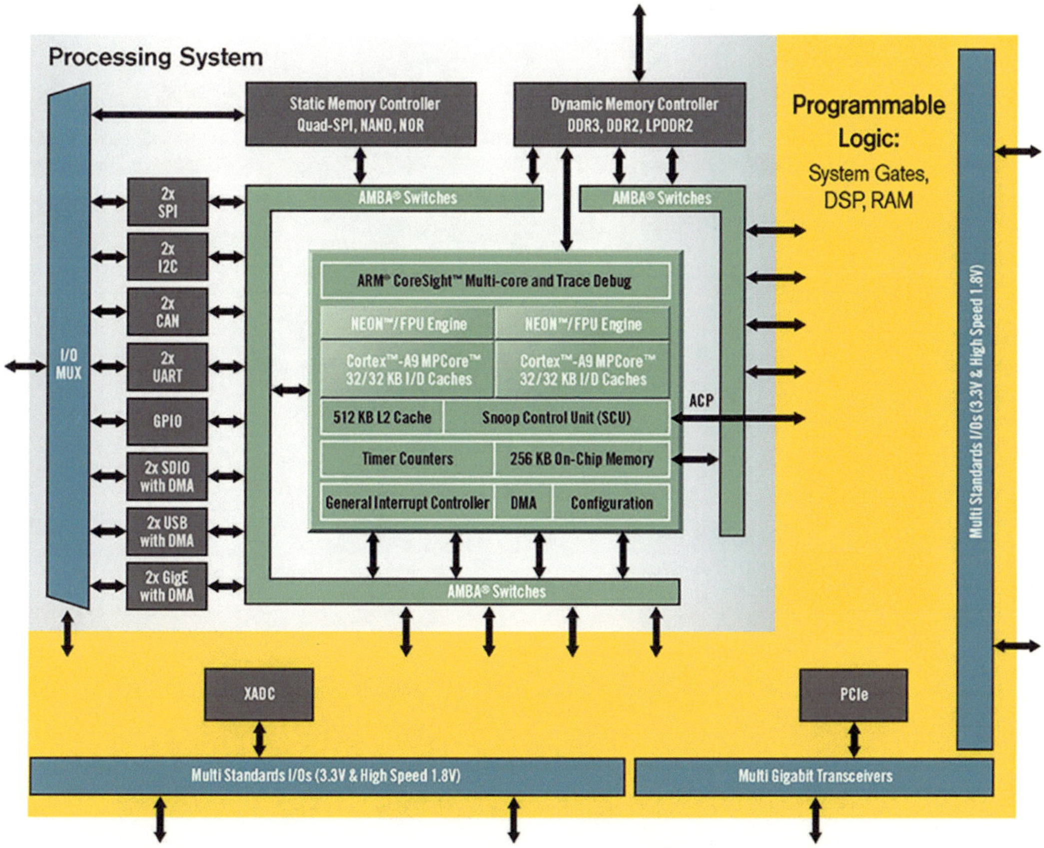

Figure 14. Architecture of an embedded system "hardware" based on ARM (picture courtesy of Xilinx).

The possible options when choosing a software microprocessor block are:

- Microprocessor core designed and customized by the user.

  Advantages: totally customizable architecture depending on the application. The source code of the microprocessor is available to perform modifications. Lower cost and commercialization possibilities.

  Disadvantages: limitations in the architecture.

- Microprocessor core designed by the manufacturer of the FPGAs.

  Advantages: the possibility of having a microprocessor utilized by a high number of users. Exchange of free-distribution programs and peripherals. Lower cost and development time.

  Disadvantages: limitations in the architecture.

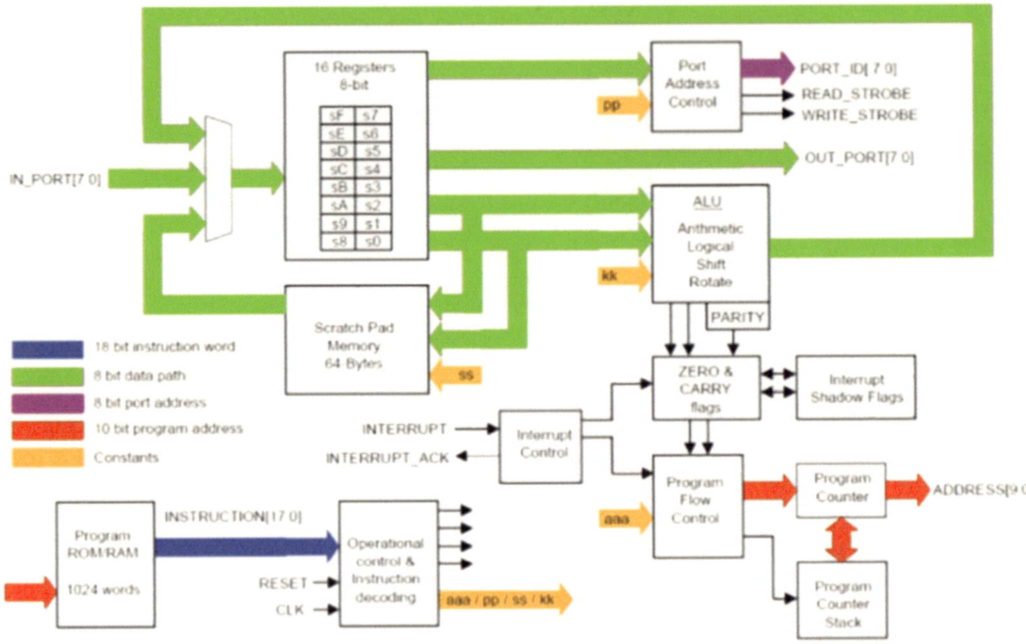

Figure 15. Architecture of an embedded system "software" based on PicoBlaze (picture courtesy of Xilinx).

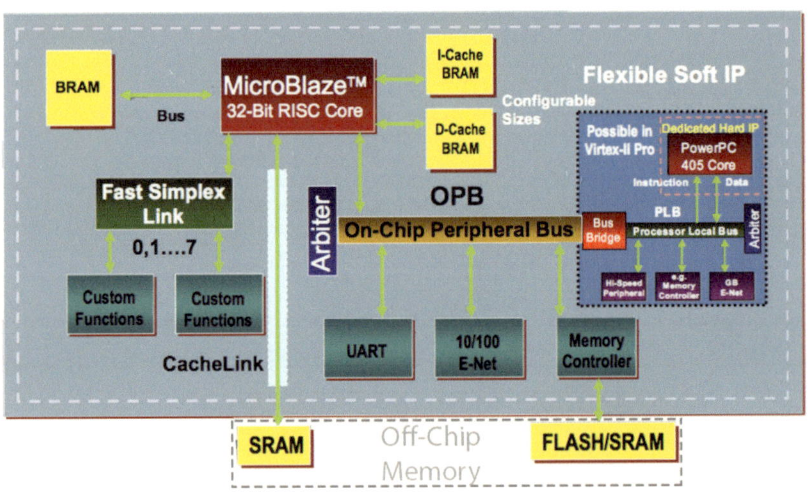

Figure 16. Architecture of an embedded system "software" based on MircoBlaze.

2.2. Codesign hardware/software

After understanding the embedded system, it is important how to stop and comprehend the design. For those designs that involve a high complexity, it is required to work on the necessary hardware, both for the microprocessor and its peripherals. Accordingly, it is necessary to know what program is going to be executed by the microprocessor and which additional components are involved in the application.

This process is usually known as codesign hardware/software [3] and basically can be defined as: the concurrent design of developers, part of the same team, to complete hardware and software tasks that are involved in an embedded system. Methodologies and tools are utilized which consider hardware/software iteration.

Figure 17 shows a basic codesign flow diagram in which the next stages are involved.

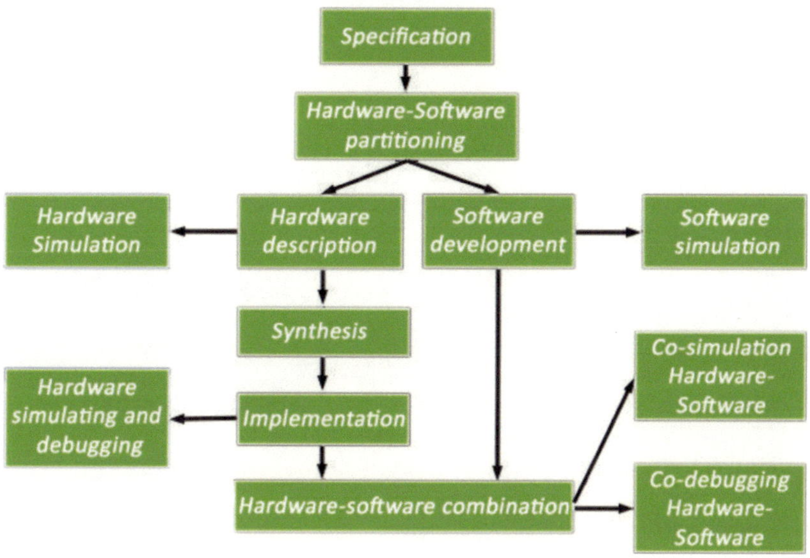

Figure 17. Processors evolution of Xilinx FPGAs (picture courtesy of AVNET).

Specifications:

Define the features of the system that is going to be developed. Hardware/software partitioning determines which functions are implemented through software routines in the embedded microprocessor and which through specific hardware circuits. In the market, certain commercial tools are available for high cost designed specifically to conduct this function.

Hardware description:

- Chose the suitable microprocessor and FPGA for the application.
- Design of the peripherals using HDL languages.

Software description:

- Development of the application in Assembler/C language. Compilation.

Hardware simulation:

- Through simulation programs that normally are part of the selected development tool.

Software simulation:

- Through simulation programs of the implemented programming language.

Co-simulation hardware/software:

- Tools such as EDK, ChipScope, synthesis, and implementation. The specific tool provided by the FPGA manufactured is employed (ISE).

A basic codesign, for example, the control of a text liquid crystal display (LCD), with two rows and 16 characters is presented in **Figure 18** which show the different options that could be implemented.

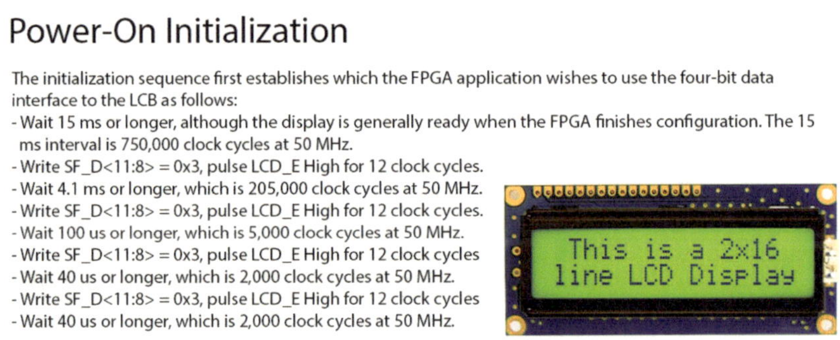

Figure 18. A basic example of codesign.

Hardware/software partitioning:

The operation slowness of the LCD and the complexity of the required control configuration sequences make the application very appropriate to be managed by a microprocessor. Nevertheless, slow operation time forces the microprocessor to attend and control the display for long periods of time, so it would be suitable release the microprocessor from that task.

First option (software):

Implement the control of the display totally by software, adding the minimum hardware. The software of the microcontroller will be in charge of sending and receiving instructions and data to the LCD (low-level routine) and of sending messages (high-level routine).

Second option (hardware):

Implement the control of the display totally using specific hardware, adding the minimum required software for the LCD. In this scenario, the device is totally managed by the hardware. In this option, a FPGA is needed to control the display, which implies higher costs and a great logic area.

Third option (codesign):

The most critical tasks, above all in time, will be implemented in specific hardware while the rest will be programmed through C in order to be executed by the microprocessor. The hardware acts as a microprocessor. In general, working with advance digital systems is the best strategy. An example of codesign is shown in **Figure 19**.

```
void XromLCDPrint2Strings(char *line1, char * line2) {
    int i=0;
        XromLCDSetLine(1);
        for(i=0; i<16; i++){
            if(line1[1])
                XromLCDPrintChar(line1[1]);
            else break;
        }
        XromLCFSetLine(2);
        for(i=0; i<16; i++)[
            if(line2[1])
                XromLCDPrintChar(line2[1]);
            else break;
        }
    return;
}
```

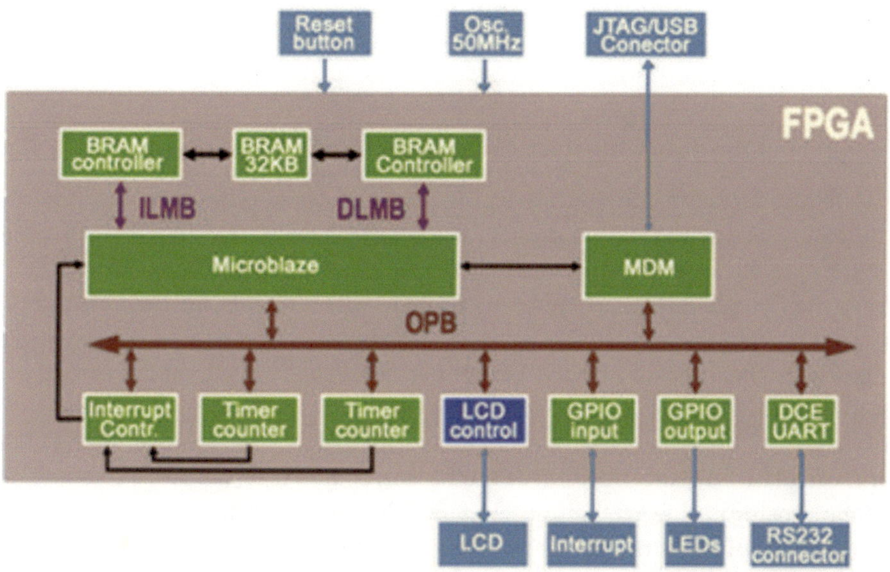

Figure 19. Example of codesign: LCD interface application.

3. Implementation of a real SoC application: description of a programmable SoC real application based on Linux with web server, database, and concurrent processing

This section is presented a real application based on a Zynq®-7000 All Programmable SoC (AP SoC) which has dual-core central processing units (CPUs) and a field programmable gate array (FPGA) in one chip, a versatile architecture that enables to lower costs and enhances the efficiency in regard to another system of analogue characteristics. The whole implemented software and hardware development that is described in this section covers from the user interface application to the free distribution operating system based on an embedded Linux [4] [5], which avoids closed source software and license costs. Additionally, extra open software is compiled to run in the specific system in order to produce a more efficient and reliable application. A web server based on Apache 2 provides the remote control and monitoring functionality, and the data storage and management is performed by a database application based on MySQL engine. Moreover, for a dynamic iteration between the web user interface and the database, a PHP server scripting language is compiled to run on the operating system.

The embedded system is designed to manage the Geowire, a novel electromechanical device that measures the temperature inside the pipes of vertical borehole heat exchangers (BHEs) throughout the thermal response test (TRT). The TRT is the standard method to quantify the thermal characteristics of borehole surrounding subsoil by measuring the temperature evolution at inlet and outlet. However, it assumes that the homogeneous isotropic subsoil calculates an average conductivity value for the overall geological domain. The Geowire provides additional information during the TRT by recording a series of depth-dependent temperature profiles during the TRT, which allow the system to identify the heterogeneity of ground stratigraphy. Thus, the minimum depth of the drilling for the maximum heat transfer can be calculated to save installation costs and build more optimized BHE.

3.1. Description of the device involved in the application

The Geowire measures the temperature inside the geothermal pipes by controlling the vertical displacement of a wired digital sensor. The sensor is connected to a cable furling inside a watertight case while a slip ring allows the transmission of the signal from the rotating structure to the Zynq board. The temperature probe goes out from the fluid output pipe connection and a servomotor rotates the furling to release or collect cable while a weight maintains it tightly by the effect of gravity. Before the cable goes outside of the case, it is guided between a roll with a magnetic encoder that transmits the signal outside the case to measure the displacement of the wire. Then, the software application is designed to control the servomotor and calculate the exact position of the sensor. In this way, a user interface makes possible to define customizable acquisition sequences by indicating the depth-dependent points of interest, the sampling time and the number of temperature samples per each point. Once the data for the acquisition process is defined, the device will begin to automatically sample and storage the data while the progress can be followed remotely in real time through the user interface.

The device was developed in such a way that it can be easily incorporated in geothermal pipes utilized during the TRT with or without water flow. The temperature inside the pipes can be measured with a maximum spatial resolution of 1 cm, a maximum temperature resolution of 0.0625°C, and an acquisition time smaller than 1 second. An electromechanical limit switch is employed to determine the starting point of the measurement path. This is activated when the weight used to sink the sensor pushes it. The provided signal from this switch is used to calibrate the measurement point every time when the sensor goes down and up. Also, it is

http://dx.doi.org/10.5772/66579

connected with the driver of the servomotor as an additional security measure to stop the motor and avoid the weight and the probe to roll inside the device enclosure. **Figure 20** shows a representation of Geowire enclosure parts.

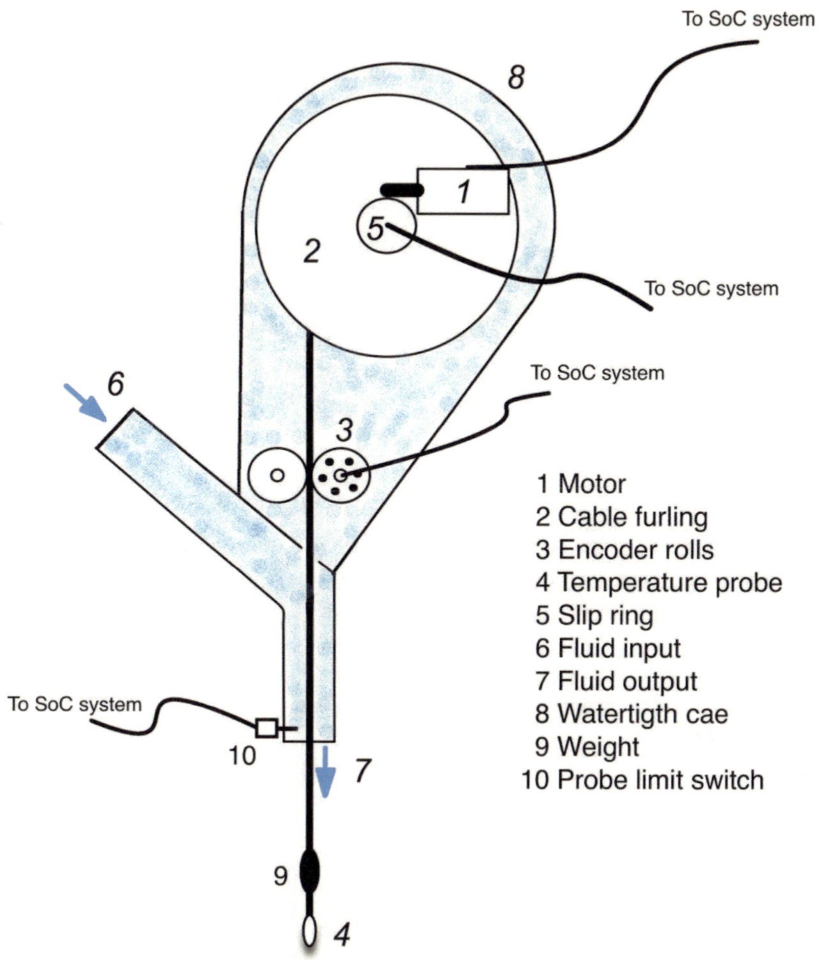

Figure 20. A representation of the different parts that comprehend the Geowire enclosure.

3.2. Application system architecture

The temperature measurement instrument described in the previous section is controlled by a μClinux OS that has been embedded in the dual-core ARM processor of the development board Digilent Zybo Zynq-7000. The board has been implemented as the CPU of the system that manages the performance of the secondary elements that compose the device through a user interface application. One of the most interesting characteristics of this kind of platforms is the possibility of using PMODs (peripheral module) through connectors with some fixed pins from the ARM and ADCs and reconfigurable pins from the programmable logic. So, eases the redesign tasks where any upgrade or modification is required. In this application, these connectors are utilized to communicate with the motor driver, read an encoder, manage a real-time clock (RTC), and control a driver to read a digital temperature sensor. Concretely, these peripherals are connected with the SoC in order to carry out the next tasks:

a. Motor driver: It is based on a one-quadrant digital servo amplifier driver and its main task is to provide protection against possible current overloads, power surge or brownout, polarity inversions and shortcuts in the motor. This driver integrates a microcontroller which follows the parameters sent by Zybo board processors. Thereby, by using an RS232, communication protocol is possible to manage the motor control settings. Thus, the CPU of the system regulates the speed either in clock wise (CW) or counter clock wise (CCW) directions and reads configuration, information or alarm registers from the driver.

b. Encoder: A magnetic encoder measures the angular movement of a roller that is rotated when the wire of the temperature sensor comes out from the Geowire to be inserted in the geothermal pipes. The wire is guided through two rollers inside the enclosure of the device to ensure the rotation by friction. When the roller rotates, the encoder generates several digital pulses that are sent to a core implemented in VHDL that measures the distance traveled by the sensor. An important advantage of this encoder is the suppression of mechanical contacts that may originate tear and wear problems. Furthermore, this technology holds the encoder electric connections shielded against water or humidity so that the water can flow inside the device enclosure without damaging the electric connections.

c. RTC: The PMOD connected to the port of the Zybo board is based on a real-time clock which contains a battery to maintain the operating system time upgraded when the system wakes up after a shut down. In this way, the RTC provides the time and date to the systems, every time it sends a request via an I^2C (inter-integrated circuit) communication protocol.

d. Temperature sensor: This application is very important to know the exact temperature inside geothermal pipes and does not lose information during the long communications between the sensor and the CPU. Thereby, a digital temperature sensor with a resolution of 0.0625°C and an acquisition time smaller than 1 second was used. The sensor implements a 1-wire communication protocol with a parasite power that derives from data line, hence only two wires are need to operate. A typical and complicated problem that may appear in this kind of communications where the cable length is quite large is related to electromagnetic interference (EMI). For that, the design of a PMOD module board to host an I^2C to 1-wire bridge device was carried on, a chip that comes with a built-in electronic configuration in order to reduce noise and perform more reliable communications.

e. Temperature probe limit switch: An electromechanical limit switch inside the pipe of the Geowire is activated when the weight used to descend the probe pushes it. Concretely, the produced signal from the limit switch is read through one of the GPIO ports of the ARM

http://dx.doi.org/10.5772/66579

in the Zybo in order to calibrate the initial position of every measuring down-up path of the sensor. The motor driver also detects this signal as a security measure that does not allow the motor to continue rolling up, avoiding the weight and the sensor to get inside the device enclosure.

Figure 21 shows the architecture application diagram and **Figure 22** shows the SoC device with peripherals connected.

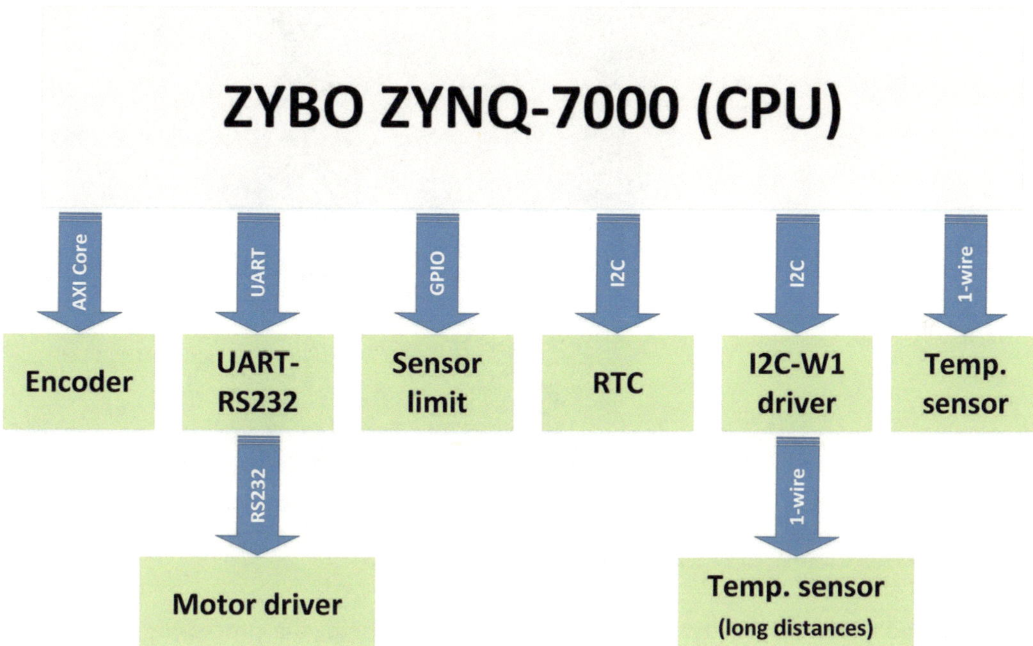

Figure 21. Implemented system architecture.

3.3. Hardware implementation

In order to implement the described application, the Zybo development board based on the Xilinx Zynq-7000 AP SoC architecture was employed. It integrates both a dual-core ARM Cortex A9 processor and Xilinx 7 series FPGA. SoCs provide a very interesting solution due to the combination of both FPGA reprogram ability and flexibility, and powerful ARM processors. Thus, the processor can run operating systems and manage parallel FPGA hardware processing.

Specifically, the processor runs with a clock of 650 MHz and the FPGA is connected to a clock core of 100 MHz. Regarding to communication between the processors, the FPGA cores, and the memories, the AXI-4 interface (Advanced eXtensible Interface) was selected. This interface belongs to the fourth generation of the ARM advanced microcontroller bus architecture (AMBA) interface specification [6]. Furthermore, AXI-4 makes easier the integration of the FPGA IPs and reduces the design effort due to it is optimized in order to achieve more

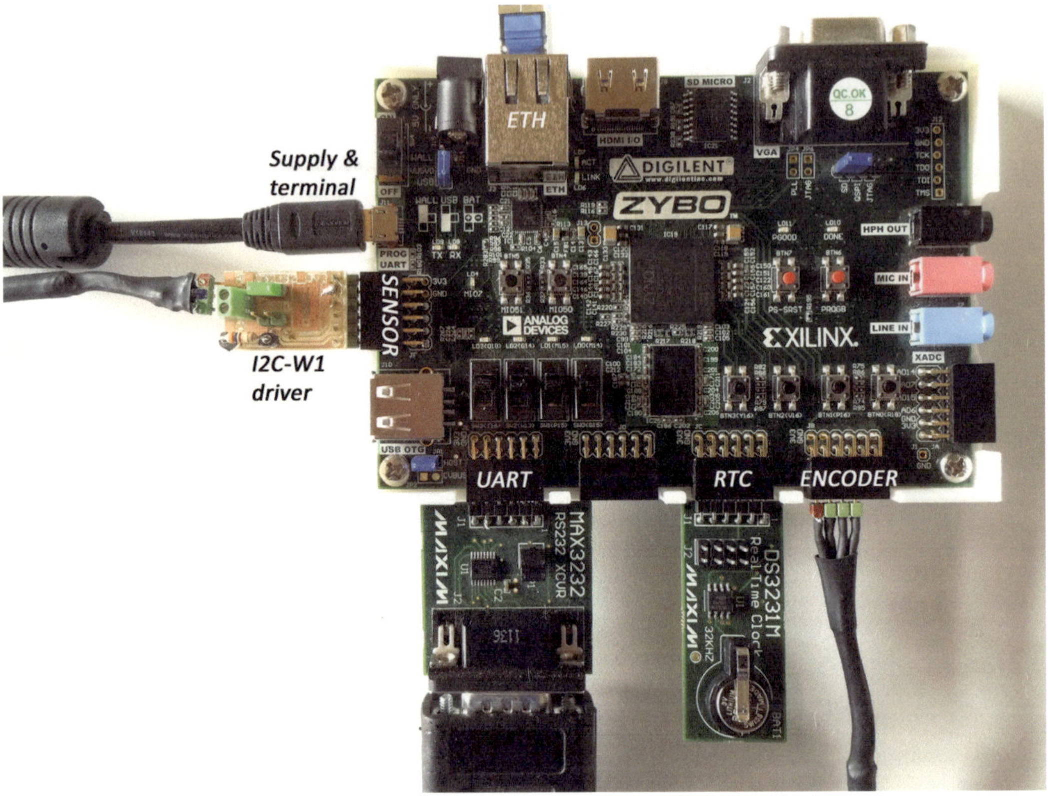

Figure 22. ZYBO board with all the peripherals connected.

flexibility and performance. The processor communicates with the processing system (PS) and programmable logic (PL) peripherals through AXI-4 bus protocol that also enables the data exchange between the external modules across the ports of the Zynq architecture. The other peripherals that compose the hardware are listed as follows:

- DDR3 volatile memory: it loads the operating system and processes.
- SPI-FLASH nonvolatile memory: it is employed to storage the FPGA hardware configuration.
- Ethernet port: it carries out TCP/IP network communications in order to access the services through the implemented web server.
- MicroSD card: this slot stores the embedded Linux kernel and its filesystem together with a copy of the hardware configuration and the bootloader.
- UART-USB port: this peripheral makes possible to communicate with the system console and perform maintenance tasks.
- UART port: this serial communication sends the commands to the servomotor in order to control it.

- I2C port: with this peripheral is managed both temperature sensor driver and the RTC device.
- GPIO port: it detects the sensor limit and communicates with the temperature sensor.
- A custom core implemented in VHDL: reads the signals from the encoder, identifies the direction of rotation, and stores the information in a BRAM register. What is more, the core communicates with the CPU of the system and sends a signal when the distance established in the main program is reached. The core runs in parallel with the ARM processors in order to guarantee a continuous detection of the pulses and avoid loss of information in faster acquisition processes

The software that was employed to carry out the hardware development is the Vivado Design Suite 4.4 of Xilinx. It is an interesting software suite that enables fast and efficient programming of the SoCs and also can automatically generate configurable cores and several interface circuits by using the block design tool.

The previous hardware and interconnections description can be checked in the block scheme which is shown in **Figure 23**.

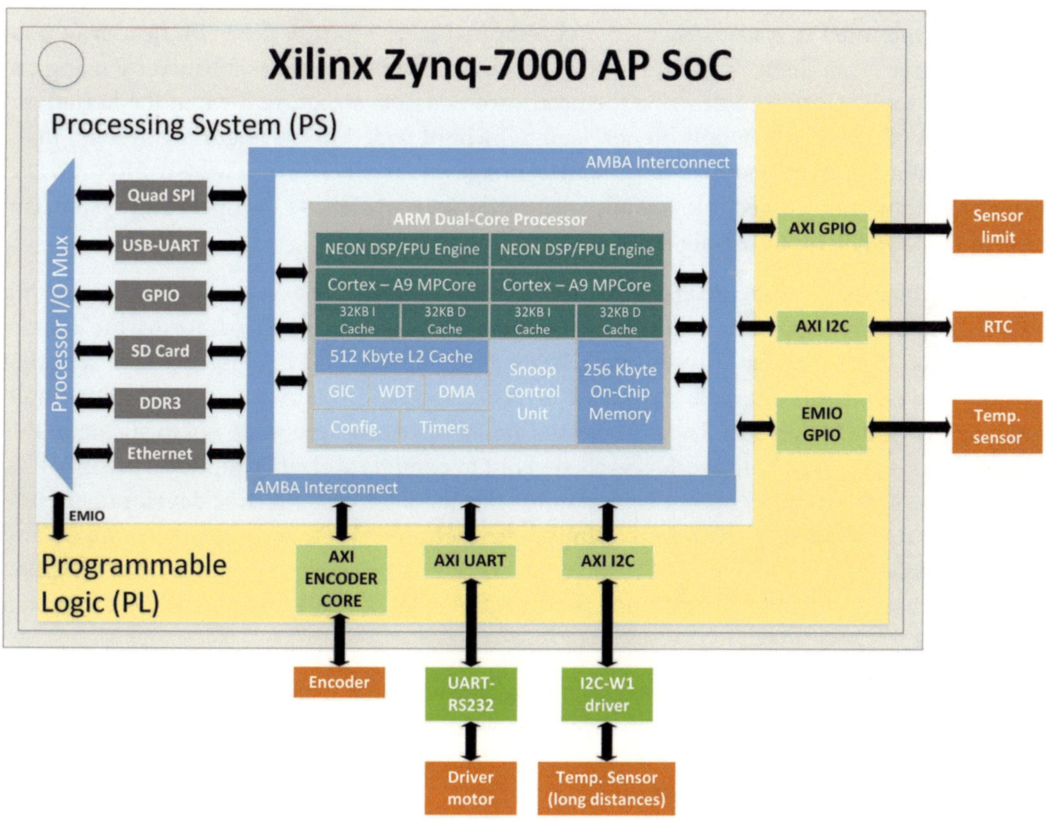

Figure 23. Implemented architecture in Zynq-7000 AP SoC chip.

3.4. Embedded operating system

The source code of an embedded Linux operating system was configured and compiled to run over the previously described hardware. This operating system is based on a modified version of the standard Linux kernel without a memory management unit (MMU), which is optimized for running in embedded systems, but keeping Linux robustness. The main advantage of this system is its well-known free open-software characteristic. However, it has another important advantage, such as all the available resources can be managed by the operating system, while the programmer is abstracted from that layer of hardware. Furthermore, there are a lot of libraries and utilities that can be integrated in Linux make easier the upgrade or development of new applications.

The operating system was implemented using the Linux version provided by the tools of uClinux distribution, which is supported by the ARM processor architecture integrated in the selected SoCs. Specifically, a stable version, in this case the version 3.14, was configured and compiled for the hardware platform in order to create the kernel image (uImage). By using Vivado Suite tool, a first stage bootloader (FSBL) for Zynq was built to set the FPGA with the previously defined hardware, load the operating system image in the DDR3 memory, and begin executing it (BOOT.bin). As the root file system, a prebuilt ramdisk image provided by Xilinx was mounted in a microSD card and wrapped in one part of the FSBL header to boot with it. Moreover, a device tree blob (DTR) project application was established by using the announced tools in order to describe the hardware in a data structure file that the kernel can understand. Thereby, the details do not need to be hard coded in the operating system, making it portable. Once the bootloader, the kernel image and the device tree were configured and generated, they were copied into the microSD card root partition, so the operating system can boot and access the mounted file system.

At last, the database, webserver, and main application processes are executed over the operating system. Looking for a powerful-reliable webserver and database, as well to improve the interaction between them, Apache 2 (webserver), MySQL (database), and PHP (server-side scripting language) were cross compiled for the ARM processor with the toolchain arm-xilinx-linux-gnueabi provided by Vivado. These processes are an open source but are not available in this kernel distribution, so a hard work was carried out to compile and run properly. Nevertheless, the benefits of incorporating them compensate with the later cost-effective development, reliability, features, and performance. In a similar manner, in order to manage the compiled and installed processes, the software of Vivado Suite was employed to program and cross compiled the executable applications. The relation between the most important parts and files is shown in **Figure 24**.

The executable applications are based on a state machine that manages the system flow that interacts with the peripherals and the database.

3.5. Webserver and database application

The main application is known as LAMP (Linux-Apache-MySQL-PHP), a combination of Linux as the operating system, Apache as the webserver, MySQL as the relational database management system, and PHP as the object-oriented scripting language. The advantages of the LAMP stack technology compared with other systems over closed system platforms are as follows:

- Open sourced.
- Highly secure.
- Highly flexibility.
- Scalability.
- Interoperability.

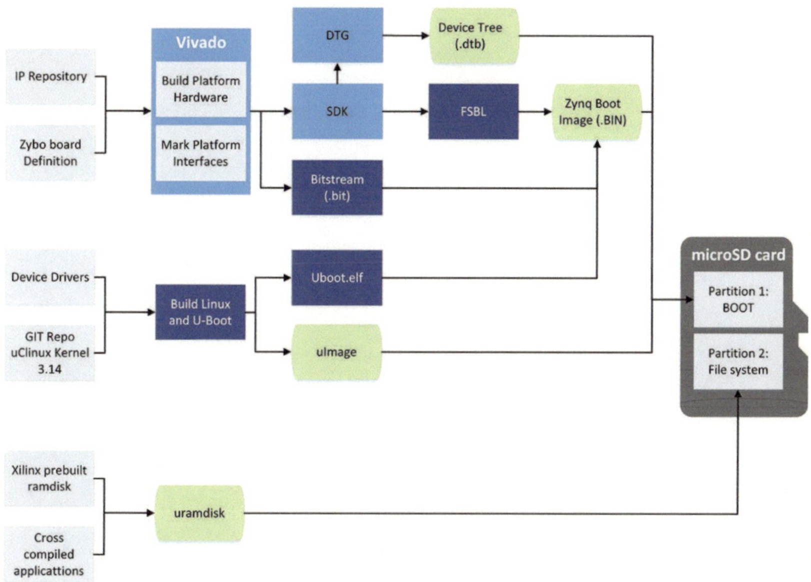

Figure 24. Diagram of files to configure and boot the system.

These characteristics help the designer to develop new applications because it is relatively easy and there is plenty of documentation available. Therefore, the LAMP stack combination makes possible to create truly database-driven and dynamic website that is easy to update and provides a lot of resources to support users. The embedded operating system layer architecture used in this application is shown in **Figure 25**.

Likewise, Apache is the dominant and most important webserver in the word due to its reliability and performance, so that combining it with a powerful database as MySQL and the possibility of using server-side scripting offer very cost-effective and versatile applications.

Commonly, SQlite [7] is implemented as database management in embedded systems because its resource optimization. Nevertheless, in this study due to the powerful processor of Zynq chip, a MySQL database was cross compiled to run in this specific architecture. MySQL offers the following advantages regarding to SQlite:

- Permits multiple queries and modifications at the same time.
- More data types.
- Better compatibility.
- Each table is in a different file (as views and objects).
- Reduces the latency in the queries.
- It is possible manage users with different levels of access.

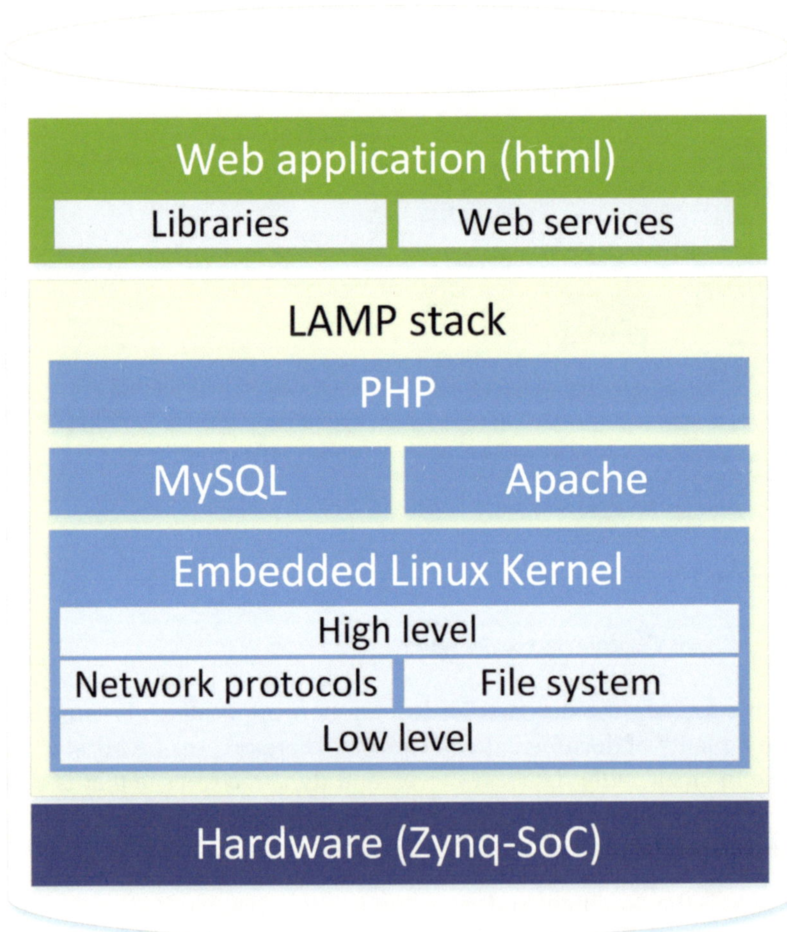

Figure 25. Embedded operating system layer architecture.

The higher software layers of the web application are divided into the following categories: HTML (HyperText Markup Language), JS (JavaScript), CSS (Cascading Style Sheets), and PHP files. This allows the user to control, configure, and monitor the performance of the

Geowire device through surfing the pages of the programmed web interface. Hence, the web interface is structured in three sections:

- Settings: It allows the user to set up motor driver parameters and visualize motor performance indicators and alerts.
- Test: It is possible configure database tables in order to begin a temperature acquisition process at preestablished sets of depth or load previously implemented acquisitions profiles. These parameters are storage in the database together with a timestamp during the acquisition process.
- Graph and charts: In this section, the user can visualize the representation of previously measured depth-dependent temperature profiles, visualize real-time acquisition processes, or download a Microsoft Excel spreadsheet with the recorded temperature profiles.

The design of the web application interface is shown in **Figure 26**.

Basically, when the client requests the application files to the web server in order to render the web interface, the PHP files are first processed by the server. Then, the output of those PHP scripts is transmitted to the client so dynamic content can be added. In this manner, when the client makes a PHP request that script is processed in the server which can manage the database or the system applications.

Thereby, it is possible to execute the server side applications that define the system operation flow from the client interface and manage the device peripherals. The implemented LAMP stack client-server communication diagram is shown in **Figure 27**.

Moreover, initially a validation is required in order to protect the web content from no authorized accesses. In this manner, even if the system is connected to Internet, the data content will not be accessible to the public if they are not accredited.

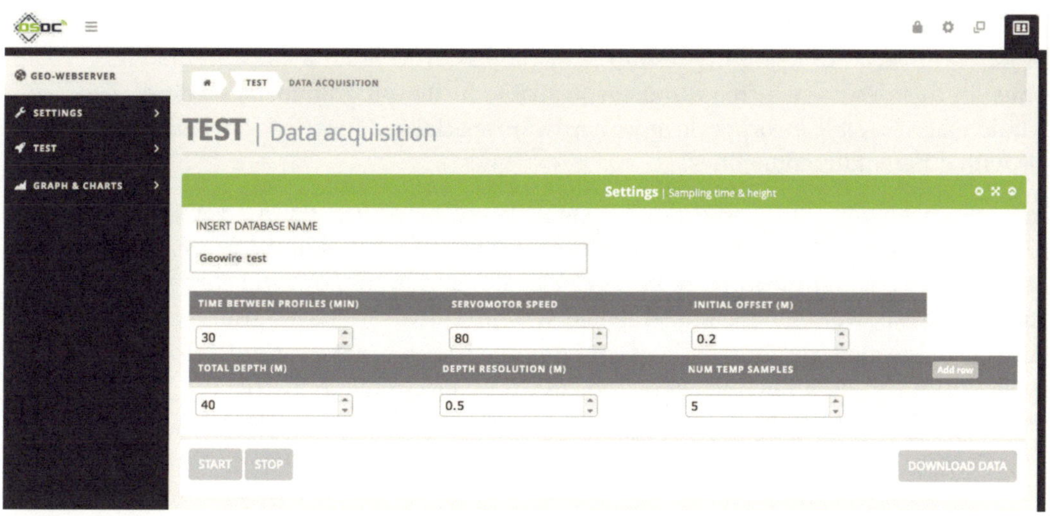

Figure 26. Web application interface for the acquisition process.

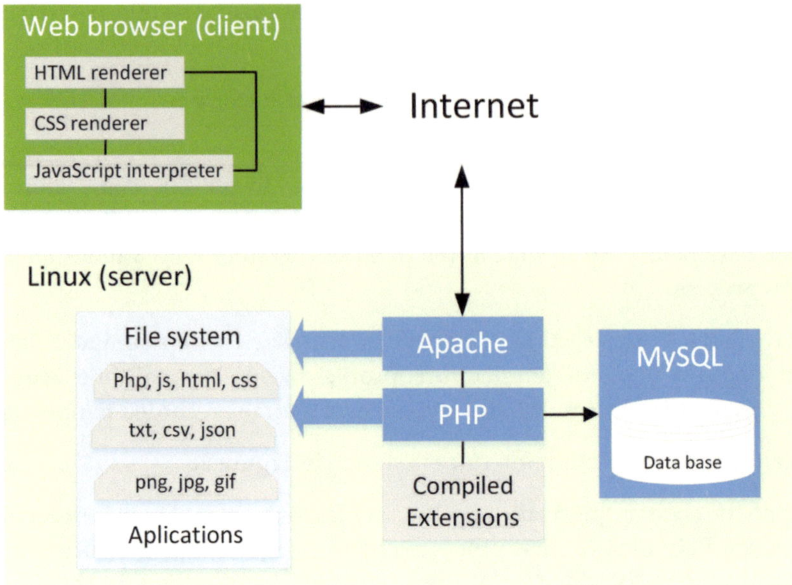

Figure 27. Implemented LAMP stack client-server communication diagram.

4. Results

4.1. System implementation and laboratory results

Figure 28 presents the obtained results after the hardware postimplementation using the Vivado tool. There are no more resources available for the mixed-mode clock manager (MMCM) and the resources for the input/output are 78%, but still there are plenty of resources for use more memory and look up table (LUT). One of the advantages of a system based on the AP SoC architecture is that it permits to implement the operating system in the ARM processor, instead of using the resources of the FPGA to build a software microcontroller. Thus, in this design, the free resources available in the programmable logic of Zynq chip would make possible incorporate more hardware specialized functions or configure peripherals with different functionalities.

In comparison with other embedded systems based on ASICs, the AP SoCs give the same function by porting the uClinux system except the performance of the processors. A design using an ASIC could incorporate faster clock rates and more powerful core processors that consume less power than the implemented system. However, the programmable logic in the AP SoC offers more flexibility. The reconfigurable IP cores make easier to reduce the design cycle which can be a key factor for the consumer electronic industry. The peripheral can be easily added or modify to the hardware with less effort according to the necessities of the customers. Additionally, the designer can implement multiprocessor by using the dual core ARM and the logic in the FPGA, which will improve the performance of the embedded system.

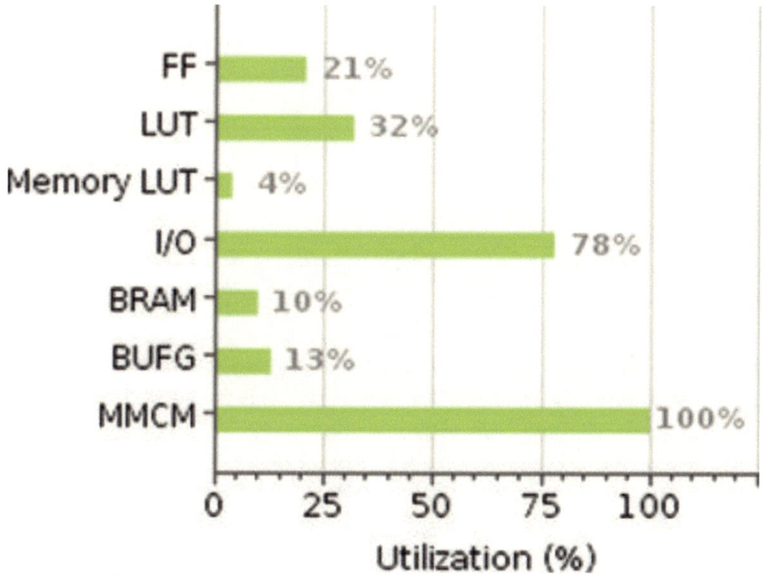

Figure 28. Resources utilization after the hardware postimplementation.

The webserver robustness was test by a concurrent access of different users to the web application. A number of 20 users from different devices were able to access the different sections of the interface without affecting the system performance. Neither the speed nor the efficiency was affected in comparison with the access of a single user.

The digital temperature sensor was calibrated using a thermal bath and accurate thermometer. The calibration was implemented using 5 points, from 0°C and by increasing 5 until 25°C, normally the temperatures during a TRT are comprehend inside that range. Then, a linear trend line was calculated to fit the data from the digital sensor with the recorded temperatures through the accurate sensor. The calibration equation is applied to the obtained data from the sensor before saving the values in the database.

The wire connected to the sensor passes through three rollers that rotate when the wire is released to lower the sensor inside the geothermal pipes. In one of these rollers, six magnets were attached along the diameter and separated the same distance between them. On the external part of the device enclosing a hall effect sensor that detects the polarity changes of the magnets. This mechanism is the encoder that measures the angular movement of the roller. Because the roller diameter, the separation between the magnets and the wire diameter is known, the system program is able to calculate the distance traveled by the sensor. The maximum spatial resolution of 1 cm was achieved after testing and calibrating the instrument by *in situ* measurements. In order to ensure the repeatability of the device for longer distance increments more trials were conducted. For a distance increment of 0.5 m, a deviation of ±1 cm was observed after 20 trials. Besides, a maximum deviation of ±5 cm was observed between distance intervals of 50 m after 20 trials (**Table 1**).

Distance interval	Deviation
0.5 m	±1 cm
50 m	±5 cm

Table 1. Geowire distance interval measurement deviation.

4.2. Experimental results throughout a distributed thermal response test in a borehole heat exchanger

In this section, an experimental test of the implemented system in one of the four BHE installed on the campus of the University of Liege (Liege, Belgium) is presented, with the aim of testing its performance and analyzing the obtained results.

The installation is equipped with double-U geothermal pipes of 100 m long, over a surface area of 32 m². Deposits of sand and gravel characterize the site geology until a depth of approximately 8 m. Then, the bedrock follows until the end of the borehole, which consists mainly of siltstone and shale inter-bedded with sandstone, while fractured zones are detected in the rock mass mostly until a depth of 35 m. In this installation, thermal behavior of fractured bedrock stratigraphy was being investigated throughout TRTs and distribution temperature sensing (DTS) technique. Hence, during the insertion of the geothermal pipes, fiber optics thermometers were tapped every 50 cm in direct contact with the outside part of the pipe wall. Given the relatively small borehole diameter (136 mm), spacers were not used during the installation and the distance between the U-legs is in the order of 3 cm. Among the four available boreholes, the test was conducted in B2, which was backfilled with a bentonite-based commercial material (Füllbinder, $\lambda = 0.95$ W/mK).

Fiber optics makes possible to obtain continuous, high-resolution temperature profiles along the pipes length by applying the DTS technique [8]. The temperature resolution of the fiber optic measurements presented in this study (standard deviation) was in the order of 0.05°C. Temperature was recorded every 20 cm (sampling interval) with a spatial resolution of 2 m.

On 15 December 2015, an enhanced TRT of a heat injection of 2 kW and a duration of approximately 7 days was conducted in one of the single U-pipes of B2 which are disposed in a parallel configuration. The fiber optic cables were installed along the single U-pipe in which the heat was injected while the Geowire was inserted in the nonheated single U-pipe (observer pipe), which was filled with water. **Figure 29** presents a cross section of the borehole with the U-pipes, the location of the temperature measurement instruments where the heat was injected.

From the user interface, the Geowire temperature acquisition sequence was settled to lower the sensor until a depth of 40 m. Then, the sensor was established to stop every 0.5 m for a period of 5 s in order to achieve a thermal stabilization and avoid the possible convective effects, produced by the moving water when lowering the sensor. After 5 s, the device was programmed to record three samples of temperature, one every second and storage the average value in the specified database.

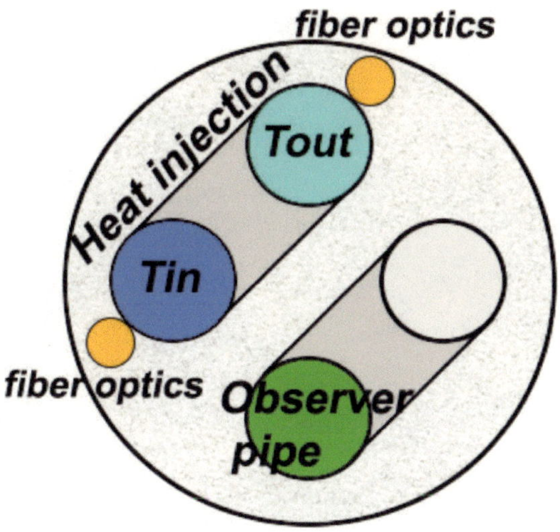

Figure 29. Cross-section of the BHE utilized in the experiment.

Figure 30 shows the Geowire adapted in a structure and the different components of the implemented system before the beginning of the test in the field.

The section of Graph & Charts of the web application allows the users to visualize the recorded data remotely during a real test. Once the device is running it is possible to observe real-time charts or load previously stored temperature profiles from the database. **Figure 31** shows the user interface for a temperature profile obtained during the announced test.

Figure 32 presents the recorded temperature profile with the proposed system inside the observer pipe, and the registered fiber optic temperature measurements (along pipe inlet and pipe outlet outer surfaces) during the heating phase of the TRT. The oscillations observed in the recorded datasets along the observed pipe may be attributed to the varying distance through depth between the observer pipe and the U-heated pipe, as well as to the ground heterogeneity. Moreover, in the first ~18 m, temperature is also affected by the air temperature as previously studied by Radioti et al. [9].

The main advantage of the recorded temperature profiles is that, by applying the proposed procedure of Aranzabal et al. [10], it can contribute to calculate a detailed depth-dependent thermal conductivity profile of the BHE subsoil surrounding layers. Basically, it consists in an iterative simulation process of a numerical model in order to fit simulation results with experimental data.

Figure 30. The implemented system before the beginning of the acquisition process in a BHE.

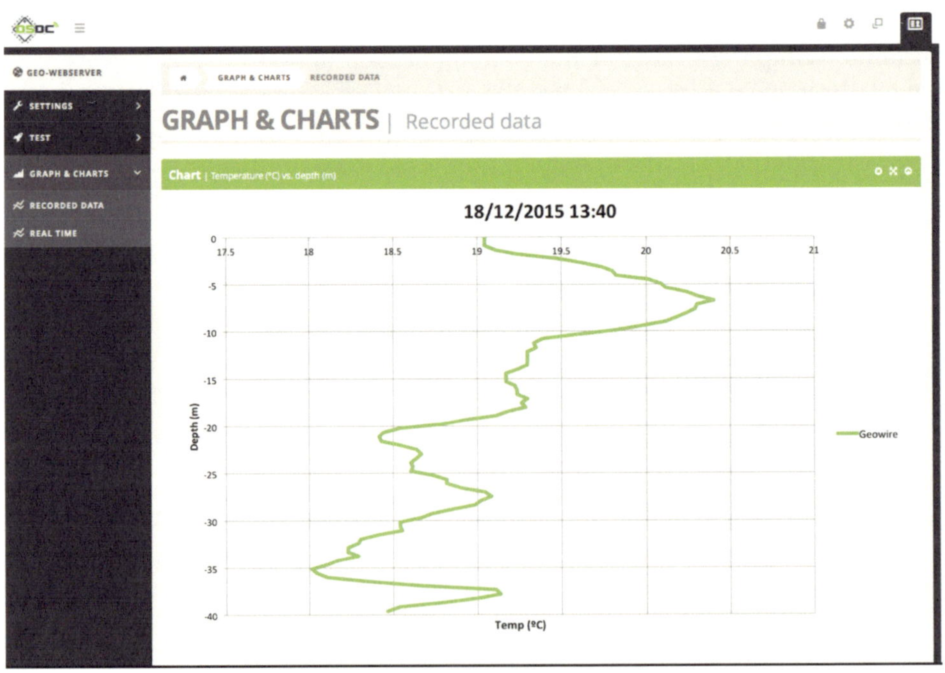

Figure 31. Recorded temperature profile along the depth through the user interface.

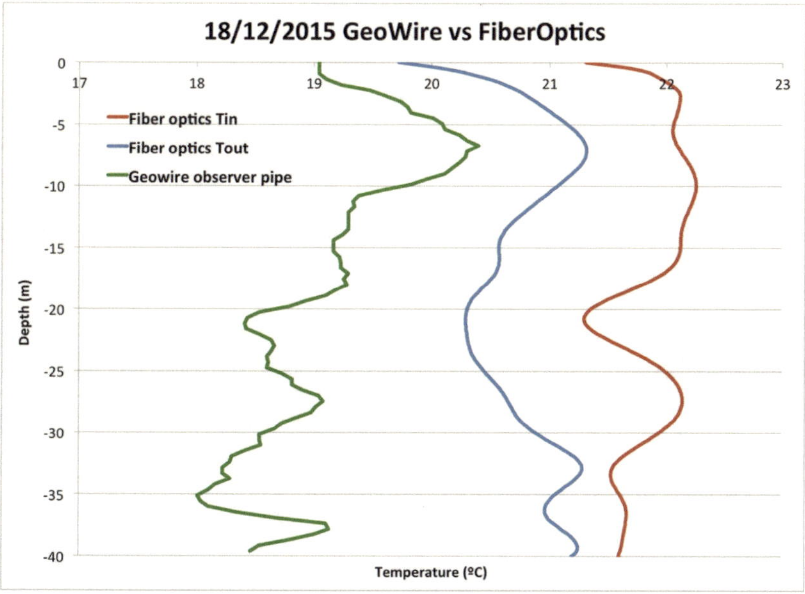

Figure 32. Recorded vertical temperature profile by the Geowire and the fiber optics for 18 of December.

5. Conclusions

The embedded systems have increase its popularity given the society evolution towards a more technology-based and automated necessities. There is no doubt about how fast this market is progressing, one reason is because more resources are being invested and second due to the constant increase in the integration capacity of the components inside the chips. For that reason, many designers need to update continuously their know-how about the methods to design SoC-based applications, which combine processing systems and programmable logic, in order to be competitive.

This chapter has been covered the evolution timeline of FPGA-based systems from its beginning until the final AP SoC chips. They are complex devices and it is necessary to have a deep understanding to utilize them in more efficient way. In this manner from all the advance digital systems, the SoCs are analyzed in more detail because they provide a solution that combines the different digital and analogic elements embedded only in one chip. That permits to develop more versatile applications but it is necessary a well-known comprehension of the analysis and design methodologies.

The introduced codesign hardware/software methodologies are very useful technique to develop time-effective applications. A good coordination between the developers of the same team is required, being the tools that can help to simulate software and hardware designs fundamental to achieve competitive results.

This study we presented the evident improvements in the SoC systems, which can run an embedded Linux for interfacing FPGA-based designs. Thus, the development of a real application has been described and the advantages over other system have been highlighted.

Author details

Nordin Aranzabal*, Adrián Suárez, José Torres, Raimundo García-Olcina, Julio Martos, Jesús Soret, Abraham Menéndez and Pedro A. Martínez

*Address all correspondence to: nordin.aranzabal@uv.es

Department of Electronic Engineering, University of Valencia, Burjassot, Valencia, Spain

References

[1] Stephen M. Three Ages of FPGAs: A Retrospective on the First Thirty Years of FPGA Technology. Proceedings of the IEEE Vol. 103, No. 3; 2015. DOI: 10.1109/JPROC.2015.2392104

[2] Sang Don K. and Seung Eun L. An ARM Cortex-M0 Based FPGA Platform in Teaching Computer Architecture. International Journal of Computer and Information Technology Vol. 04 – Issue 06; 2015

[3] Schaumont P. A. Practical Introduction to Hardware/Software Codesign. In: Springer; 2013; ISBN: 978-1-4614-3737-6

[4] Tyson S. H, James O. H. Using an FPGA Processor Core and Embedded Linux for Senior Design Projects. IEEE International Conference on Microelectronic Systems Education; 2007

[5] Szabolcs H and Sándor-Tihamér B. Implementation of Embedded Linux Systems on FPGA Based Circuits for Real Time Process Control. International Conference on Recent Achievements in Mechatronics, Automation, Computer Science and Robotics; 2015

[6] Young-Taek K, Taehun K, Youngduk K, Chulho S, Eui-Young C, Kyu-Myung C, Jeong-Taek K and Soo-Kwan E. Fast and Accurate Transaction Level Modeling of an Extended AMBA2.0 Bus Architecture. Proceedings of the conference on Design, Automation and Test in Europe Vol. 3; 2005

[7] Kun Y, Linying J, Liu Y and Heming P. Research of Embedded Database SQLite. Application in Intelligent Remote. International Forum on Information Technology and Applications; 2010

[8] Hermans T, Nguyen F, Robert T. and Revil A. Geophysical methods for monitoring temperature changes in shallow low enthalpy geothermal systems. Energies; 2014. DOI: 7 (8): 5083–5118

[9] Radioti G, Delvoie S, Radu J. P, Nguyen F, and Charlier R. Fractured bedrock investigation by using high-resolution borehole images and the Distributed Temperature Sensing technique. In: Canada. ISRM Congress 2015 Proceedings - Int'l Symposium on Rock Mechanics; 2015

[10] Aranzabal N, Martos J, Montero A, Monreal L, Soret J, Torres J, García-Olcina R.. Extraction of thermal characteristics of surrounding geological layers of a geothermal heat exchanger by 3D numerical simulations. In: Applied Thermal Engineering; 2016. DOI: 99: 92–102

Chapter 7

The Use of FPGA in Drift Chambers for High Energy Physics Experiments

Gianluigi Chiarello, Claudio Chiri,
Giuseppe Cocciolo, Alessandro Corvaglia,
Francesco Grancagnolo, Marco Panareo,
Aurora Pepino and Giovanni Francesco Tassielli

Additional information is available at the end of the chapter

http://dx.doi.org/10.5772/66853

Abstract

In this chapter, we describe the design of a field programmable gate array (FPGA) board capable of acquiring the information coming from a fast digitization of the signals generated in a drift chambers. The digitized signals are analyzed using an *ad hoc* real-time algorithm implemented in the FPGA in order to reduce the data throughput coming from the particle detector.

Keywords: drift chamber, FPGA, CluTim, impact parameter

1. Introduction

A drift chamber (DC) is a detector used in high energy physics experiments for determining charged particle trajectories. It consists of a gas volume and of an array of thin wires at high voltages generating high electric fields. Charged particles passing through the gas ionize it creating electron/ion pairs along their path [1], which, accelerated by the electric fields, produce signal pulses on the wires. The signal pulses from all the wires are then collected and the particle trajectory is tracked assuming that the distances of closest approach (the impact parameter) between the particle trajectory and the wires coincide with the distance between the closest ion cluster and the corresponding nearest wire [1, 2]. The widespread use of light, helium-based gas mixtures, aimed at minimizing the multiple scattering contribution to the momentum measurement for low momentum particles, produces, as a consequence, a low ionization clusters density (12 cluster/cm in a 90/10 helium/isobutane mixture) [3], thus introducing a sensible bias in the impact parameter assumption, particularly for short impact parameters and small drift cell [4]. Recently, it has been proposed an alternative track reconstruction (cluster counting/timing) technique, which consists in measuring the arrival times on the wires of each individual ionization cluster and combining these times to get a bias-free estimate of the impact parameter [5, 6]. Typical time separations between consecutive ionization acts, in a helium-based gas mixture, range from a few ns, at small impact parameters, to a few tens of ns, at large impact parameters [7, 8]. Therefore, in order to efficiently applying the cluster timing technique, it is necessary to have readout interfaces [9, 10] capable of processing high speed signals, in which one can easily isolate pulses due to different ionization cluster.

1.1. Hardware

The wire signals generated by the drift chamber, before being processed, are converted from analog to digital with the use of an analog-to-digital converter (ADC). Requirements on drift chamber performance impose the conversions at sampling frequencies of at least 1 GS/s with at least 8-bit resolution. These constraints, together with the maximum drift times, usually of the order of 1 microsecond, and with the large number of acquisition channels, typically of the order of tens of thousands, mandate some sizeable data reduction, which, however, must preserve all the relevant information. Identifying both the amplitude and the arrival time of each peak associated with each individual ionization cluster is the minimum requirement on the data transfer for storage [5, 6].

This chapter deals with the possibility of using FPGAs for the real-time analysis of the data generated by a drift chamber [11] and successively converted by an ADC (**Figure 1**).

More specifically, a fast readout algorithm (CluTim [12]) for identifying, in the digitized drift chamber signals, the individual ionization pulse peaks and recording their time and amplitude has been developed as VHDL/Verilog code implemented on a Xilinx ML605 Evaluation board [13] shown in **Figure 2**, making use of a Virtex 6 FPGA.

In particular, the used device is a Virtex-6 XC6VLX240T-1 FFG1156 [14] that allows for a maximum input/output clock switching frequency of 710 MHz. The hardware setup includes also an evaluation board AD9625-2.0EBZ [15] shown in **Figure 3** with a pipeline ADC.

The analog-to-digital converter (ADC) used is an AD9625 [16], a 12-bit monolithic sampling ADC that operates at conversion rates of up to 2.0 GSPS, with 3.48 W power dissipation.

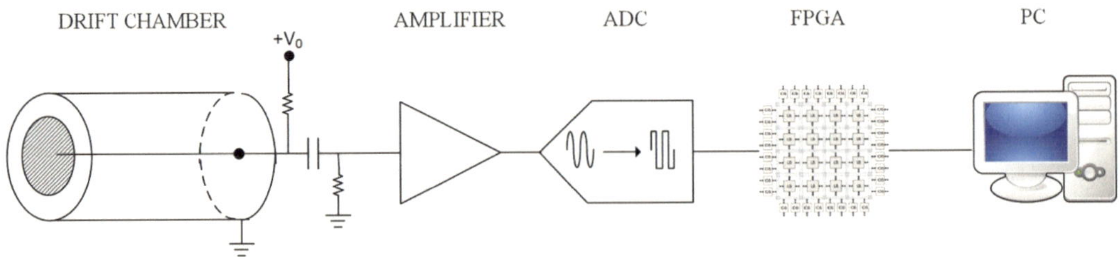

Figure 1. Channel setup.

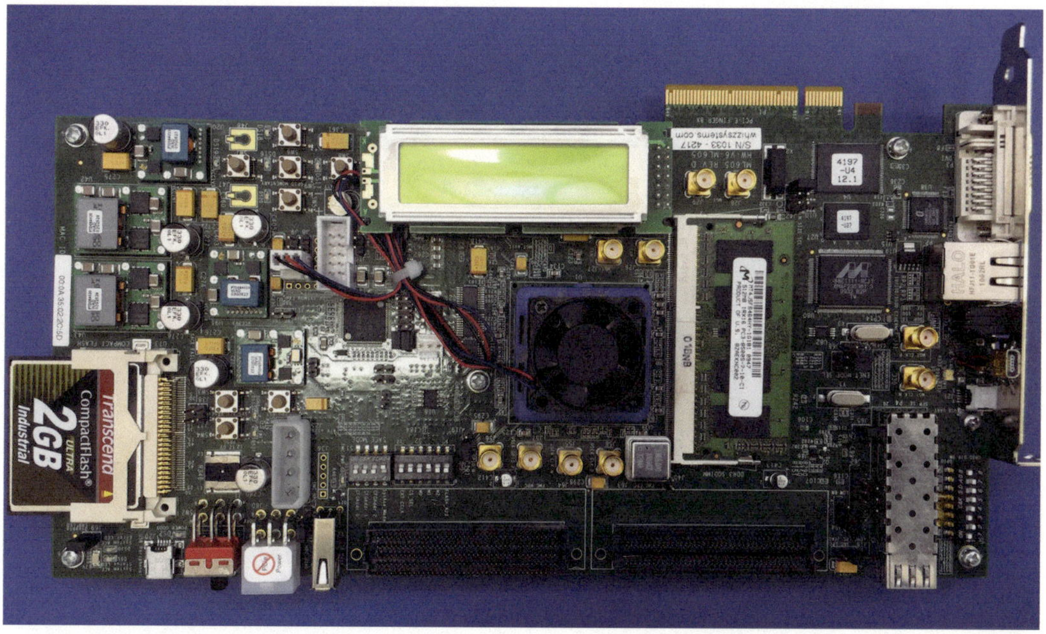

Figure 2. Xilinx ML605 Evaluation Board.

This product is designed for sampling wide bandwidth analog signals up to the second Nyquist zone. The combination of wide input bandwidth, high sampling rate, and excellent linearity of the AD9625 is ideally suited for data acquisition systems, and for the purpose of the experiment.

The analog input clock signals are differential. The standard output used is JESD204B-based high speed serialized output [17] that is configurable in a variety of one-, two-, four-, six-, or eight-lane configurations.

The ADC configuration (sampling frequency, number output lines, power control, etc.) is set with three single-ended lines (clock, date, and enable) that configure the internal register of an SPI. The signals on these lines are managed from the VIRTEX, where a VHDL script generating the bits stream to configure the SPI register, the enable signal and the clock is implemented, and sent to the ADC.

The AD9625 digital output complies with the JEDEC Standard No. JESD204B, serial interface for data converters.

JESD204B is a protocol linking the AD9625 to a digital processing device over a serial interface up to link speeds in excess of 6.5 Gbps. The benefits of the JESD204B interface over LVDS include a reduction in the required board area for data interface routing and enabling smaller packages for converters and logic devices.

The JESD204B data transmit block assembles the parallel data from the ADC into frames and uses 8-bit/10-bit encoding as well as optional scrambling to form serial output data. Lane

Figure 3. AD9625-2.0EBZ Evaluation Board.

synchronization is supported using special characters during the initial establishment of the link. Additional data that are used to maintain synchronization are embedded in the data stream thereafter. A JESD204B receiver (the FPGA) is required to complete the serial link.

The AD9625 JESD204B transmits block maps to two digital down converters for the outputs of the ADC over a link.

A link can be configured to use up to eight JESD204B lanes. The JESD204B specification refers to a number of parameters to define the link, and these parameters must match between the JESD204B transmitter and receiver.

The JESD204B protocol stack consists of seven functional blocks in the transmit path and seven functional blocks in the receive path, as shown in **Figure 4**.

Xilinx offers a complete working solution that simplifies the adoption of JESD204B [17], including best-in-class serial transceivers, IP, design tools, reference designs, and ecosystem partners.

- **GTH transceiver architecture:** Reducing the BER in serial transceiver transmissions directly correlates to the peak bandwidth that can be achieved through the serial link. Xilinx FPGAs offer industry-leading jitter performance, achieved through a unique capability called adaptive receiver equalization. This improves the performance of the decision feedback equalizer (DFE) block, significantly reducing BER.
- **Xilinx JESD204B IP core:** Xilinx offers the industry's first fully JESD204B-compliant IP core for programmable logic. This core supports the full JESD204B bandwidth specification of 12.5 Gbps over one to eight lanes. It can be configured as both a transmitter for interfacing to a DAC and as a receiver for interfacing to an ADC. The core also includes support for scrambling and initial lane alignment.
- **JESD204B reference designs:** Complete JESD204B reference designs are provided for Xilinx development boards by a variety of third-party analog vendors, such as analog devices.

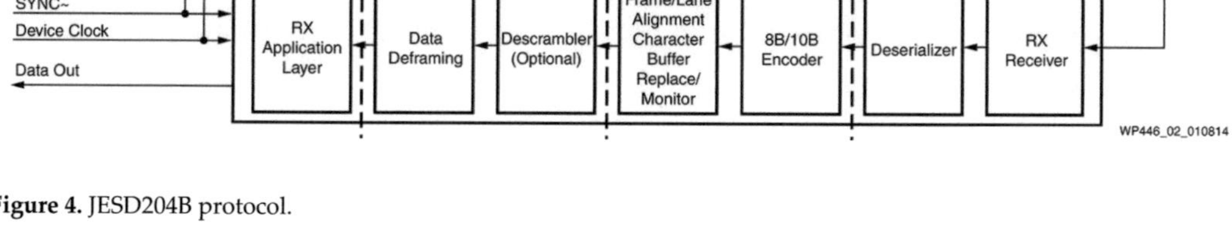

Figure 4. JESD204B protocol.

The two evaluation boards are connected by a high-speed VITA 57 Mezzanine Connector in order to limit parasitic effects due to pin couplings.

The FPGA Mezzanine Card (FMC) standard has proven to be highly popular with over 100 total FMC cards now available from a variety of partners. Over 30 of these FMCs specifically support high-speed data converters. The FMC provides a way for customers to quickly configure their standard Xilinx development boards with real-world analog interfaces.

Xilinx partners have been providing many easy-to-use (and to re-use) reference designs that save customers weeks or even months of development time. Building on this success are the first high-speed analog FMC cards supporting JESD204B from industry-leading analog providers such as analog devices, IDT, 4DSP, NXP, and others.

1.2. Software

The Xilinx ISE 14.5 software has been used to design, develop, and test the CluTim algorithm. It allows for the analysis and synthesis of source code written in a hardware description language (HDL) such as Verilog and VHDL, provides designs, and is also able:

- **To perform timing analysis:** To provide a detailed analysis of the FPGA design. This ensures that the specified timing constraints are properly passed to the implementation tools.
- **To examine RTL diagrams:** After the HDL synthesis phase of the synthesis process, it is possible to show a schematic representation of the synthesized source file. This schematic shows a representation of the preoptimized design in terms of generic symbols, allowing for control of design issues early in the design process.
- **To simulate a design:** ISE simulator (ISim) provides a complete, full-featured HDL simulator integrated within ISE, with which is possible to perform waveform tracing, waveform viewing, and HDL source debugging.
- **To configure the target device**: It is possible to program the device with a JTAG programmer.
- **To provide IP core**: IP (intellectual property) core is a block of logic or data that is used in an FPGA. It is part of the growing electronic design automation (EDA) industry trend toward repeated use of previously designed components. Ideally, an IP core should be entirely portable—that is, must be easy to insert into any vendor technology or design methodology.

The real data can be stored and visualized using the Chip Scope PRO software. It inserts logic analyzer, system analyzer, and virtual I/O low-profile software cores directly into design, allowing them to view any internal signal or node, including embedded hard or soft processors. Signals are captured in the system at the speed of operation and brought out through the programming interface, freeing up pins for the design. Captured signals are then displayed and analyzed using the ChipScope Pro analyzer tool. These signals can be also saved to be processed with other tools, i.e., MATLAB.

http://dx.doi.org/10.5772/66853

2. Algorithm description and FPGA implementation

The chosen hardware involves the use of an FPGA device in which 20 transceiver are implemented. This means that one can use a single FPGA to connect multiple ADC in parallel, up to three, if one adopts a configuration which uses six out of the eight lines at high speeds, configurations with fewer lines at the maximum sampling frequency are also possible, but this requires a digital-down-conversion (DDC) of the signal, implemented internally to the ADC, which cannot be applied in the described case, since it acts like a filter to the signal which, if not limited to a narrow band, rather than harnessing the entire band up to 1 GHz (as in the described case), results in loss of information.

The ability to connect multiple devices in parallel has the effect of having a very large number of data to store; if the data from the ADC were stored without preprocessing to perform a reduction and store only those useful for experimental purposes would involve several disadvantages, as follows:

- A time window of observation of the event greatly reduced, because the internal memory to the FPGA would be filled very quickly.
- A very long data transfer time from FPGA to PC for their postprocessing, due to the very large amount of data to be transferred.
- A very long time to postprocess data.

All of these issues require a real-time data preprocessing to store only useful data.

The CluTim algorithm [12], here described, is able to process the data in real time. In particular, it

- identifies, in the digitized signal, the peaks corresponding to the different ionization electrons.
- stores each peak amplitude and timing in an internal memory.
- sends the data stored to an external device when specific trigger signals occur.

The ability of an FPGA to perform multiple operations in parallel turns out to be useful having to manage a large amount of data coming from the ADC at a very high rate.

In fact, in this way, one can execute on different data, at the same time, multiple instructions performing the same function.

Before the written algorithm starts to work properly, the ADC is configured by loading the SPI internal registers with appropriate values.

The clock handling the loading of the SPI has a frequency of 12 MHz, supplied by FPGA by means of the IP core Clocking Wizard 3.6 [18], able to generate a number of clocks shifted in a phase by predetermined values and with frequencies selected from a specific range, starting from the same master clock signal frequency.

In the case of the FPGA used, this range is from a minimum value of 10 MHz to a maximum of 700 MHz (the maximum frequency value sustainable by the FPGA). The master clock used has a frequency of 66 MHz and is generated by a MBH2100H—66 MHz oscillator, which is mounted on the demo board.

After properly configuring the ADC, it begins the first phase of communication between the JESD204B transmitter implemented on the ADC and the receiver implemented on, called synchronization group code (CGS).

In this phase, the receiver finds the boundaries between the 10-bit characters in the data stream. During the CGS phase, the JESD204B transmit block transmits a known sequence of characters *K*. The receiver must locate this *K* characters in its input data stream using clock and data recovery (CDR) techniques. The receiver issues a synchronization request by activating the SYNCINB± pins of the ADC. The JESD204B Tx begins sending *K* characters until the next clock boundary. When the receiver has synchronized, it waits for the correct reception of at least four consecutive *K* characters. It then deactivates SYNCINB±. The ADC then transmits an initial lane alignment sequence (ILAS) on the following clock boundary.

The ILAS phase follows the CGS phase and starts on the next clock boundary. The ILAS consists of four multiframes, with *R* known characters marking the beginning and *A* known characters marking the end. The ILAS begins by sending *R* characters followed by 0–255 ramp data for one multiframe. On the second multiframe, the link configuration data are sent starting with the third character. The second character is *Q* known characters to confirm that the link configuration data follows. All undefined data slots are filled with ramp data. The 3 and 4 multiframe are the same as multiframe 1.

After the initial lane alignment sequence is completed, the user data is sent. In a usual frame, all characters are user data.

The synchronization clock signal comes from the ADC with a frequency 1\4 of the sampling frequency, 500 MHz. This external clock is used as a reference clock of the RX PLL implemented in each transceiver. The transceiver gives in output a clock at a half frequency of 250 MHz.

The schematic representation of the connection is shown in **Figure 5**.

From such a signal, using a block called advanced mixed-mode clock manager (MMCM_ADV) [19] provided by ISE, all the clock signals necessary to the management of the transceiver modules and of the module JESD204B provided by XILINX are generated.

The ADC communicates with the FPGA through the transceivers. Each transceiver has a high-speed (up to 6.5 Gbps) serial data line as input and a 32-bit word with a frequency of 125 MHz as output, all 32-bit words (which may be from 1 to 8 according to the number of lines used) are passed simultaneously to the JESD204B block. Here, they are recombined to provide at its output the correct information consisting of 16 words of 12 bits (the ADC resolution) in output at a frequency of 125 MHz (sampling frequency = 16 * 125 MHz = 2 GHz).

The algorithm, if it were to carry out its function to each data serially, must have an execution frequency of 2 GHz in order to be able to process all the information before it is overwritten

http://dx.doi.org/10.5772/66853

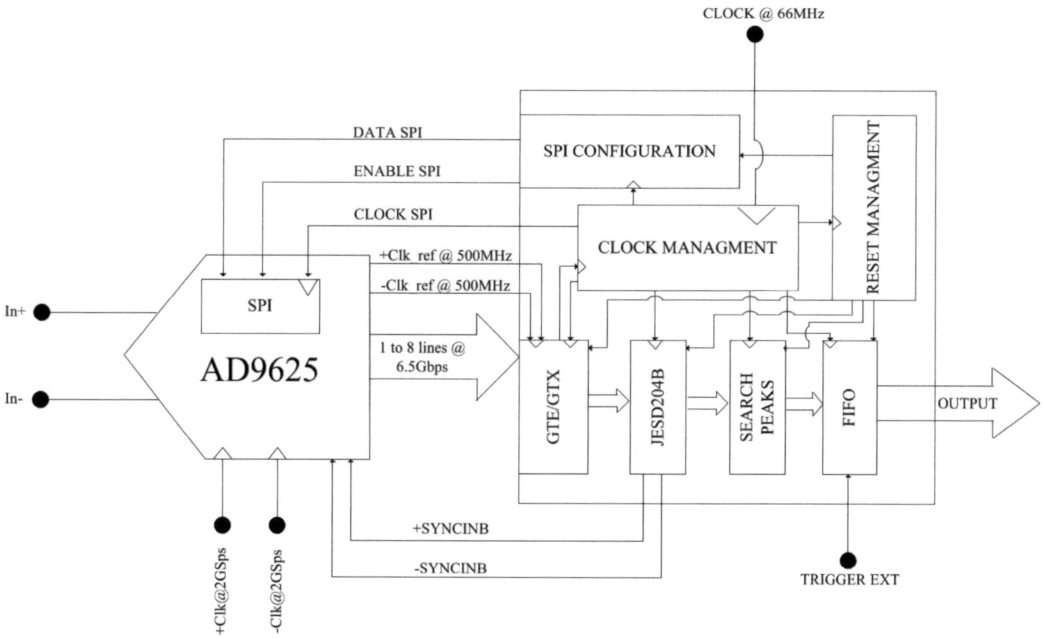

Figure 5. Schematic representation.

by new data. This is physically impossible because the maximum operating frequency of the used FPGA is 700 MHz. To overcome this problem, one has to exploit the ability to process more data in parallel, thus reducing the operating frequency of the FPGA with the advantage of relaxing the time constraints by avoiding the introduction of time delays inside the device.

At the beginning of the signal processing procedure, a counter starts to count providing the timing information related to the event under scrutiny. The determination of a peak is done by relating the *i*th sampled bin to a number *n* of preceding bins, where *n* is directly proportional to the rise times of the signal peak.

The value of *n* has been chosen to be 2. Supposing a 1 ns rise time for the signal, which is sampled at a rate of 2 GS/s, two maxima must be separated by at least three samples to be associated with two distinct peaks.

The implemented algorithm is shown schematically in **Figure 6**.

Among the 16 samples $S_{K,X}$ where *K* is the sample number among those available and *X* is the time at which they are present, the functions $D1_{K,X}$ and $D2_{K,X}$ are calculated according to the relations of Eqs. (1) and (2), respectively (step 1).

$$D1_{K,X} = \left(\frac{2 * S_{K,X} - S_{K-1,X} - S_{K-2,X}}{16} * 3 \right) \tag{1}$$

$$D2_{K,X} = \left(\frac{2 * S_{K,X} - S_{K-2,X} - S_{K-3,X}}{16} * 5 \right) \tag{2}$$

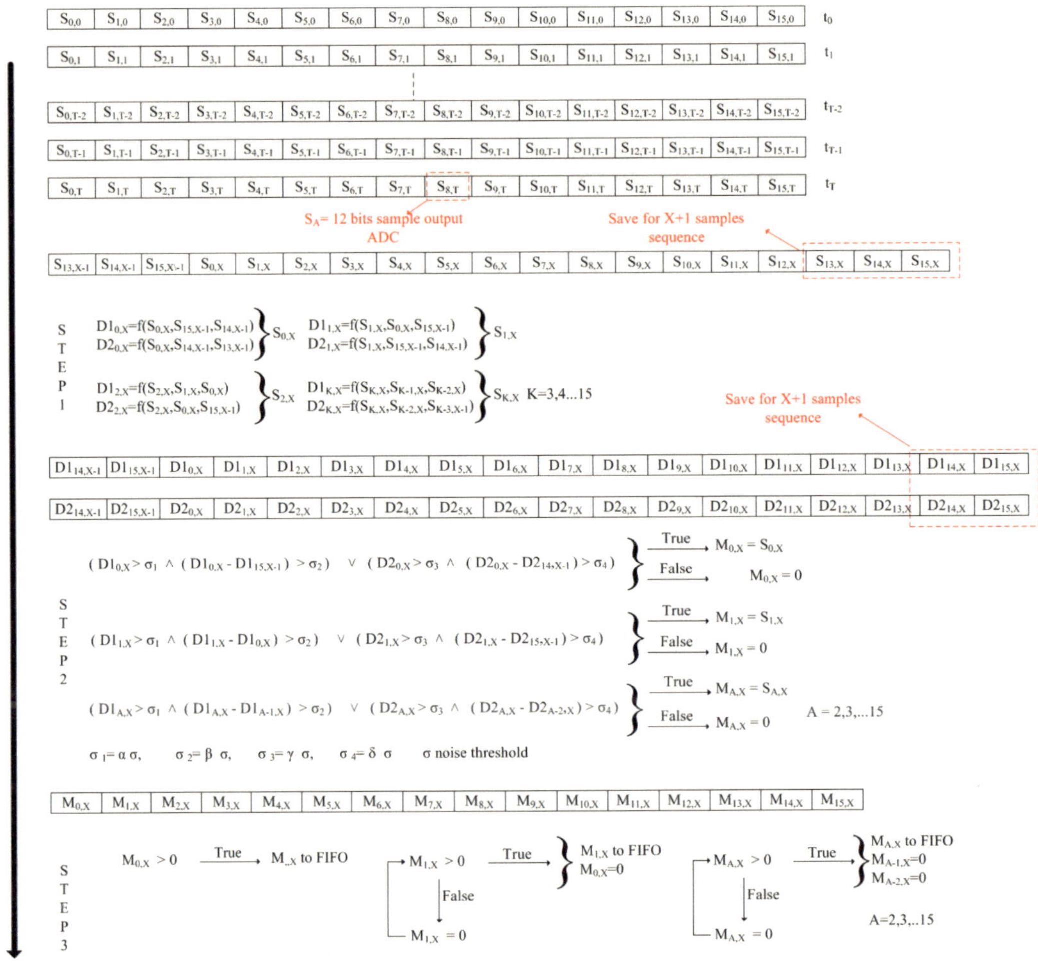

Figure 6. Algorithm implemented.

The value of $D1_{K,X}$ function provides an estimate of the variation of the amplitude of the *i*th sample compared to the (*i*-1)-th and (*i*-2)-th samples. Likewise, the $D2_{K,X}$ function as far as the (*i*-2)-th and (*i*-3)-th samples are concerned.

Figure 7 shows the input signal to the ADC, the peaks found, and the values of the functions and of their differences. As can be noticed, the functions assume their maximum values in correspondence with the signal peaks.

The values of the $D1_{K,X}$ and $D2_{K,X}$ functions are stored in a 16-element vector. For the first three samples in the series of 16 input words, they make use of the last three corresponding samples of the previous 16 input words. A temporary storage is, therefore, used to this purpose. Likewise, in order to be able to calculate the differences $D1_{K,X}$ and $D2_{K,X}$ in the samples head, the values of the functions $D1_{K,X-1}$ and $D2_{K,X-1}$, calculated at the previous time and relating to the samples tail must be temporarily stored.

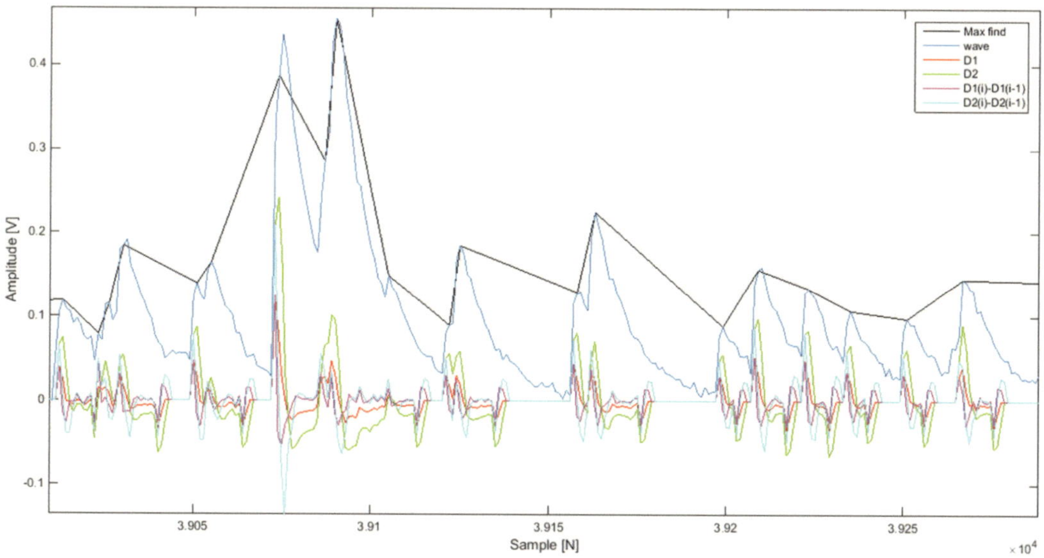

Figure 7. Input signal, found peaks, and discrimination functions.

The values of $D1_{K,X}$ and $D2_{K,X}$ and the differences between $D1_{K,X}$ and $D1_{K\text{-}1,X}$ and between $D2_{K,X}$ and $D2_{K\text{-}1,X}$ are compared with respect to thresholds proportional to the level of noise present in the input signal (step 2).

If the imposed conditions are met, the test sample is identified as a possible maximum and its value is temporarily stored in $M_{K,X}$.

In order to transfer the data to the memory, the last step consists in checking that, according to the conditions imposed on the signal rise time, there are no adjacent peaks (step 3).

To this purpose, a check is performed on the last maximum. If its corresponding location contains a nonzero value, this is transferred into the memory and the two preceding locations are assigned a zero value. The procedure continues by scrolling all locations and sending to memory all effective maxima. The time information is provided with a counter, the value of which is stored in an FIFO every time a peak is found. The counter is clocked at 125 MHz clock frequency and, therefore, when a peak is found, the time to be stored is multiplied by 16 to take into account the correct ADC sampling rate and added to the sample number corresponding to the maximum found, which is illustrated in **Figure 8**.

The memories are continuously filled as new peaks are found. When a trigger signal occurs at time *t*0, indicating with *T* the observed time window, which coincides with the maximum drift time, the reading procedure is enabled and only the data related to the found peaks in the [*t*0, *t*0 + *T*] time interval are transferred to an external device. This results in data reduction factors of more than one order of magnitude.

For storing data, ISE provides several types of IP cores. The one used is the FIFO generator 9.3 [8]. Within this IP, one can choose between various options, each one supporting different

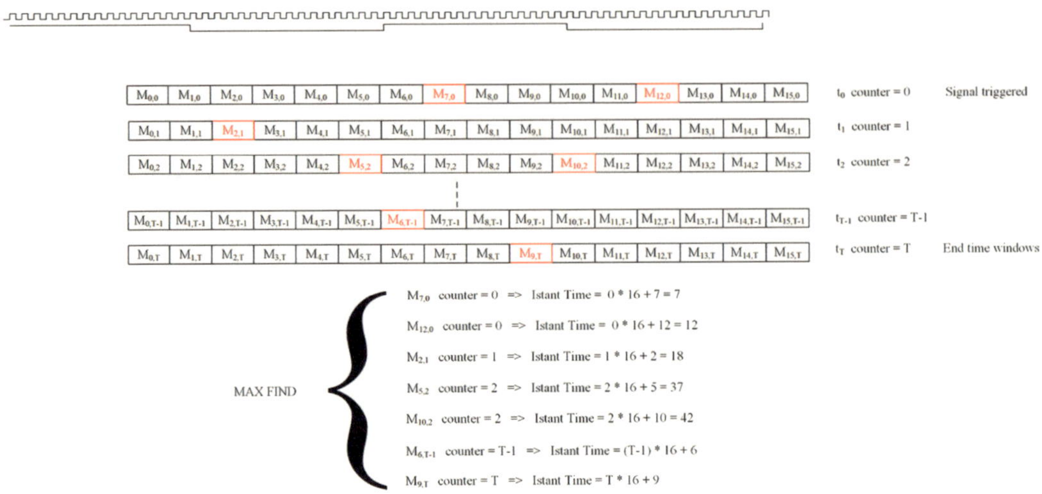

Figure 8. Time information algorithm.

features. For these purposes, an FIFO common clock has been chosen, the depth of which depends on the observed time window *T* of the event, while the width has been chosen of 8 bits. This is linked to the way in which the data is sent to the external device.

ISE provides several methods of data exchange, such as communication through UART and Ethernet.

A UART-type interface has been used. It requires 1 start bit, 8 data bits, and 1 stop bit to communicate.

The dates are sent with a baud rate of 115,200 bps. The clock UART (16 times baud rate) is obtained from a clock of 50 MHz obtained from the Clock wizard IP core.

However, since its transmission speed is very low, the time required to transfer data stored in the FIFO to the external device is very long. To considerably reduce this time, an Ethernet module, having significantly faster transfer times, can be implemented.

The implemented algorithm has been executed both in MATLAB and VHDL, using a test signal with a defined noise of 0.5 mV rms, and a number of peaks. A comparison of the obtained results can be used to derive an index of the algorithm performance. By comparing the results obtained from the number of real and fake peaks found, one can assess if any error is due to the algorithm itself or due to approximations in the VHDL implementation, when the function $D1_{K,X}$ and $D2_{K,X}$ are rounded to integer.

Figures 9 (MATLAB) and **10 (VHDL)** show the algorithm efficiency, calculated using Eq. (3), and the percentage of fake peaks, calculated using Eq. (4)

$$\text{Eff}[\%] = 100^* \frac{Pf - \text{Pfake}}{Pr} \tag{3}$$

$$Pfake[\%] = 100^* \frac{\text{Pfake}}{Pf} \tag{4}$$

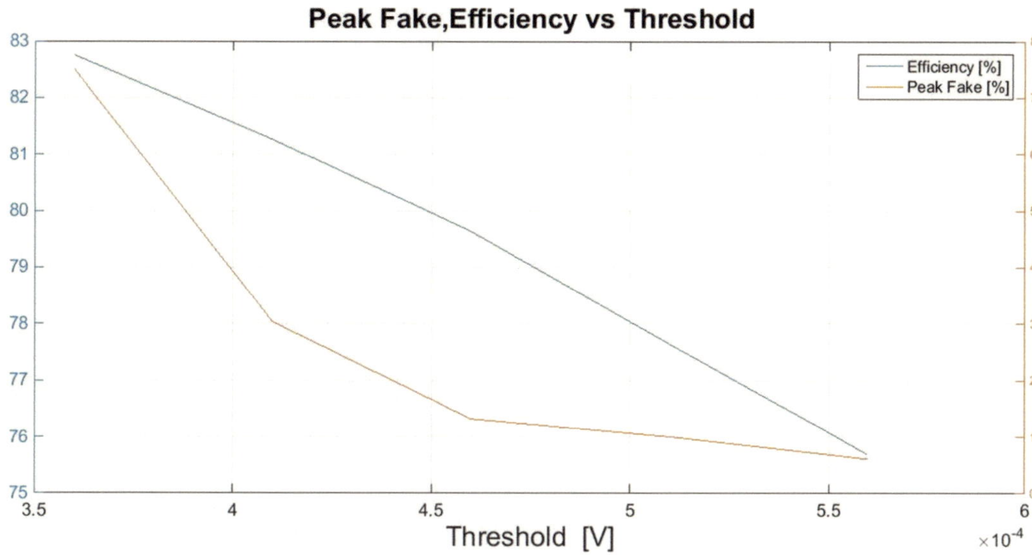

Figure 9. Algorithm results with MATLAB.

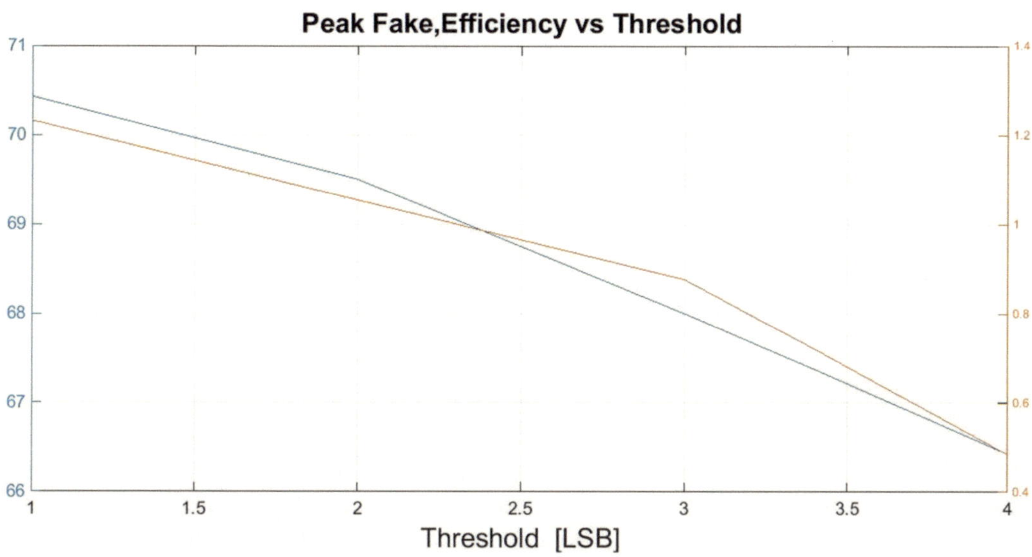

Figure 10. Algorithm results with VHDL.

where *Pf* is the number of peaks found in the signal, *Pr* is the number of real peaks, and *P*fake is the number of fake (equal to *Pf* – *Pr*).

The results were obtained by varying the proportionality factors used to calculate the thresholds to which the $D1_{K,X}$ and $D2_{K,X}$ functions and their differences are compared.

As expected, in both cases, increasing the thresholds not only decreases the efficiency but also reduces the number of false peaks.

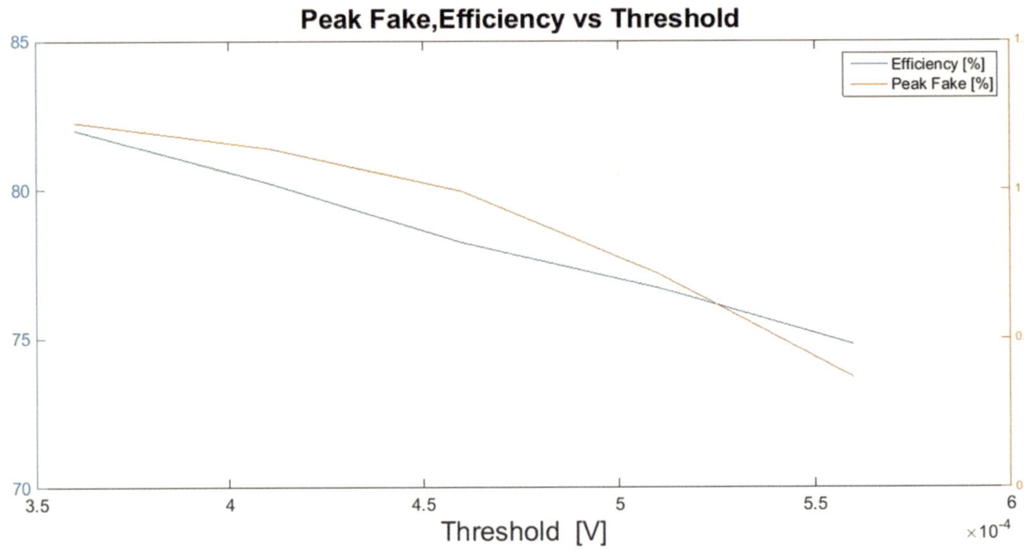

Figure 11. Algorithm results with MATLAB without noise.

The algorithm executed with MATLAB performs slightly better than VHDL, for example, for a rate of fake peaks of 1%, one has an efficiency of 76% in the case of MATLAB and an efficiency of 69% for VHDL.

A further test was performed by implementing the algorithm in MATLAB on the signals previously analyzed without the noise contribution. The results obtained are shown in **Figure 11**.

By comparing **Figures 9** and **11**, it can be seen how the presence of noise induces a reduction in efficiency and an increase of false peaks, indicating that a part of the errors of the algorithm may actually be due to the characteristics of the signal at input.

This problem can be mitigated somehow by trying to increase the signal-to noise-ratio by filtering the input signal to the ADC.

The use of the algorithm described results in data reduction factors of more than one order of magnitude. It is executed for each readout channel and, considering the high number of I/O FPGA pins, it is possible to process multiple channels corresponding to different drift chamber signals with a single device.

3. Conclusion

A key role in the algorithm is represented by the parallel data processing combined with the use of a pipeline structure, which allows managing more input data at the same time, thus relaxing the speed specifications.

http://dx.doi.org/10.5772/66853

The choice of an FPGA device turns out to be the best among those available, since it allows many advantages as follows:

- Adaptability: The possibility of being programmed to perform *"ad hoc"* functions, optimizing its performance in relation to the task to be performed.
- Parallelization: The possibility of executing the same instructions on multiple data simultaneously. In the written HDL code, one can declare multiple processes, which are executed in parallel sheets, while the internal instructions are executed serially.
- Diffusion: The FPGA devices are widely used in various application areas, this has led several manufacturers of electronics to develop scripts in HDL that can be implemented in FPGA to provide the user devices a starting code to be modified according to the needs to create a device-FPGA communication link.
- Management of the links: The user during the programming phase can create appropriate constraints to manage internal links to the FPGA, thus being able to optimize the time delays of the lines, the most critical connections, with a rooting favoring closer.
- IP core: The FPGA programming tool (in this case ISE) makes available the intellectual property, a block of logic data having multiple features, such as the clock management, the management of interfacing with various external devices in the various possible standards, the management of the storage of the information with various types of implemented memories, and many others.

Examining how the performance of FPGA devices has evolved over the years, by taking into consideration factors such as the data storage capacity and maximum operating frequency, one can note that these are strongly linked to the progressive miniaturization of the technology with which they are made.

They follow the technological scaling, which led to the advent of FinFET technology to reduce the size of the channel length to less than 16 nm. The effect of technological scaling leads to different types of advantages as follows:

- A reduction of the length of the connections, thus allowing for a lower propagation delay between the various cells and for a higher operating frequency.
- Being able to integrate, for equal occupied area, more logic cells, thus increasing the computational capacity and the ability to store data.
- A reduction in size, and hence a greater integration of multiple devices in a smaller area.
- A reduction in the power consumption.

The adoption of an FPGA, as the main block of data management and communication between various devices, appears to be winning, not only relying on the existing technology and on the various application tools created by the different manufacturers, but also thinking about the future, that is, how much they can be improved in terms of performance with improved technologies and how fast they can be improved.

Author details

Gianluigi Chiarello[1, 2], Claudio Chiri[2], Giuseppe Cocciolo[1, 2], Alessandro Corvaglia[2], Francesco Grancagnolo[2*], Marco Panareo[1, 2], Aurora Pepino[1, 2] and Giovanni Francesco Tassielli[1, 2]

*Address all correspondence to: franco.grancagnolo@le.infn.it

1Department of Mathematics and Physics "Ennio De Giorgi" – Salento University, Lecce, Italy

2 INFN (Istituto Nazionale Fisica Nucleare), Lecce, Italy

References

[1] Blum W., Riegler W., Rolandi Li. Particle Detection with Drift Chambers. 2nd ed. Springer-Verlag, Berlin, Heidelberg; 2008. pp. 448. DOI: 10.1007/978-3-540-76684-1

[2] Sauli F., Principles of operation of multiwire proportional and drift chambers. In: Experimental Techniques in High Energy Physics. 2nd ed. Addison-Wesley; 1987. pp. 79–188. DOI: 10.5170/CERN-1977-009

[3] Cataldi G., Grancagnolo F., Spagnolo S. Cluster counting in helium based gas mixtures. Nuclear Instruments and Methods in Physics Research Section A: Accelerators Spectrometers Detectors and Associated Equipment. 1997;**386**:485–469. DOI: 10.1016/S0168-9002(96)01164-3

[4] Tassielli G.F., Grancagnolo F., Spagnolo S. Improving spatial resolution and particle identification. Nuclear Instruments and Methods in Physics Research, Section A: Accelerators, Spectrometers, Detectors and Associated Equipment. 2007;**572**(1 SPEC. ISS.):198–200. DOI: 10.1016/j.nima.2006.10.300

[5] Cascella M., Grancagnolo F., Tassielli G. Cluster Counting/Timing Techniques for Drift Chambers. In: Francesco Grancagnolo and Marco Panareo, editors. 1st Conference on Charged Lepton Flavor Violation; May 2013; Lecce (Italy). Elsevier; 2014. pp. 127–130. DOI: 10.1016/j.nuclphysbps.2014.02.025

[6] Signorelli G., Donofrio A., Venturini M. A novel method to estimate the impact parameter on a drift cell by using the information of single ionization clusters. Nuclear Instruments and Methods in Physics Research, Section A: Accelerators, Spectrometers, Detectors and Associated Equipment. 2016;**824**:581–583. DOI: 10.1016/j.nima.2015.11.028

[7] Grancagnolo F. The ultimate resolution drift chamber. Nuclear Physics B—Proceedings Supplements. 2007;**172**:25–27. DOI: 10.1016/j.nuclphysbps.2007.07.033

[8] Baldini A.M., Baracchini E., Cavoto G., Cascella M., Cei F., Chiappini M., Chiarello G., Chiri C., Dussoni S., Galli L., Grancagnolo F., Grassi M., Martinelli V., Nicol D., Panareo M., Pepino A., Piredda G., Renga F., Ripiccini E., Signorelli G., Tassielli G.F., Tenchini F., Venturinia M., C. Voena. Single-hit resolution measurement with MEG II drift chamber prototypes. Journal of Instrumentation. 2016;**11**:18. DOI: 10.1088/1748-0221/11/07/P07011

[9] Chiarello G., Corvaglia A., Grancagnolo F., Panareo M., Pepino A., Primiceri P., Tassielli G. A Full Front End Chain for Drift Chambers. In: 1st Conference on Charged Lepton Flavor Violation; May 2013; Lecce. Elsevier; 2014. pp. 140–142. DOI: 10.1016/j.nuclphysbps.2014.02.029

[10] Chiarello G., Corvaglia A., Grancagnolo F., Panareo M., Pepino A., Pinto C., Tassielli G. A high performance front end for MEG II tracker. In: 6th IEEE International Workshop on Advances in Sensors and Interfaces (IWASI), 18–19 June 2015; Gallipoli. IEEE; 2015. DOI: 10.1109/IWASI.2015.7184937

[11] Krzysztof T. P. FPGA-based, specialized trigger and data acquisition systems for high-energy physics experiments. Measurement Science and Technology. 2010;**21**(6):17. DOI: 10.1088/0957-0233/21/6/062002

[12] Cappelli L., Creti P., Grancagnolo F., Pepino A., Tassielli G. A fast readout algorithm for cluster counting/timing drift chambers on a FPGA board. Nuclear Instruments and Methods in Physics Research Section A: Accelerators, Spectrometers, Detectors and Associated Equipment. 2013;**718**:226–228. DOI: http://dx.doi.org/10.1016/j.nima.2012.10.087

[13] Xilinx. ML605 Hardware User Guide [Internet]. 2012 . Available from: http://www.xilinx.com/support/documentation/boards_and_kits/ug534.pdf [Accessed: 2016-08-10]

[14] Xilinx. Virtex-6 Family Overview [Internet]. 2015. Available from: http://www.xilinx.com/support/documentation/data_sheets/ds150.pdf [Accessed: 2016-08-10]

[15] Analog-Devices. Evaluating the AD9625 Analog-to-Digital Converter [Internet]. [Updated: 2015]. Available from: https://wiki.analog.com/eval/ad9625 [Accessed: 2016-08-11]

[16] Analog-Devices. 12-Bit, 2.6 GSPS/2.5 GSPS/2.0 GSPS, 1.3 V/2.5 V Analog-to-Digital Converter [Internet]. 2014. Available from: http://www.analog.com/media/en/technical-documentation/data-sheets/AD9625.pdf [Accessed: 2016-08-14]

[17] Anolog-Device. JESD204B Survival Guide [Internet]. 2014. Available from: http://www.analog.com/media/en/technical-documentation/technical-articles/JESD204B-Survival-Guide.pdf

[18] Xilinx. LogiCORE IP Clocking Wizard 3.6 (ISE) / 4.2 (Vivado) [Internet]. 2012. Available from: http://www.xilinx.com/support/documentation/ip_documentation/clk_wiz/v4_2/pg065-clk-wiz.pdf [Accessed: 2016-08-11]

[19] Xilinx. Mixed-Mode Clock Manager (MMCM) Module (v1.00a) [Internet]. 2009. Available from: https://www.xilinx.com/support/documentation/ip_documentation/mmcm_module.pdf [Accessed: 2016-08-11]

Chapter 8

Real-Time Adaptive Optic System Using FPGAs

Steffen Mauch and Johann Reger

Additional information is available at the end of the chapter

http://dx.doi.org/10.5772/66184

Abstract

For "adaptive optics" (AO) that are used in a control loop, sensing of the wavefront is essential for achieving a good performance. One facet in this context is the delay introduced by the wavefront evaluation. This delay should be kept to a minimum. Since the problem can be split into multiple subproblems, field-programmable gate arrays (FPGAs) may beneficially be employed in view of the FPGAs' power to compute many tasks in parallel. The evaluation of, e.g., a Shack-Hartmann wavefront sensor (SHWFS) may simply be seen as the evaluation of an image. Therefore, in general, image processing methods may be split into multiple assignments such as connected component labeling (CCL). In this chapter, a new method for real-time evaluation of an SHWFS is introduced. The method is presented in combination with a rapid-control prototyping (RCP) system that is based on real-time Linux operating system.

Keywords: SHWFS, FPGA, real-time Linux, RTAI, CCL, RCP, PCIe, adaptive optics

1. Introduction

"Adaptive optics" (AO) have been successfully utilized for more than one decade to improve the image quality of optical imaging systems. One reason for the high popularity originates from the fact that the image quality may be improved without mechanical adjustment, for example, the lenses. Additionally, the technological progress with respect to the manufacturing of deformable mirrors, an increase of computational power, and new approaches for controlling and sensing the wavefront allows broadening the scope of AO to new application fields, e.g., additive laser manufacturing, general beam shaping, and laser link communication [1].

In **Figure 1**, the general AO principle is illustrated within the context of controlling the wavefront. It is clear that besides good performance with respect to the stroke and the dynamic response of the deformable mirror, the wavefront needs to be measured accurately. To compensate for wavefront distortions, e.g., time-varying disturbances with or without a stochastical and/or dynamical model, the disturbance has to be measured with adequate precision. For the quasi-continuous measurement of the wavefront in AO systems, Shack-Hartmann wavefront sensors (SHWFSs) have widely been employed for measuring the wavefront, thus, the phase of the electromagnetic wave [2–5].

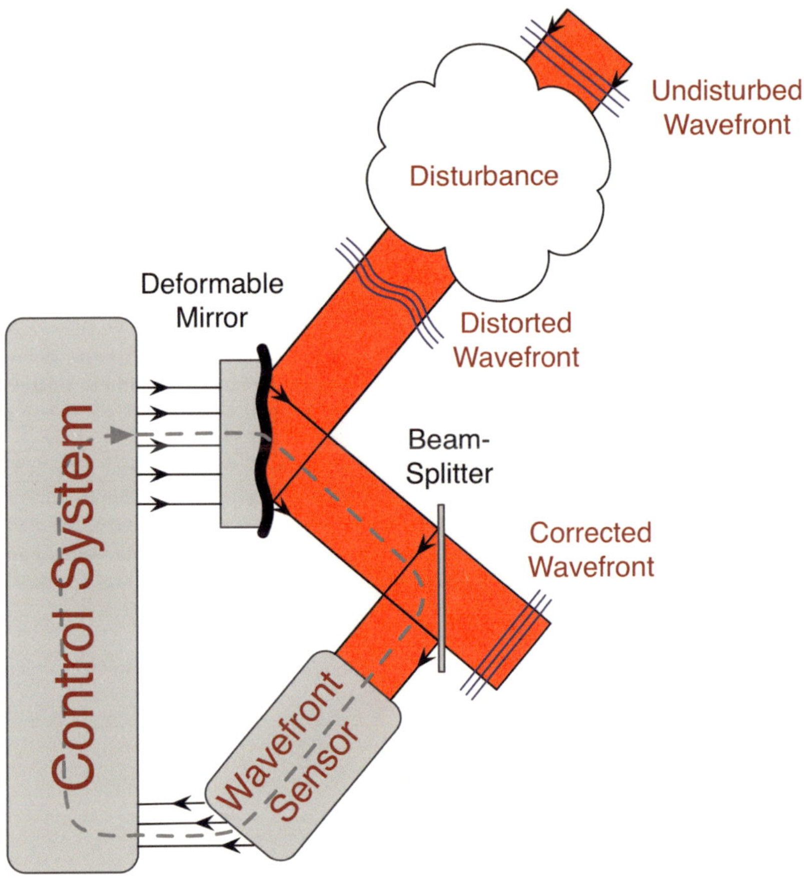

Figure 1. General scheme of adaptive optics, consisting of deformable mirror, wavefront sensor, and control system for closed-loop operation [9].

The SHWFS has shown some performance benefits when compared to interferometers as the SHWFS does not require a reference wave during the measurement process. Furthermore, the measurement sensitivity of an SHWFS primarily depends on the read-out noise of the detector, the luminosity of the wavefront and, hence, the intensity of the spots, and on the algorithm to find and assign the centroids, respectively [6–8].

The most commonly used wavefront measurement sensors, together with their advantages and disadvantages are discussed in Ref. [10]. The SHWFS itself relies decisively on the

http://dx.doi.org/10.5772/66184

determination of the centroids, i.e., on the image-processing techniques being applied. The different approaches are elaborated in Ref. [11]. As the computational performance and the dynamic behavior of the deformable mirrors are improving continuously, the sensing of the wavefront should also be accelerated which results in the demand of a low-latency and very large frame-rate. A straightforward attempt is to accelerate the image processing by utilizing parallel approaches; e.g., graphics processing units (GPUs) or field-programmable gate arrays (FPGAs).

The bandwidth demand of closed-loop AO systems is continuously increasing, see Ref. [12] or the report of the European Southern Observatory (ESO) ([13], ch. 7.9), to name but a few. In this regard, the application of GPUs is not as promising as FPGAs for evaluation of the wavefront because the GPU requires the use of the central processing unit (CPU) for data management whereas an FPGA may directly access the image sensor (typically a CMOS or CCD image sensor), that is, the pixel information. This allows parallelism with a low latency and thus a low delay. The problem with the delay is that even just a few milliseconds induced by the wavefront sensor may tend to ruin the overall performance of the closed-loop system as long as no adequate disturbance model is known, see e.g., the Xinetics AO system in Ref. [14]. FPGAs show some flexibility in interfacing to a standard computer, e.g., by using the PCIe interface or Universal Serial Bus 3.0 (USB3.0). Furthermore, the FPGA may be used to perform more tasks, for example, performing the computation for closed-loop operation or interfacing the digital to analog converter (DAC) for controlling the actuators of a deformable mirror without additional expensive cards from the hardware manufacturer.

In the last years, FPGAs became more common in academia but also in the industry due to their enormous capabilities regarding parallelism capability, achievable clocking frequency, and wide logic resources. In this course, FPGAs have been introduced as means for SHWFS evaluation. For instance, in Ref. [15], an FPGA solution is implemented under the assumption that spots cannot leave the associated subapertures.

In this chapter, we present a recently developed rapid-control prototyping (RCP) system that is based on an FPGA, mounted on a hard real-time Linux computer. Using a novel implementation, the evaluation of the SHWFS is performed on the FPGA directly. The implementation guarantees minimum delay during the evaluation of the wavefront and an enhanced dynamic range. We illustrate the algorithm for the spot detection and their ordering. Furthermore, we explain the code generation from a MATLAB/Simulink model to the hard real-time Linux system and the FPGA implementation of the PCIe interface.

2. FPGA-based SHWFS evaluation

For controlling the wavefront in an AO system, the wavefront itself has to be measured in an appropriate way. Several methods have been developed for that purpose, e.g., Pyramid, Shack-Hartmann (SHWFS), Curvature, or Holographic wavefront sensors [3, 16, 17]. Until now, an SHWFS is typically used for this objective as it may offer the best trade-off between performance, flexibility, and price. Since the SHWFS is based on capturing the intensities on an image plane (in general, a complementary metal-oxide semiconductor (CMOS) or charge-coupled device (CCD) image sensor), the evaluation of the SHWFS may be seen as some kind of image processing, calculating image moments. **Figure 2** depicts the basic principle of an SHWFS. Generally, an SHWFS will consist of an array of these lenses, called lenslet array.

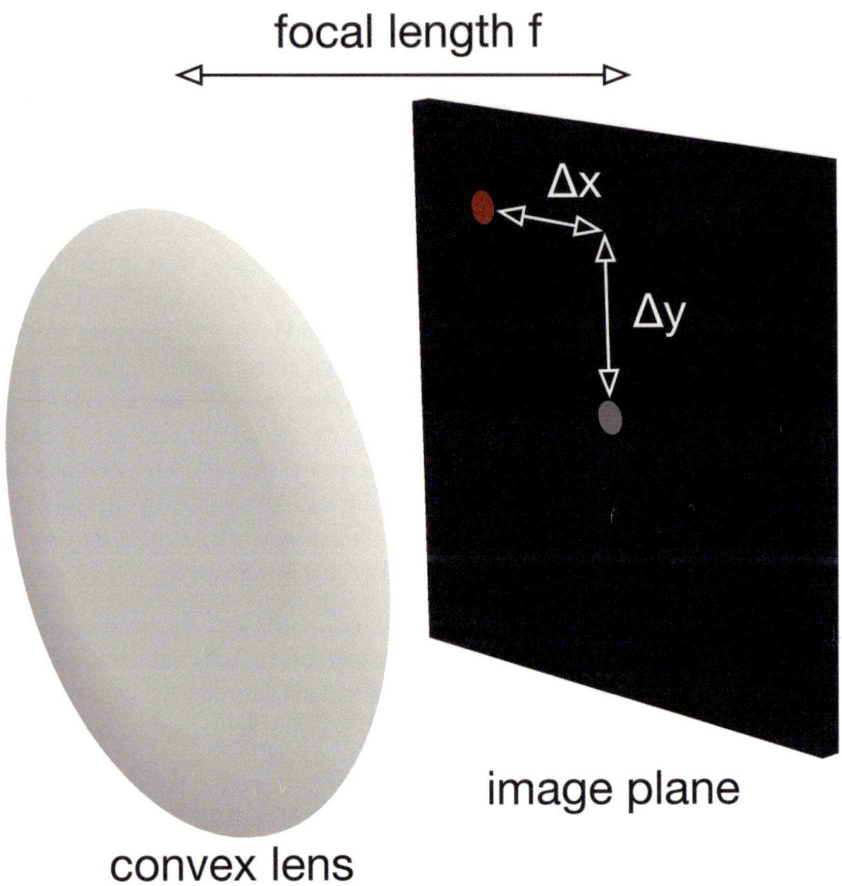

Figure 2. Single convex lens; the gray dot marks the incident flat wavefront, the red dot marks the tilted incident wavefront.

The lenslet array is positioned in parallel to the image plane with a distance of the focal length of the lenses such that the focal point is on the image plane. If a flat wavefront is incident on the lenslet, the spot lies in the projected center of the convex lens in the image plane (marked with the gray dot in **Figure 2**). Due to the nature of the convex lens, the partial derivative of the wavefront with respect to x- and y-direction is averaged over the area of the lens. Thus, the deviation of the focal spot on the image plane with respect to the projected center of the convex lens denotes the local derivative of the wavefront. If a lenslet array is used then one would define a given area on the image plane in which the spot must lie. This may limit, of course, the possible dynamic range because the steepness of the partially tilted wavefront is limited by the area of the image plane and the focal length of the lenslet array.

The task is to determine the position or the deviation Δx and Δy of the spots, as shown in **Figure 2**. For a given area with a predetermined number of pixels, this can be formulated as

$$x_c = \frac{\sum_{i=0}^{M-1}\sum_{j=0}^{N-1} x_i I(x_i, y_j)}{\sum_{i=0}^{M-1}\sum_{j=0}^{N-1} I(x_i, y_j)} \tag{1}$$

$$y_c = \frac{\sum_{i=0}^{M-1}\sum_{j=0}^{N-1} y_i I(x_i, y_j)}{\sum_{i=0}^{M-1}\sum_{j=0}^{N-1} I(x_i, y_j)} \tag{2}$$

where M and N denote the number of pixel in x- and y-direction, respectively. $I(x_i, y_j)$ is the intensity of the pixel at the coordinate (x_i, y_j).

In the past, several methods have been developed for extending the dynamic range of the SHWFS, such as hardware modification, tracking, similarity approaches, to name but a few. We may now calculate $\Delta x = x_c - X_{pc}$, where X_{pc} is the x-coordinate of the projected center of the convex lens. The corresponding calculation can be done for Δy too. For the work presented here, the SHWFS with its CCD camera is directly connected to an FPGA to the end of lowest possible latency.

Evaluating the pixel information of the SHWFS for determining the phase of the wavefront may be divided into two problems: First, determine the individual centroids, i.e., calculate the centroid of the connected areas and, second, the ordering of the centroids to the lenslet for calculating the deviation with respect to the default position and, thus, computing the local derivatives of the wavefront.

As mentioned previously, as long as a predefined area is given in which the spots have to stay, the dynamic range of the SHWFS is limited. Since the approach of the connected areas does not use any predefined area, the restriction is no longer prevalent. However, to be fair, the default algorithm performs the determination and ordering of the centroids in a single step whereas the ordering is a subsequent step which is discussed in the following.

The determination of the connected areas for calculation of the centroids may be based on different methods. These methods mainly differ in their ability for online calculation of the connected areas, meaning that the pixel stream is processed sequentially at the end of it. Such methods are called single-pass algorithms, emphasizing that only a single pass is required without necessarily storing the complete pixel information. The methods have extensively been studied, e.g., in road sign detection or line tracking systems for lane assistance. The general name for these algorithms is connected-component labeling (CCL). CCL—also denoted by connected-component analysis, region labeling—is an algorithmic application of graph theory. The subsets of connected components are often denoted as "blobs." Blobs are uniquely labeled, based on a predetermined heuristic, mostly along the neighbor relationship.

For this approach, the labeling used for the CCL is based on an eight-point neighborhood system, see **Figure 4**. Another popular neighborhood system is the four-point neighborhood system which is presented in **Figure 3**. In **Figures 3** and **4**, the symbol "*s*" marks the actual pixel and the corresponding neighbors of pixel *s* are marked in gray.

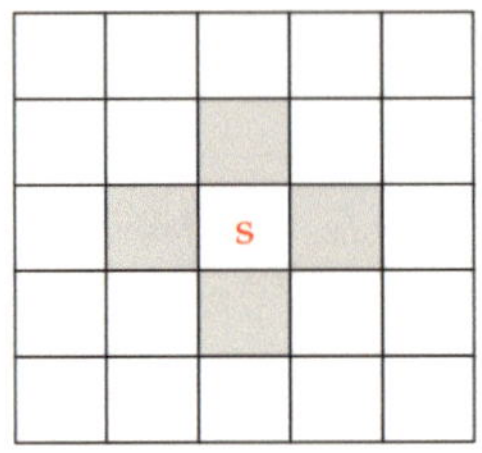

Figure 3. Four-point neighborhood system.

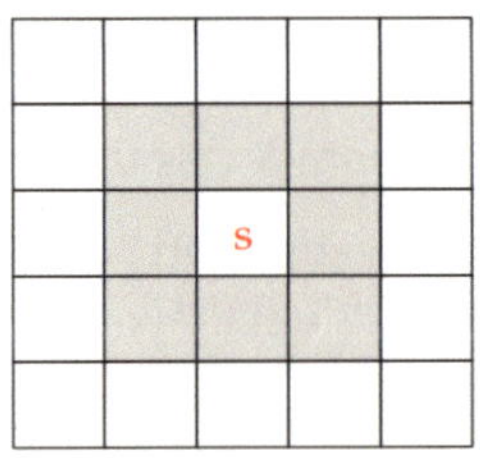

Figure 4. Eight-point neighborhood system.

The procedure for CCL is straightforward. The pixels (the intensity information for each pixel) are streamed sequentially, typically from left to right and top to bottom. If the intensity information is larger than a threshold value, the pixel is assumed to be "1" else "0." This step is called binarization. In **Figure 5**, the boxes with a gray background have already been processed and the actual pixel carries the symbol "?."

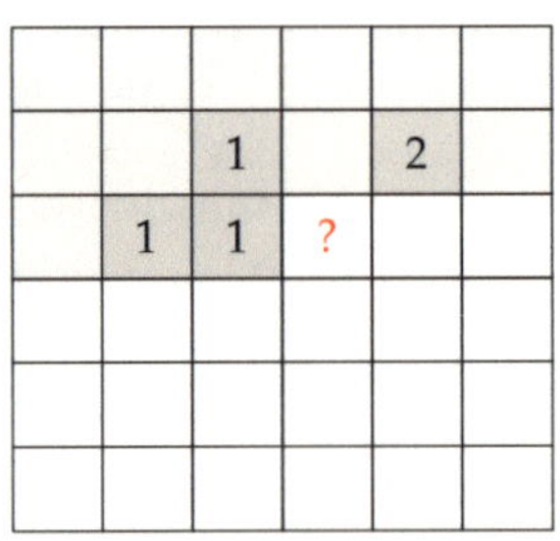

Figure 5. Label collision due to nonconvex blob.

In the case when two sets are connected, but due to the sequential processing receive two different numbers, a label collision may occur. As long as the blobs are convex sets, a label collision cannot occur when using an eight-point neighborhood. Experiments have shown that the assumption of convex blobs is not valid for the typical application scenario of an SHWFS. This may be caused by a disturbed pixel intensity information, recording the noise of the camera sensor, the photon noise, nonperfect lenses, and other effects. Due to thresholding with a fixed value, a single count in terms of the digitalized intensity information can lead to nonconvex blobs, see e.g., **Figure 6**.

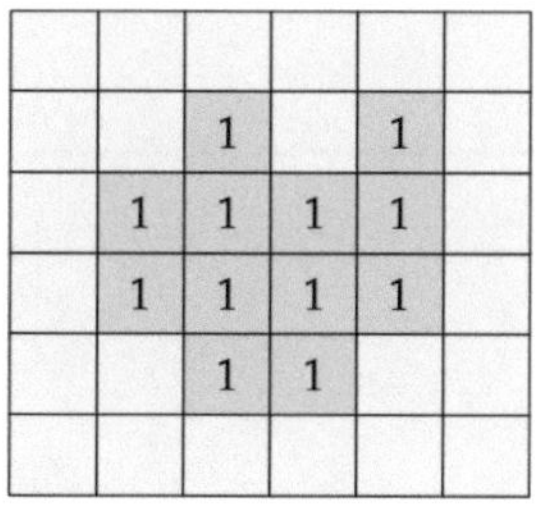

Figure 6. Resolved label collision due to blob merging.

Application of morphology methods, such as dilation or erosion, is not possible without storing large parts of the image, thus are not single-pass compliant. Additionally, morphology methods significantly increase the delay.

The handling of label collisions can be accomplished by using a label stack which allows label reusing after a label collision has occurred. By means of label reusing the number of provided labels can be kept to a minimum; otherwise, under some circumstances, twice the number or even more labels must be provided. More information is given in Refs. [9, 18, 19].

After having determined the blobs, thus the connected areas, the division of the numerator and denominator may be performed for each valid blob found. The numerator and denominator have to be stored separately as the division step can only be performed when the connected set is maximum. In the block diagram given in **Figure 7**, this step is done in the "centroid calculation/feature extraction" block which also performs the assignment of the centroids to the lenslets.

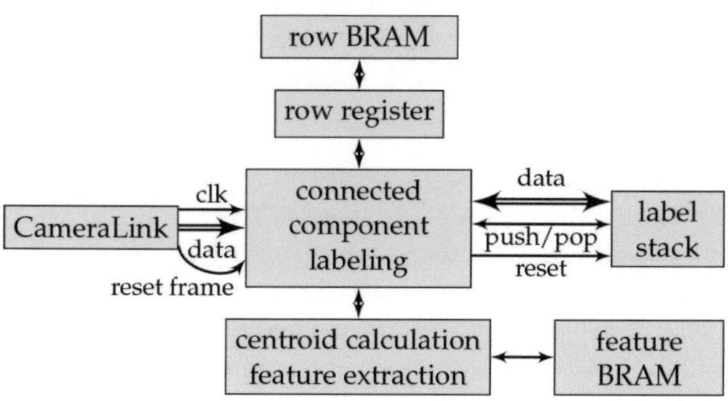

Figure 7. Block diagram of the complete CCL FPGA implementation [18].

One of the key elements of the drafted implementation in **Figure 7** is that solely the former line of the pixel stream has to be stored, not the whole pixel stream. For the applied camera, this results in storing 224 pixels where each pixel is one bit wide because only the binarized value has to be stored. Furthermore, only parts of the former line have to be accessed in parallel such that a small row register is sufficient which is automatically loaded from a Block RAM (BRAM). Using a BRAM has the advantage that the consumption of logic cells is reduced as the BRAM is a dedicated peripheral offered by most FPGAs. The overall logic consumption can be kept at a minimum [18].

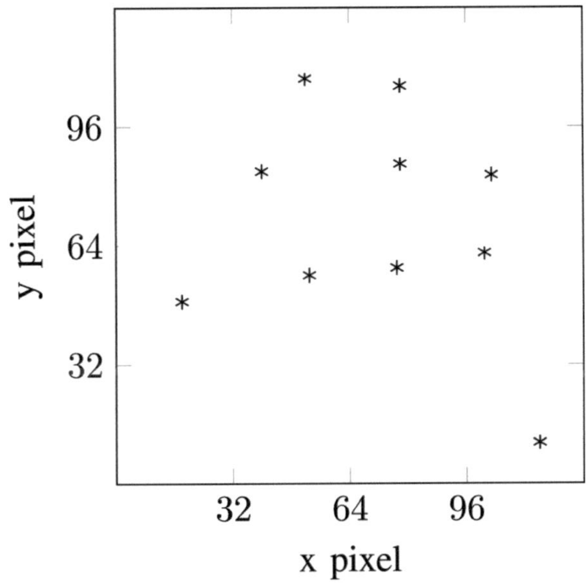

Figure 8. Exemplary centroids for a 4 × 4 lenslet array; some spots are missing.

The assignment or segmentation of the centroids is visualized in **Figures 8** and **9**. This idea has been presented in Ref. [18] and behaves similar to the standard approach for the regular case, that is, the wavefront is not strongly disturbed. However, the advantage of this approach appears whenever a large defocus is present in the wavefront to be measured since shrinking or increasing the overall distance between two neighbored centroids is not a problem for the segmentation method.

The fundamental principle is that the centroids are ordered in parallel with respect to their x- and y-value such that two separately ordered lists exist. Then, straight lines are used to segment the centroids in x- and y-direction by using their distance between each other. As **Figure 9** illustrates, this method is working well also for the case when some centroids are missing due to shadowing or insufficient light intensities. When a very large shearing occurs, however, the method will not be ideal because straight vertical lines are used. But if this problem appears, the standard approach is also not applicable anymore. This algorithm is called simple straight line segmentation.

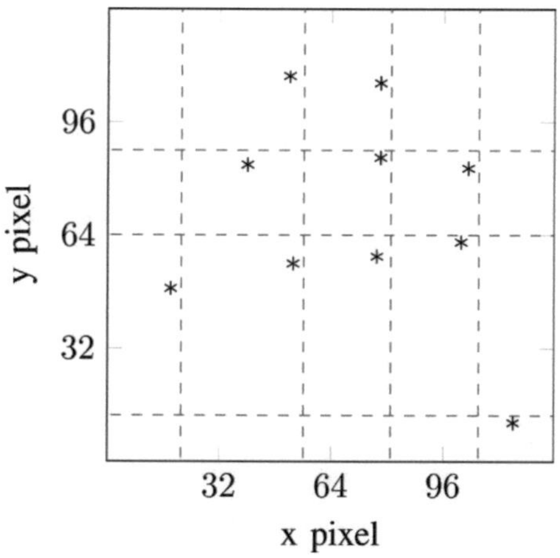

Figure 9. Segmented centroids applying the method presented in [18].

The described algorithm is very simple and straightforward. In Ref. [19] the so-called spiral method has been extended to be deterministic and real-time capable using the centroids gathered by employing CCL. It is obvious that depending on the specific application other methods may be better suited. The CCL may be enhanced by making the thresholding adaptive to compensate the natural intensity inhomogeneity [9, 20]. Another enhancement is the adaptive positioning which for most cases may solve the problem when the number of rows and columns after assignment of the centroids are not the same as with the lenslet array. This circumstance, in general, will lead to ambiguity of the assignment. The adaptive positioning, however, uses an approach based on the similarity of the shape of the segmented centroids and minimizes the shift. Based on this information, the assignment is shifted by one row or column to reduce the offset.

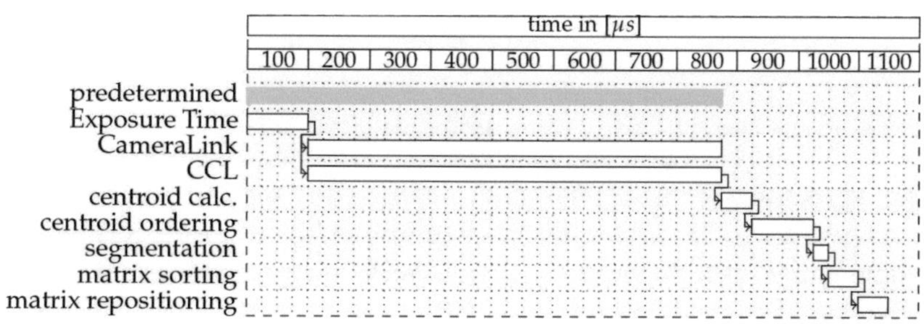

Figure 10. Evolution of processing time, each step rounded up to 25 μs, during SHWFS evaluation applying the FPGA approach.

Figure 10 holds the timeline for the whole evaluation of the SHWFS beginning with the exposure until the assignment of the centroids to the lenslets. The "Imperx ICL-B0620M" camera, on which the "Imagine Optics HASO™ 3 Fast" wavefront sensor is based, is used for this setup. The camera has a maximum frame rate of approximately 900 Hz at 224 × 224 pixel which corresponds to 1111 ms. Thus, the proposed method has a delay equal or less than one single frame.

3. FPGA PCIe integration into the real-time Linux system

The evaluation of the SHWFS is only one part of the AO system since the partial derivatives of the wavefront must be either used for reconstruction of the wavefront and/or used for controlling a deformable mirror (DM) in closed-loop operation. A simple, basic AO concept is used for the work presented in this text, see **Figure 11**. In the experimental setup, the FPGA, besides the evaluation of the SHWFS, is also used for interfacing the digital-to-analog converter (DAC) card. The benefit is that the FPGA can easily guarantee a true parallel output (same guaranteed phase) for all analog outputs even if multiple DACs have to be used.

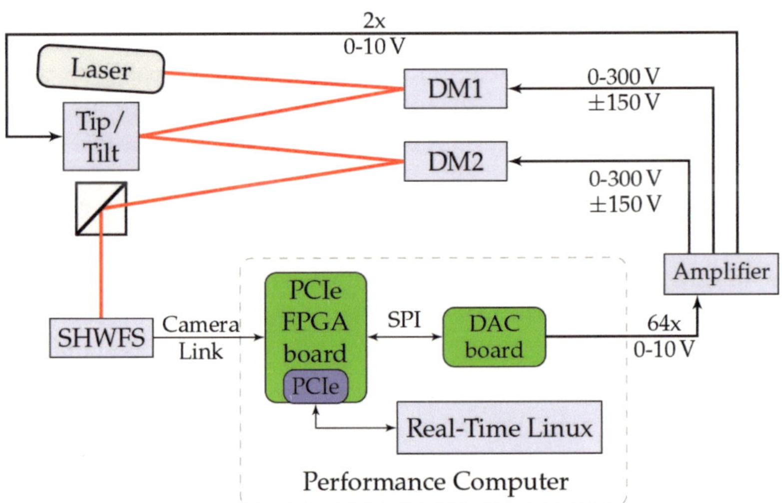

Figure 11. Overview over the basic AO concept; for the detailed concept see [21].

The subsequent processing of the SHWFS data is carried out by a performance computer using state-of-the-art hardware. On this performance computer, the control algorithm is running on a hard real-time Linux operating system (OS). This OS in combination with the performance computer offers rapid-control prototyping (RCP) capabilities in view of the direct MATLAB/Simulink interface. Such an RCP system reduces the implementation effort drastically when different control schemes and approaches need to be tested or compared with each other.

The PCIe FPGA card, see **Figure 12**, is a self-developed card based on the Xilinx Kintex-7 FPGA module TE0741 from Trenz Electronics. The PCIe FPGA card offers more connectivity than only the CameraLink interface. Nevertheless, in this context, only CameraLink, PCIe, and the Serial Peripheral Interface (SPI) are used. The other interfaces are neglected in this context but are presented in detail in Refs. [9, 21].

Figure 12. Developed PCIe FPGA board based on a TE0741 (Xilinx Kintex-7) module from Trenz Electronics.

The integration of the FPGA card is realized via the PCIe interface. Thus, almost any modern computer can be used for interfacing the PCIe FPGA card. The SHWFS, more exactly the CCD camera, is connected with the CameraLink interface to the card. Additionally, two separate DAC boards are installed where each DAC board offers 32 analog channels.

The outputs of the DAC cards are fed into an amplifier which amplifies the small signals to drive, for example, the piezoelectric actuators that are part of the DM. In the setup, two DMs have been applied. This circumstance allows the feature that one DM may be used for an artificial, but realistic disturbance generation, whereas the other compensates for such disturbance. In principle, the disturbance may also be virtually induced by adding some signal to the output of the SHWFS; however, a meaningful emulation can be rather involved. This may limit the performance of the system. For this reason, a real disturbance has been incorporated. The amplifier offers the feature to switch between regular and symmetric voltage by modifying the reference ground. Here, the benefit of the symmetric voltage is that the stroke is symmetric as well. Due to the creeping behavior of the piezoelectric actuators, simply applying an offset of [+150] V is not the same as symmetric operation.

For integrating the PCIe FPGA card into the Linux kernel, a kernel driver has to be developed. So as to integrate data acquisition cards, Linux offers a special interface called comedi (control and measurement device interface). Using this interface is very comfortable because the core functionality is already implemented and only low-level driver modules have to be developed for supporting a new data acquisition card. In addition, a user-space library called "comedilib" is available which allows the utilization of user-space to access the functionality of the data acquisition card (**Figure 14**).

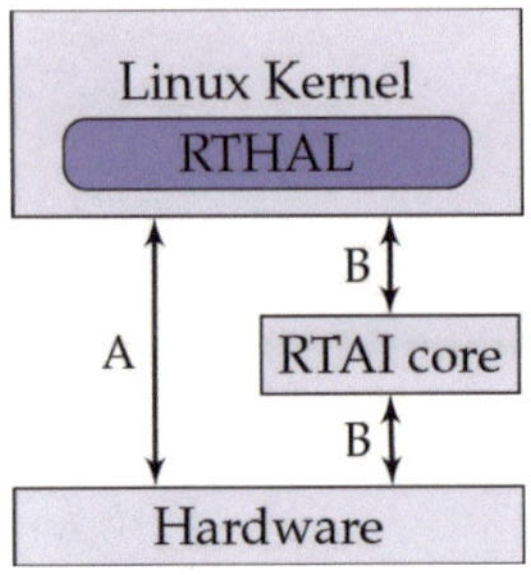

Figure 13. RTAI principle for RTAI-core active or inactive [21].

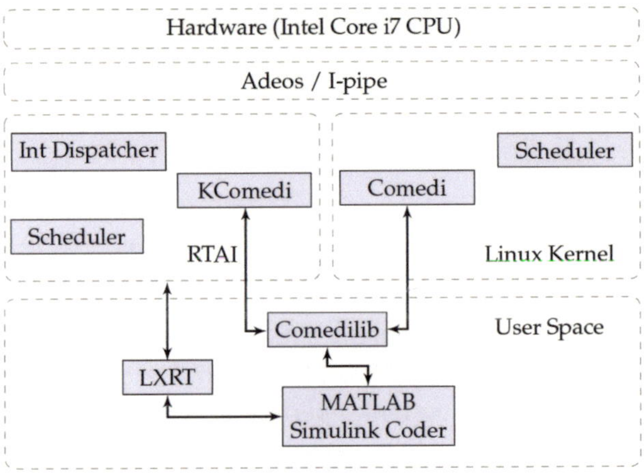

Figure 14. Block diagram of the different abstraction layers used in RTAI/LXRT [9].

The Linux kernel is patched with the RTAI (real-time application interface) [22] patch which itself is based on Adeos. The purpose of the Adeos project is to offer an environment so as to allow sharing of hardware resources among multiple operating systems. RTAI uses that approach (shown in **Figure 13**) for scheduling Linux in the hard real-time support. If RTAI is loaded then case B is active, otherwise case A.

Furthermore, RTAI supports comedi without disturbing the hard real-time behavior. RTAI has the LXRT extension that offers the feature to run real-time applications as user-space programs, see **Figure 14**. Additionally, a MATLAB/Simulink target is available which uses the Simulink Coder for C/C++ code generation [23]. Based on these prerequisites it is easy to extend the given code generation to support more comedi implemented features such as block memory reads or trigger commands.

The PCIe implementation is based on the Xilinx 7 Series Gen2 Integrated Block for PCI Express IP-core which has been extended to support Direct Memory Access (DMA). This way, the FPGA may write the assigned centroids into the main memory of the computer without involving the CPU, see **Figure 15**. PCIe is based on sending and receiving Transaction Layer Packets

http://dx.doi.org/10.5772/66184

(TLPs). The block "COMEDI_SHWFS_READ," see **Figure 16**, performs a blocking read request on the memory destination also being used for the DMA transfer. Behind these Simulink blocks, we have predefined *s*-functions which are based on the functionality provided by comedi. The "COMEDI_SHWFS_TRIGGER" triggers the start of the frame capture; thus, the image acquisition is synchronous to the real-time application which is essential for guaranteeing a deterministic behavior. As shown in the timeline in **Figure 10**, after approximately 1050 μs the data is transferred via DMA to the main memory of the computer.

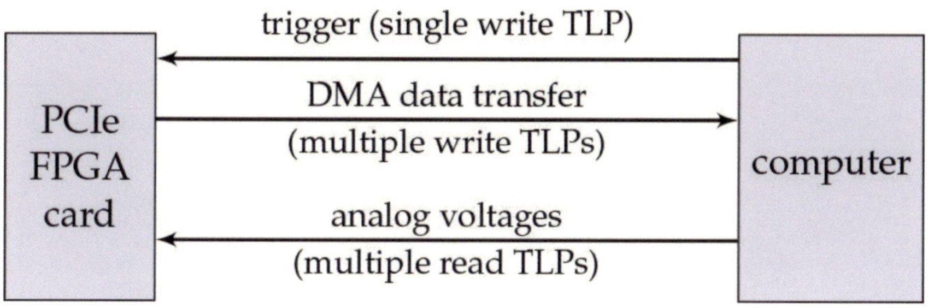

Figure 15. Communication using PCIe interface between FPGA and computer.

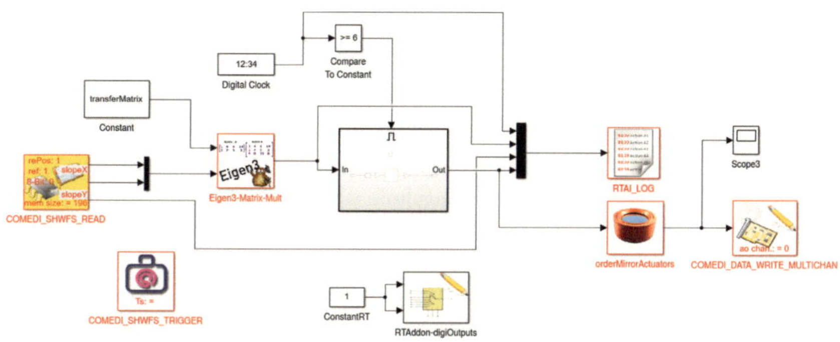

Figure 16. Simulink model used for code-generation and based on Simulink Coder.

The captured data, e.g., from the SHWFS as well as the control output and error values are fed into the "RTAI_LOG" block. This module creates an interface with which another user-space program may record the data and write it either to the main memory or the hard disk.

4. Results from the adaptive optics setup

For closing the loop of the AO setup, a stabilizing controller is required. In Ref. [9], an $\mathcal{H}_\infty$ controller has been synthesized which robustly stabilizes the AO system. In the past, in general, PI(D) controllers have been used which often were tuned by hand. It would go too far to present the method for controller synthesis. So, this is not exposed in this survey. Since the AO setup uses an RCP approach, it is not very time consuming to test other control schemes, as only the Simulink model has to be adjusted accordingly. Of course, the design of a stabilizing controller which is also robust against a set of uncertainties may be rather complicated and time consuming.

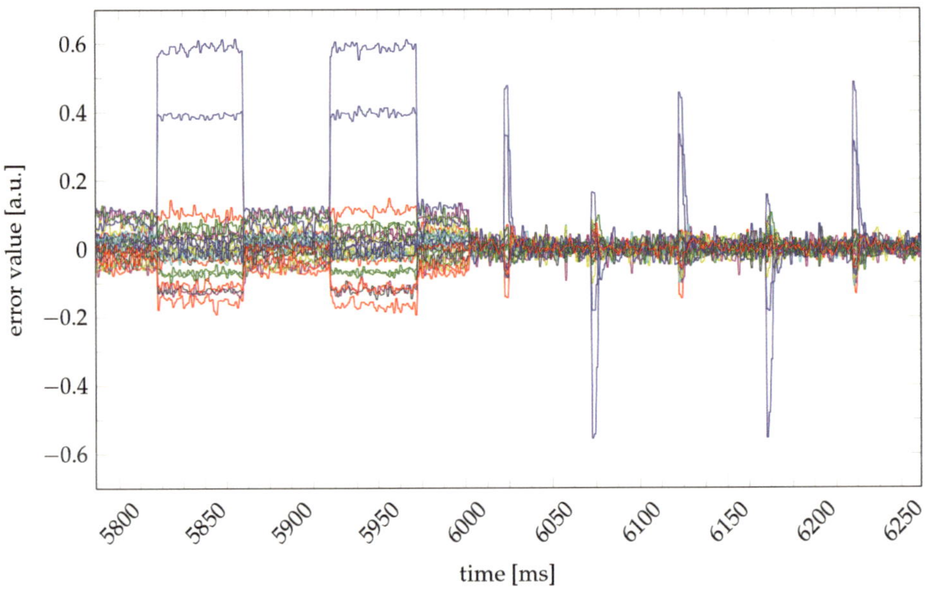

Figure 17. Application of a 10 Hz step disturbance while controller is switched on at $t = 6$ s; $f_{\text{controller}} = 1600$ Hz and $f_{\text{shwfs}} = 800$ Hz [9].

To validate the applicability of the presented approaches, **Figure 17** shows some recorded data. The controller has been switched on at time instance 6000 ms. The disturbance is a 10 Hz rectangular offset that is applied to one actuator of DM1. As the actuator patterns of the DM2 and DM1 are not the same and, additionally, do not have the same number of actuators, the result is that multiple actuators are required for compensating the disturbance. A rectangular disturbance is ideally suited to visualize the power of the controller as the steady-state error as well as the time required for compensating the disturbance may be analyzed.

The error value is obtained after the multiplication of the control matrix with the centroids (in **Figure 16** the signal after the "Eigen3-Matrix-Mult" block). The control matrix itself is the pseudo-inverse of the actuator influence function [8, 9].

The dimension of the error value is the same as the number of actuators. Nevertheless, the error value itself does not give a direct insight on how the wavefront looks like. Therefore, **Figure 16** visualizes the reconstructed wavefront as a 3D surface. The respective error values are depicted in **Figure 18** separately. To calculate the Strehl values based on the reconstructed wavefront in **Figure 20**, the wavefront at time instance 6320.63 ms has been used as reference;

thus, showing a Strehl value of exactly one. Both the three-dimensional representation and the error value visualize that it took 3–4 frames to reject the disturbance.

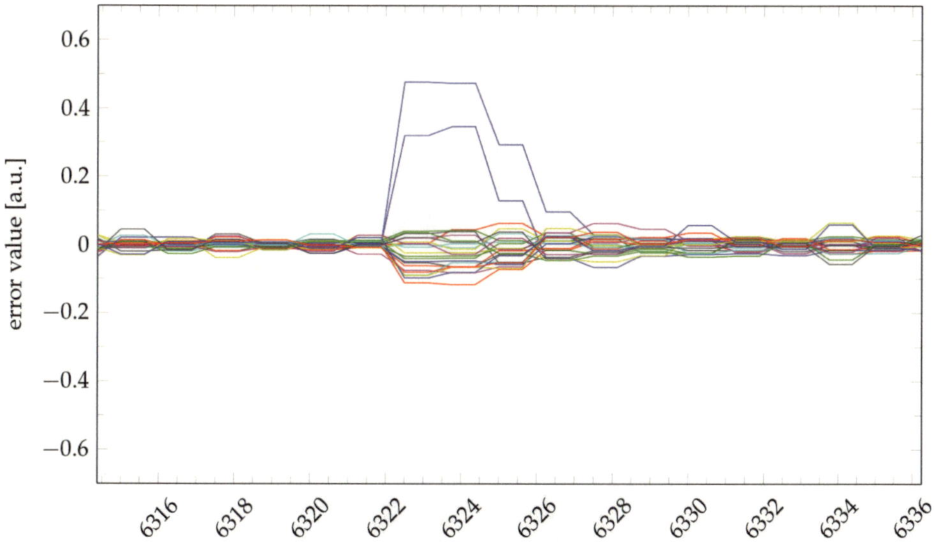

Figure 18. Experimental data with zoomed *x*-axis to highlight the control behavior after a step disturbance, $f_{\text{controller}} = 1600\text{Hz}$ and $f_{\text{shwfs}} = 800\text{Hz}$ [9].

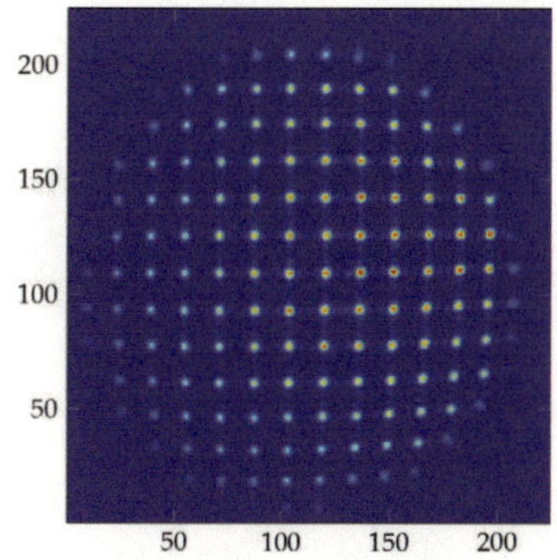

Figure 19. Captured image of the SHWFS, having a lenslet array of 14×14, during experiments.

Finally, **Figure 19** shows some captured camera image of the SHWFS. The colors have been adjusted for better visualization. The SHWFS has a lenslet array of 14 × 14 while the pixel area is 224 × 224 pixels. The standard approach cannot be used here as the spots are leaving the area on the image sensor. The area on the image sensor for each lens would be 16 × 16 pixels.

However, the new approach has no problem and correctly assigns the spots to the lenses and thus correctly measures the wavefront (**Figure 20**).

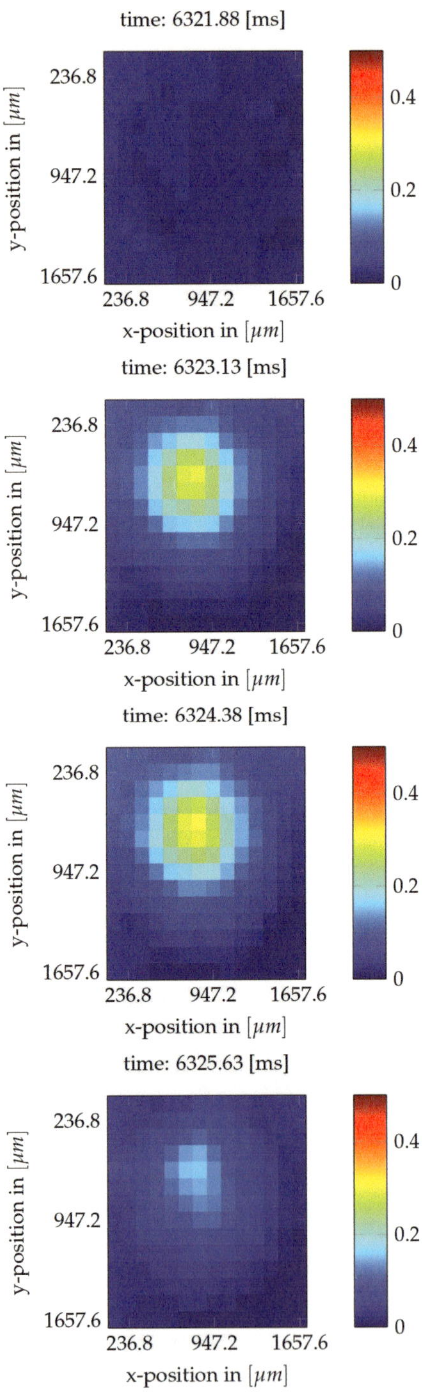

http://dx.doi.org/10.5772/66184

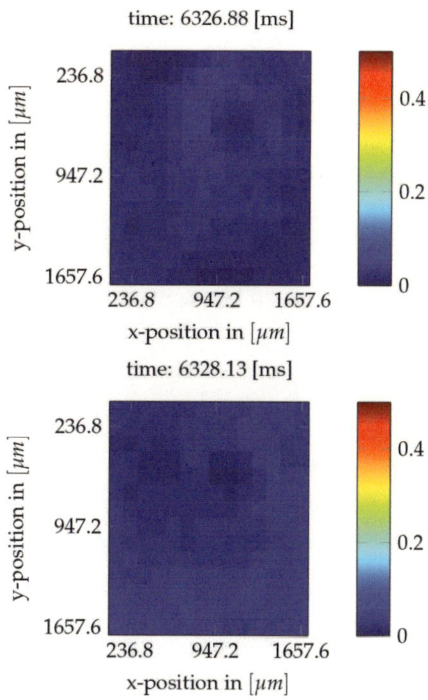

Figure 20. Reconstructed wavefront, rejecting a step disturbance; same data as in **Figure 18** [9].

5. Conclusion

The use of FPGAs in the context of AO has proven to be very beneficial with respect to the achievable performance, especially in closed-loop operation. One positive aspect is the direct evaluation of the SHWFS on the FPGA which allows to minimize the delay and to increase the throughput. For the evaluation of the SHWFS, new approaches have been presented which surpass or considerably extend existing methods. However, they have not reached possible limits so far, particularly, in terms of the achievable dynamic range of the SHWFS.

The practical applicability of the method has been demonstrated in various experiments paired with extensions such as the adaptive repositioning and thresholding [9, 18–20].

For designing AO setups and optimizing its performance, interdisciplinary groups are indispensable. In this context, the control engineers may synthesize their simulation models directly in code for closed-loop operation. Such an RCP system may also be a commercial solution such as dSpace. Yet, the presented RTAI-based hard real-time Linux system has the important benefit to be of far lower initial cost with respect to hardware while granting higher flexibility and ease of customization.

Author details

Steffen Mauch[1*] and Johann Reger[2]

*Address all correspondence to: steffen.mauch@tu-ilmenau.de

1 Ingenieurbüro Mauch, Unorthodox Solutions, Germany

2 Technische Universität Ilmenau, Ilmenau, Germany

References

[1] M. Knapek. Adaptive Optics for the Mitigation of Atmospheric Effects in Laser Satellite-To-Ground Communications. *Dissertation, Technische Universität Munchen*, 2011.

[2] P.L. Wizinowich. Adaptive optics and keck observatory. *IEEE Instrumentation Measurement Magazine*, 8(2):12–19, 2005. ISSN 1094-6969. doi: 10.1109/MIM.2005. 1405918.

[3] R.K. Tyson and B.W. Frazier. *Field Guide to Adaptive Optics, 2nd Ed (SPIE Field Guide Vol. FG24)*. SPIE Press, 4, 2012. ISBN 9780819490179.

[4] L.C. Andrews. *Field Guide to Atmospheric Optics (SPIE Vol. FG02)*. SPIE Publications, 1, 2004. ISBN 9780819453181.

[5] P. Hickson. *Fundamentals of Atmospheric and Adaptive Optics*. Homepage University of British Columbia, 2008.

[6] R. Irwan and R. G. Lane. Analysis of optimal centroid estimation applied to Shack-Hartmann sensing. *Applied Optics*, 38(32):6737–6743, 1999. doi: 10.1364/AO.38.006737.

[7] Z. Jiang, S. Gong, and Y. Dai. Monte-Carlo analysis of centroid detected accuracy for wavefront sensor. *Optics & Laser Technology*, 37(7):541–546, 2005. doi: 10.1016/j.optlastec. 2004.08.009.

[8] R.K. Tyson. *Principles of Adaptive Optics, Fourth Edition*. CRC Press}, ISBN: 9781482252330, 11, 2015.

[9] S. Mauch. *Robust Control of an Adaptive Optics System*. Dissertation, Technische Universität Ilmenau, 2016.

[10] O. Guyon. Limits of adaptive optics for high-contrast imaging. *The Astrophysical Journal*, 629(1):592, 2005. doi: 10.1117/12.789849.

[11] A.M. Nightingale and S. Gordeyev. Shack-Hartmann wavefront sensor image analysis: a comparison of centroiding methods and image-processing techniques. *Optical Engineering*, 52(7):071413–071413, 2013. doi: 10.1117/1.OE.52.7.071413.

[12] A. Basden, D. Geng, R. Myers, and E. Younger. Durham adaptive optics real-time controller. *Applied Optics*, 49(32):6354–6363, 2010. doi: 10.1364/AO.49.006354.

[13] K. Kepa, D. Coburn, J.C. Dainty, and F. Morgan. Re-baselining the ESO ELT project. *Analysis and Roadmap from the ELT Adaptive Optics Working Group*, 2006.

[14] R.M. Rennie, D.A. Duffin, and E.J. Jumper. Characterization and aero-optic correction of a forced two-dimensional weakly compressible shear layer. *AIAA Journal*, 46(11): 2787–2795, 2008. doi: 10.2514/1.35290.

[15] K. Kepa, D. Coburn, J.C. Dainty, and F. Morgan. High speed optical wavefront sensing with low cost FPGAs. *Measurement Science Review*, 8:87–93, 2008. doi: 10.2478/ v10048-008-0021-z.

[16] J.M. Geary. *Introduction to Wavefront Sensors (Tutorial Texts in Optical Engineering)*. Society of Photo Optical, 5, 1995. ISBN 9780819417015.

[17] U. Schnars, C. Falldorf, J. Watson, and W. Jüptner. *Digital Holography and Wavefront Sensing: Principles, Techniques and Applications*. Springer, 2nd ed., 2015 edition, 7, 2014. ISBN 9783662446928.

[18] S. Mauch and J. Reger. Real-time spot detection and ordering for a Shack-Hartmann wavefront sensor with a low-cost FPGA. *IEEE Transactions on Instrumentation and Measurement*, 63(10):2379–2386, 2014. ISSN 0018-9456. doi: 10.1109/TIM.2014.2310616.

[19] S. Mauch and J. Reger. Real-time implementation of the spiral algorithm for Shack-Hartmann wavefront sensor pattern sorting on an FPGA. *Measurement*, 92:63–69, 2016. ISSN 0263-2241. doi: 10.1016/j.measurement.2016.06.004.

[20] S. Mauch, A. Barth, J. Reger, N. Leonhard, and C. Reinlein. Improved thresholding and ordering for Shack-Hartmann wavefront sensors implemented on an FPGA. *International Workshop on Adaptive Optics for Industry and Medicine (AOIM)*, 2015.

[21] S. Mauch, J. Reger, C. Reinlein, M. Appelfelder, M. Goy, E. Beckert, and T. Tünnermann. FPGA-accelerated adaptive optics wavefront control. *Proceedings of SPIE*, 8978:1–12, 2014. doi: 10.1117/12.2038910.

[22] P. Mantegazza, E. L. Dozio, and S. Papacharalambous. Rtai: Real time application interface. *Linux Journal*, 2000(72es), April 2000. ISSN 1075-3583.

[23] R. Bucher and S. Balemi. Rapid controller prototyping with matlab/simulink and linux. *Control Engineering Practice*, 14(2):185–192, 2006. ISSN 0967-0661. doi: 10.1016/j.conengprac.2004.09.009. Special Section on Advances in Control Education Symposium.

Chapter 9

FPGA-SRAM Soft Error Radiation Hardening

Gabriel Torrens

Additional information is available at the end of the chapter

http://dx.doi.org/10.5772/66195

Abstract

Due to integrated circuit technology scaling, a type of radiation effects called single event upsets (SEUs) has become a major concern for static random access memories (SRAMs) and thus for SRAM-based field programmable gate arrays (FPGAs). These radiation effects are characterized by altering data stored in SRAM cells without permanently damaging them. However, SEUs can lead to unpredictable behavior in SRAM-based FPGAs. A new hardening technique compatible with the current FPGA design workflows is presented. The technique works at the cell design level, and it is based on the modulation of cell transistor channel width. Experimental results show that to properly harden an SRAM cell, only some transistors have to be increased in size, while others need to be minimum sized. So, with this technique, area can be used in the most efficient way to harden SRAMs against radiation. Experimental results on a 65-nm complementary metal-oxide-semiconductor (CMOS) SRAM demonstrate that the number of SEU events can be roughly reduced to 50% with adequate transitory sizing, while area is kept constant or slightly increased.

Keywords: SRAM, FPGA, Radiation, single event upset, hardening

1. Introduction

The dimensions of integrated circuit devices decreased in each successive technology generation. The goal of this scaling is, on the one hand, to improve the performance of integrated circuits and, on the other hand, to integrate a greater number of devices per unit area. Static random access memories (SRAMs) are not an exception to this evolution, the dimensions of the transistors forming memory cells decreased roughly following Moore's Law. Consequently, the area occupied by each cell decreased from generation to generation [1]. Current technological processes used to manufacture complementary metal-oxide-semiconductor (CMOS) SRAM memories are in the nanometer region, since the nominal characteristic dimensions of the transistors forming each cell are of the order of tens of nanometers.

The supply voltage of SRAMs has also been reduced. However, this decrease did not follow the predictions of the International Technology Roadmap for Semiconductors (ITRS), in fact, it was more moderate. This is mainly due to the limitation imposed on the transistors threshold voltage scaling to avoid an excessive increase of leakage current [2]. To meet performance demands of current electronic systems, large capacity integrated SRAMs are usually needed, and in fact, FPGA-SRAMs are not an exception. This requirement results in a large proportion of area dedicated to SRAM memory. Forecasts indicate that in the coming years this figure may reach 90% [3]. Of course, integrating large memories has an adverse impact on circuit area, which in turn results into higher costs. For this reason, designers try to integrate the largest possible number of SRAM cells per unit area. This leads to cells designs with small sizes to squeeze the full potential of technology. SRAMs are usually designed with transistors close to the minimum possible size, and arranged with the highest possible density. In addition, to reduce power consumption, voltages are kept as low as possible. Although, as mentioned before, the expected voltage reduction have not been fully implemented in real technology.

As a result of the decrease in device dimensions and of the reduction of supply voltage in successive technology generation, designing SRAM faces two major challenges: the first one is related to the stability of the cells and second one has to do with their susceptibility to radiation-induced transient events. This chapter focuses on the second challenge, the CMOS SRAM radiation problem. However, SRAM stability issues are also discussed.

SRAMs are one of the most sensitive to radiation parts of a circuit. They are especially sensitive to those effects caused by a single energetic particle. These effects are the so-called single event upsets (SEUs). They are considered soft errors (SE) because they trigger an error without permanently damaging the circuit. This chapter focuses on six-transistor (6T) CMOS SRAM SEUs and on a technique to mitigate its effects, which is easily implementable in current FPGA design workflows. The architecture of 6T RAMS cell is described in Section 2.

Regarding the process that generate SEUs, the interaction of an energetic particle creates electron-hole pairs, so that part of this deposited electric charge can be collected by a sensitive node affecting its voltage. If this node is the node of an SRAM and the perturbation is high enough, it can flip the cell state altering data stored in it, and thus generating an error. These errors are not necessarily destructive. In particular, in an SRAM, a particle is capable of modifying data stored in one or more memory cells without damaging them. This means that cells can be rewritten and operate normally. Nevertheless, cell data has been corrupted, and if the cell is read before a new write occurs, a read error will be produced.

The problem of radiation effects in integrated circuits is not new. It has been studied and taken into account for decades by designers in areas such as the aerospace industry and, since the mid-1990s, also by the aeronautics manufacturers [4]. This is due to the high flow of energetic particles that devices operating in these high-altitude environments are exposed to. The atmosphere shields part of the energetic particles that come from outside the Earth, so that, the higher the altitude, the higher the particle flux. To mitigate these effects, radiation shields, redundant components, techniques of error detection and correction and radiation tolerant elements are used. The implementation of these measures ranges from technological aspects of architecture to system level. Many of these measures increase costs and negatively impact circuit performance. There exist many well-known techniques to mitigate SEU effects, such as triple modular redundancy (TMR), which can be suitable for certain applications. However, most of them involve high penalties in terms of cost, power, or performance, which can be affordable for the space industry but could be non-acceptable for other FPGA fields of application.

In addition, due to technology scaling, SEUs are becoming a major reliability concern for electronic devices in general and SRAMs in particular, not only in harsh radiation environments but also at ground level, where radiation fluxes are low. In the case of SRAMs, this is due to the fact that the number of errors per time unit in SRAM memories due to radiation-induced transient events has increased with technology scaling [3, 5]. This fact has two main causes. The first cause has to do with both reducing the dimensions of the transistors forming the cells and with decreasing the supply voltage. Both factors contribute to reduce the amount of electrical charge used by a cell to store one bit of information. Thus, it is easier that the charge induced by the interaction of a particle upsets the cell content. The second cause includes three related factors: the increase in the number of cells integrating SRAMs, the higher density of cells, and the amount of chip area occupied by SRAM cells. All of them contribute to increase the probability that an energetic particle interacts with a sensitive area of a memory causing a transient event that leads to cell data corruption. In a FPGA, this can be a serious problem, since SRAM-based FPGAs rely on SRAMs to store configuration bits. An SEU affecting one of those bits can produce an unpredictable behavior or even a complete system failure.

To conclude, SEU effects are not a new problem and the space industry has developed specialized techniques to deal with them for decades. However, FPGAs are used in a broad range of applications, and in many of them circuits are not subject to high radiation fluxes. Nevertheless, due to technology scaling, they are becoming sensitive to radiation either from the environment or from the circuit materials. For this reason, it is necessary to implement some radiation hardening techniques, especially if the circuit is operated in critical systems. Traditional aerospace techniques are not suitable for most SRAM-based FPGA applications, since they involve high costs or significant performance degradation, which cannot be assumed. One of the most paradigmatic examples is commercial electronics or any other FPGA application field where FPGAs are attractive due to its fast time to market, flexibility, and reprogrammability, which reduce costs while keeping good performance. Thus, the aim of this chapter is to present a technique that fills this gap and can be used as a suitable technique to improve radiation reliability in a broad range of FPGA-SRAMs applications. More specifically, the technique works at the cell design level, and its goal is to enable the design of intrinsically more robust cells. In addition, the technique is also attractive because it is compatible with current memory compilers, since it does not change SRAMs cell architecture.

2. Radiation impact on SRAMs

The analysis of radiation impact on integrated circuits is difficult and is typically performed by experimental tests or using device-level simulations. However, the critical charge (Q_{crit}) is a parameter usually used as a standardized methodology to analyze the circuit-level impact of radiation on SRAMs [6, 7]. One of the main advantages of this parameter is that it can be obtained by electrical simulations, which are cheaper than experimentation and less time consuming than device-level simulations. In addition, it helps to understand how SEUs are produced.

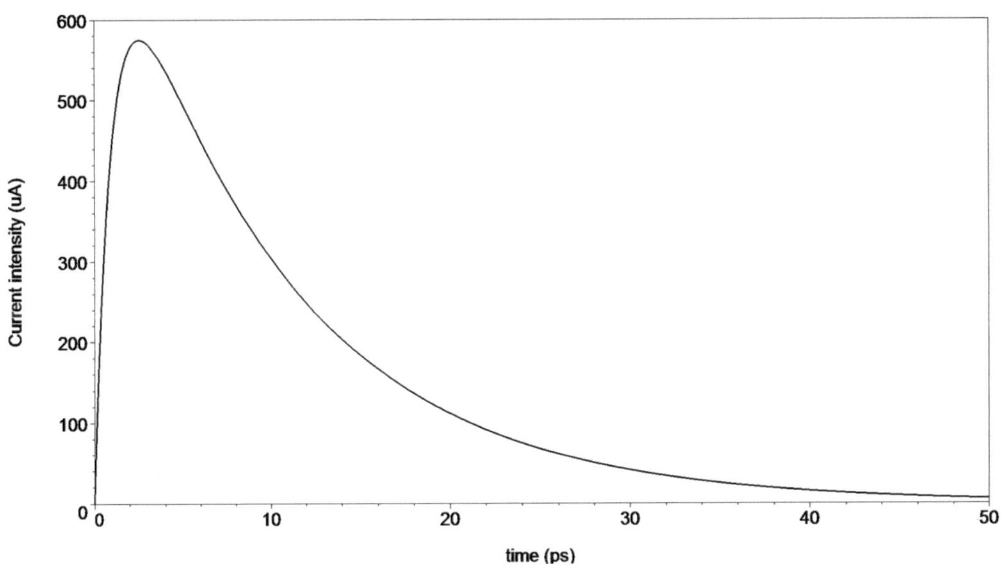

Figure 1. Example of a double exponential current pulse.

When an energetic particle impacts a CMOS circuit substrate, it induces a charge track due to electron-hole pair generation. This deposited charge can be collected by a sensitive node—typically the drain of an off transistor—which is near to the ionization track [4]. This results in a transient current pulse at the node. A sufficiently strong current pulse will modify data stored in the cell (cell flip). If this occurs, an SEU is produced. The word "Single" means that the cell upset is caused by a single energetic particle. The parameter used to quantify the minimum amount of charge collected by a memory element node that changes its state is the critical charge. Typically, Q_{crit} is determined by electrical simulation analyzing how a given memory cell flips under current pulses having different shapes and intensities. It has been reported that energetic particle strikes lead to current transients with varying pulse durations (pulse width), and that the Q_{crit} value of a node is a function of the waveform shape [8, 9]. For this reason, a proper choice of current waveforms to estimate the critical charge is important. In this chapter, we will use the well-known double-exponential current source model given by

http://dx.doi.org/10.5772/66195

$$i(t) = i_0 \left(e^{-\frac{t-t_0}{\tau_1}} - e^{-\frac{t-t_0}{\tau_2}} \right) \quad (1)$$

where $i(t)$ is the current intensity at time t, i_0 is a parameter that scales the current intensity, τ_1 determines the current fall-time, τ_2 its rise time, and t_0 is the time at which the current peak is initiated. The total charge injected in the node is the area under the $i(t)$ curve. The shape of one of these curves is represented in **Figure 1**.

Figure 2 depicts a 6-transistor SRAM (6T-SRAM) cell configuration. It has two cross-coupled inverters which form the two internal cell nodes (LN and RN). In addition, it has two access transistors, which are used to reach the internal nodes from outside the cell in the read and write operations.

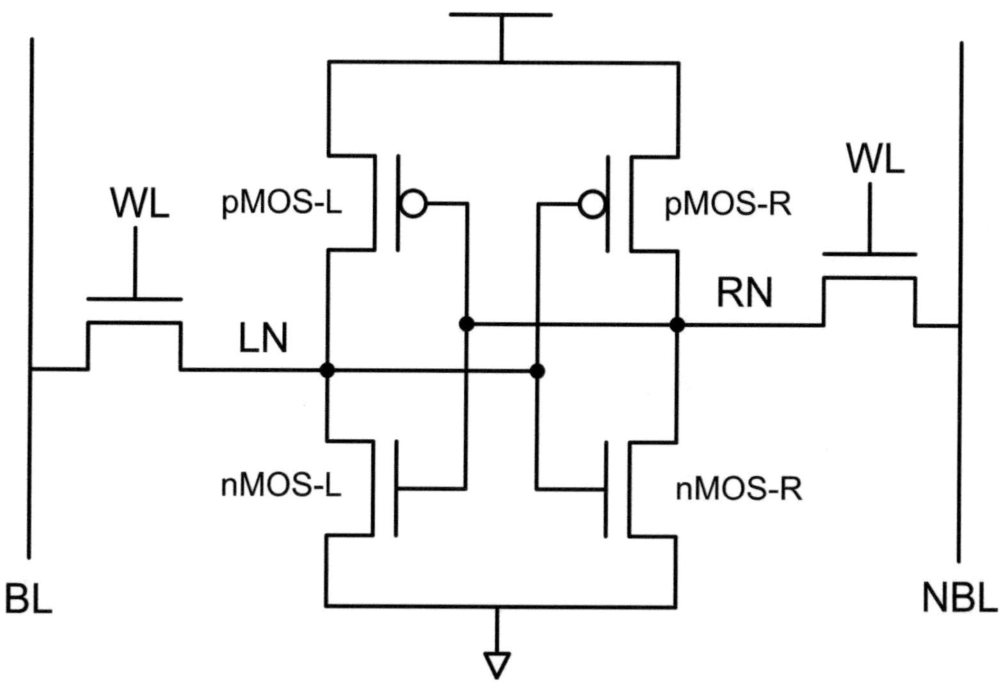

Figure 2. 6T-SRAM cell schematic.

Figure 3 shows the current sources scheme used to simulate SEUs. In particular, it is necessary to investigate two types of SEUs: a 0-to-1 SEU, where the impacted node is at 0 level, and a 1-to-0 SEU, where the impacted node is at 1 level. Due to cell symmetry, only two configurations cover all possibilities of memory cell perturbation. **Figure 3** also shows that a charge injection on a node which is at 0 requires the nMOS transistor to drain the collected charge due to the particle hit. Conversely, when a particle hits a node which is at 1, the pMOS transistor maintains the stored value by providing the current needed to hold the node electrical value.

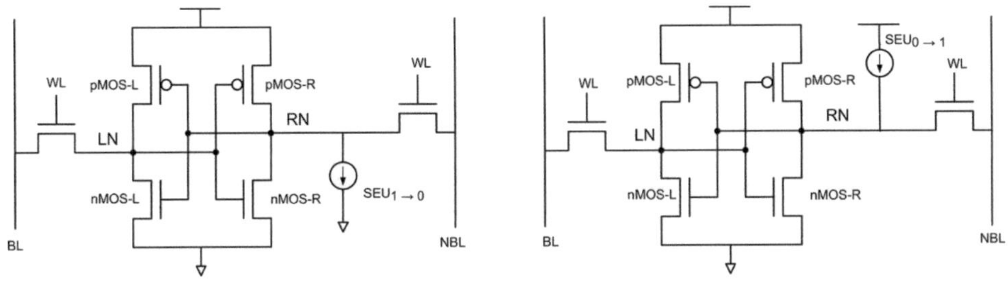

Figure 3. 6T-SRAM cell schematics for simulating a 1 to 0 SEU (left) and a 0 to 1 SEU (right).

This chapter deals only with 6T SRAM cells, although there are other SRAM which are specially designed to deal with radiation issues. In general, they are hardened SRAM cells that maintain their stored data even if the electrical state of some of their nodes is flipped by a particle strike, some of them are described in [10, 11]. The main drawbacks of them are the increase in cell transistor count with the consequent area increase. In addition, in these cells, it is difficult to implement regular cell layouts, which, as it will be described in Section 3, is a useful method for parameter variation minimization. Furthermore, nonstandard cells complicate the possibility to include them in current SRAM and FPGA design flows. Conversely, the technique that will be described in Section 3 is fully compatible with SRAM memory compilers and easily adaptable to current FPGA designs. Other mitigation techniques, such as supply voltage increase [9], are not suitable to be implemented in many applications due to their impact on power consumption and long-term reliability.

3. SRAM cell transistors channel width modulation technique

Memories are usually structures in which the maximum density of integration is requested. Therefore, the transistors forming memory cells are usually close to the minimum dimensions enabled by technology. Nevertheless, this section describes how it is possible to achieve more robust SRAM cells by varying the channel width of some of the cell transistors. This technique has a clear impact on the area occupied by each cell and, therefore, in the total memory area. For this reason, we will study how to use the area increase in the most efficient way, that is, how to get some gain in critical charge with minimum additional area. Moreover, the impact of this technique in terms of power consumption, stability, and access time is characterized in Section. 4.4.

Designing SRAMs is a challenge as technology scales down mainly due to parameter variations. There are two main causes of mismatch between the cell cross-coupled inverters: polysilicon and diffusion critical dimensions, as well as implant variations [12]. The use of subwavelength lithography and reactive ion etching are two of the main causes that converts the drawn polygon corners on the layout mask into rounded shapes on the manufactured circuits. Although proper optical proximity corrections can minimize those distortions, these techniques alone cannot compensate all the distortions, especially as the lithography gap is increasing with each successive technology node [12]. As a result, traditional cell designs are very sensitive to misalignment because they include transistor diffusion width changes. These changes in width produce bends and steps in the diffusion regions, which in turn, cause small variations of the poly placement that lead to significant poly-diffusion overlay misalignment. This variability impacts directly on transistor matching, which can compromise cell stability and functionality.

The so-called regular cell layouts (**Figure 4**) have shown to be more tolerant to parameter variations due to several factors: all poly lines are drawn in the same direction, poly lines are aligned facilitating better polysilicon critical dimension control, and helping phase shift masking techniques [13]. In addition, when a cell is inside the SRAM array, all transistors see the same polysilicon patterns, thus minimizing poly proximity issues [12]. Finally, regular cells have straight diffusions and, therefore, are much less sensitive to misalignments [14, 15].

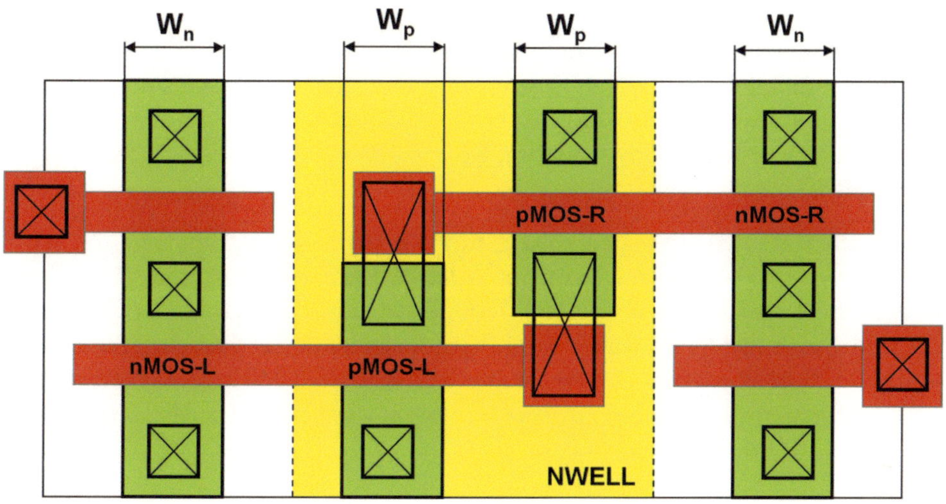

Figure 4. 6-T SRAM regular layout.

Parameter variation has become a key factor in SRAM memory design. For this reason, the regular layout is the one that is considered in this chapter. Using regular layouts imposes geometrical restrictions, for example, as previously mentioned, it is necessary to orientate all polysilicon lines in the same direction and keep them aligned. However, the determining factor that mainly affects the transistor channel width modulation technique is the impossibility to introduce steps and bends in the diffusion areas. This means that the designer will be unable to freely change SRAM transistors channel widths.

The formation of bends in the diffusion regions of a cell, like the one considered in **Figure 4**, can be avoided if all nMOS transistors channel width (W_n) is the same, as well as all pMOS transistors channel width (W_p) is also the same. In **Figure 4**, it can be seen that this way the diffusion areas (colored in green) remain straight. If we consider as a reference a cell where channel width of all transistors is the minimum (W_{min}), the restriction is expressed as

$$W_n = r_n \cdot W_{min}$$
$$W_p = r_p \cdot W_{min} \quad (2)$$

With these two restrictions, the nMOS channel width can vary independently from the pMOS channel width. This implies that the designer has two degrees of freedom.

3.1. Critical charge results

As it was mentioned before, the behavior of the cell undergoing a current injection due to an energetic particle impact depends on the duration of the pulse (pulse width); for this reason, it is interesting to use it as a parameter to explore.

Pulse widths of current transients are highly variable and depend on multiple parameters, but several studies show that they are between a few picoseconds and hundreds of picoseconds [6]. 3D simulations also show that short pulses correspond to ionization events whose track crosses the drain of a cut-off transistor, while long ones are the result of events whose track does not pass through the drain [9]. It is necessary to consider both cases, since the location of the trace ionization with respect to drain is a random parameter. For this reason, to characterize the behavior of the cell, simulations with pulse widths ranging between 20 and 200 ps have been performed.

In addition, there are two different critical charges depending on which node (the one at 0 or the one at 1) receives the collected charge modeled by the current injection. The collection of electrons by the drain junction of an nMOS in OFF state results in a current pulse that upsets the affected node from 1 to 0, so this critical charge is named $Q_{crit,e}$. Similarly, the collection of holes by a pMOS drain junction upsets the affected node from 0 to 1, so this critical charge is called $Q_{crit,h}$. If both critical charges are represented as a function of pulse width, **Figure 5** is obtained.

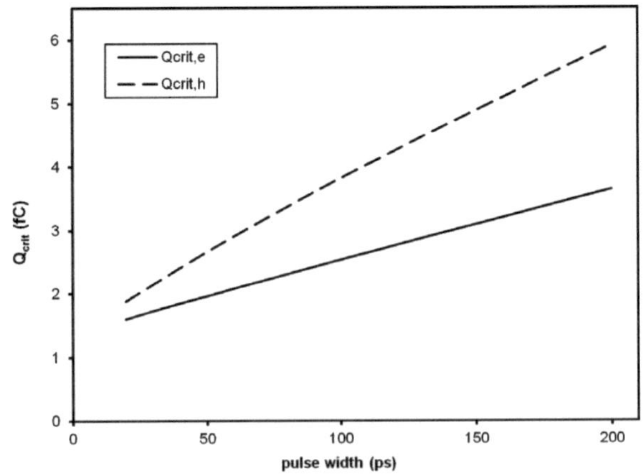

Figure 5. Critical charge for electrons and holes of a minimum-sized 6T-SRAM ($r_n = r_p = 1$) as a function of pulse width.

It can be observed that $Q_{crit,e}$ is lower than $Q_{crit,h}$. Therefore, it is normally considered that the cell-flip process is dominated by $Q_{crit,e}$, and sometimes $Q_{crit,h}$ is neglected. However, accurate models need to include both critical charges, as it will be shown in Section 4.4.

In addition, critical charges for a 6T cell for various combinations of W_p, W_n were calculated. **Figure 6** shows the results in a graph where the independent variables are r_p, r_n. Results are shown for two different pulse widths and only for $Q_{crit,e}$, since $Q_{crit,h}$ show similar results.

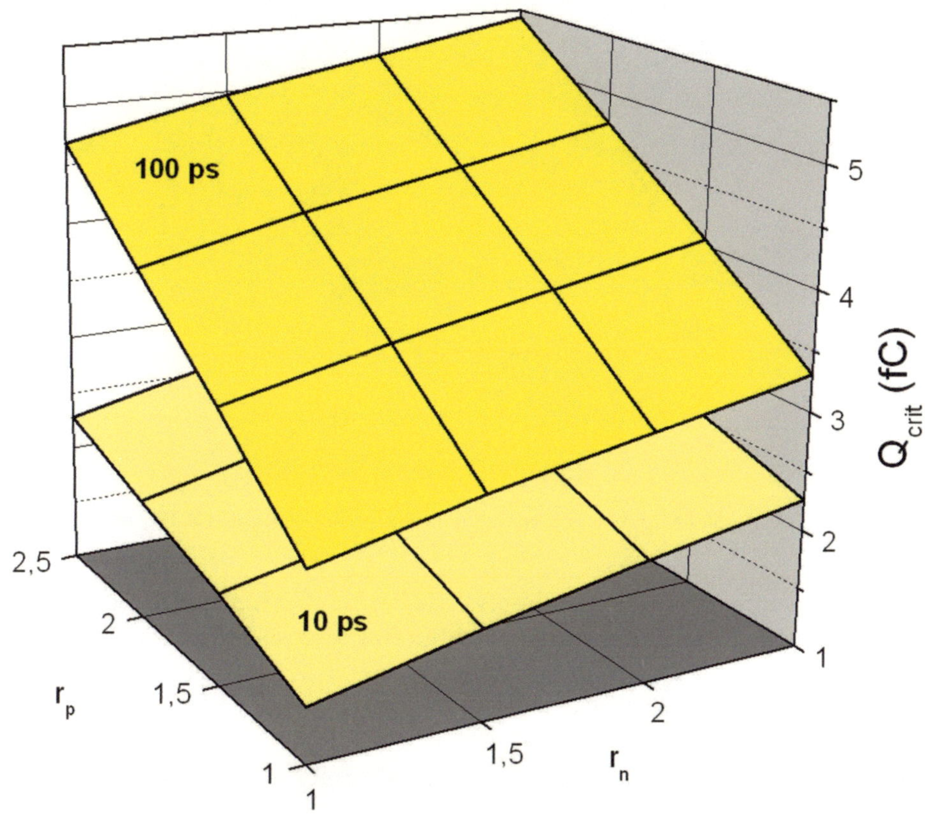

Figure 6. Critical charge ($Q_{crit,e}$) as a function of r_n and r_p and for two different pulse widths.

Figure 6 shows how the cell is more robust as the transistors channel width is increased. However, increasing the channel width of transistors produces a clear and undesired impact on the area of each cell and, therefore, on the total memory area. For this reason, it is necessary to establish a trade-off between the increased radiation robustness and the additional area used. Moreover, it is convenient to use the additional area in the most efficient possible way. This is discussed in the following subsection.

It has also been studied how the supply voltage affects cell robustness. **Figure 7** shows the results of critical charge for a typical alpha-particle pulse width of 30 ps [6] as a function of r_p, r_n for two different supply voltages.

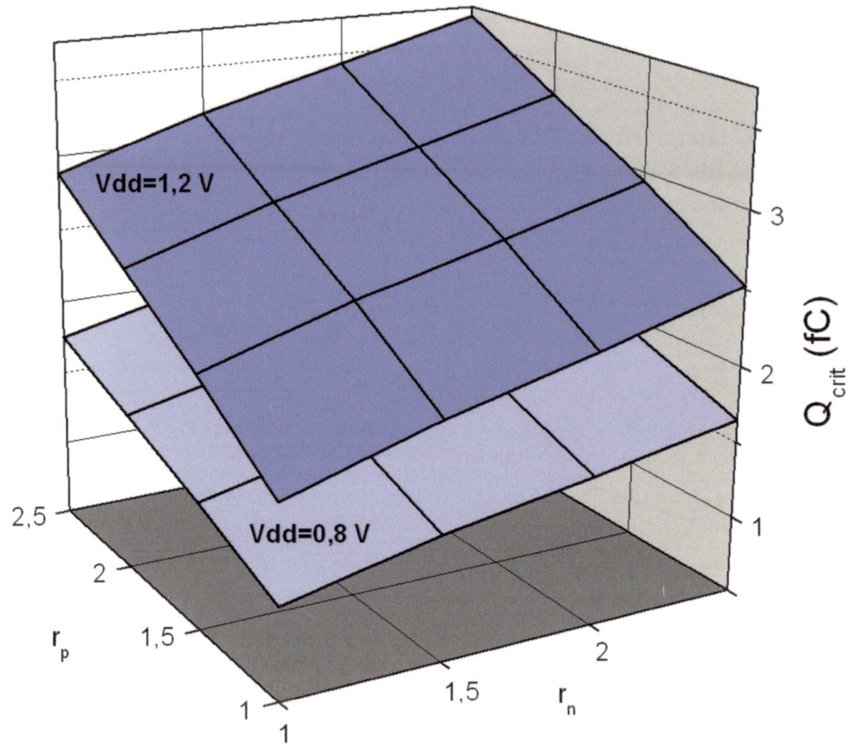

Figure 7. Critical charge ($Q_{crit,e}$) as a function of r_n and r_p and for two different supply voltages and for a 30 ps pulse width.

As it can be observed, a decrease in the supply voltage causes a reduction in the critical charge for all combinations of transistors channel widths. This result is in line with the previously mentioned fact that a cell with reduced voltage supply uses less charge to store data and, therefore, it is easier to change its stored value.

3.2. Additional area optimization to harden the SRAM cell

Due to the almost linear behavior of the graph in **Figure 6**, the following coefficients can be defined and are virtually independent of W_p and W_n:

$$\chi_p = \frac{\partial Q_{crit}}{\partial W_p} \chi_n = \frac{\partial Q_{crit}}{\partial W_n} \quad (3)$$

These two coefficients represent the efficiency, in terms of critical charge, of a certain increase in the transistors channel width (pMOS in the case of χ_p, and nMOS in the case of χ_n).Geometrically, these coefficients represent the slopes in the two horizontal directions of the planes of **Figure 6**. These slopes vary as a function of the different pulse widths; therefore, coefficients are a function of the considered pulse width. If this dependence is plotted, **Figure 8** is obtained.

http://dx.doi.org/10.5772/66195

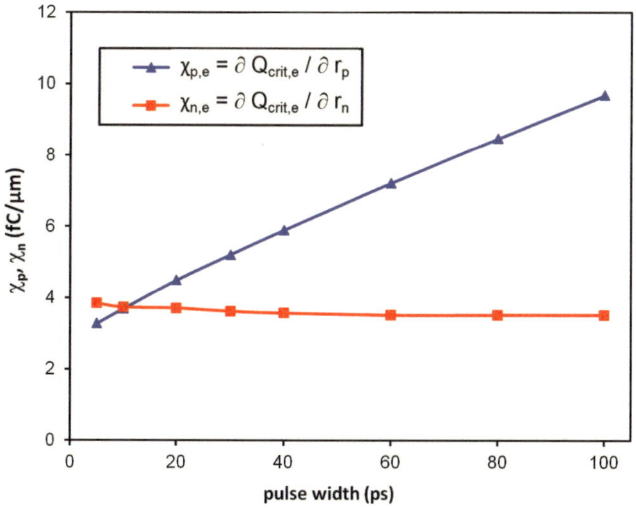

Figure 8. Dependence of $\chi_{p,e}$ and $\chi_{n,e}$ with pulse width for nominal supply voltage (1.2 V).

Figure 8 shows that, in general, χ_p is larger than χ_n, only for very short pulses χ_n tends to equal or even exceed the value of χ_p. This means that for pulses longer than about 10 ps, increasing only pMOS transistors width (W_p) is more efficient than increasing nMOS transistors (W_n). As it has been mentioned before, the widths of the current pulses generated by SEU vary. However, for alpha particles, a typical pulse width is about 30 ps [6]. For this typical pulse width, increasing W_p is more efficient than increasing W_n.

Same simulations were repeated for 0.8 V supply voltage, the results are shown in **Figure 9**.

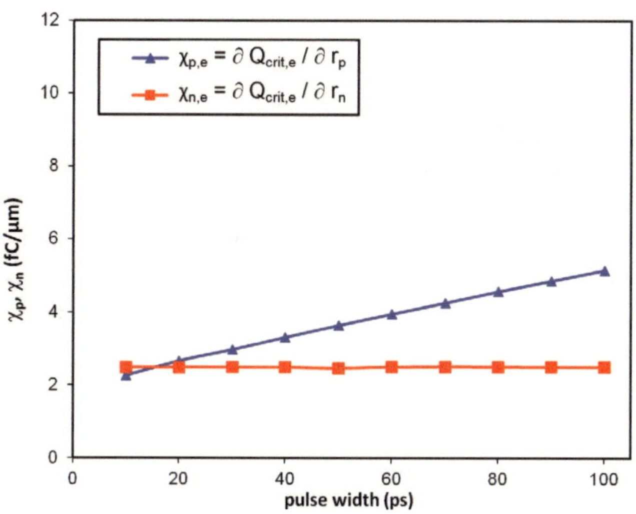

Figure 9. Dependence of $\chi_{p,e}$ and $\chi_{n,e}$ with pulse width for 0.8 V supply voltage.

The results obtained are analogous to those of **Figure 8**. However, the values of χ_p and χ_n at 0.8 V are lower than at 1.2 V (note that the graphs in **Figures 8** and **9** are represented at the same scale). This means that reducing the supply voltage not only reduces the critical charge but also reduces the efficiency in terms of critical charge to make wider pMOS transistors.

Finally, **Figure 10** plots χ_p as a function of the pulse width and supply voltage in a surface plot and as a family of curves generated by the supply voltage parameter.

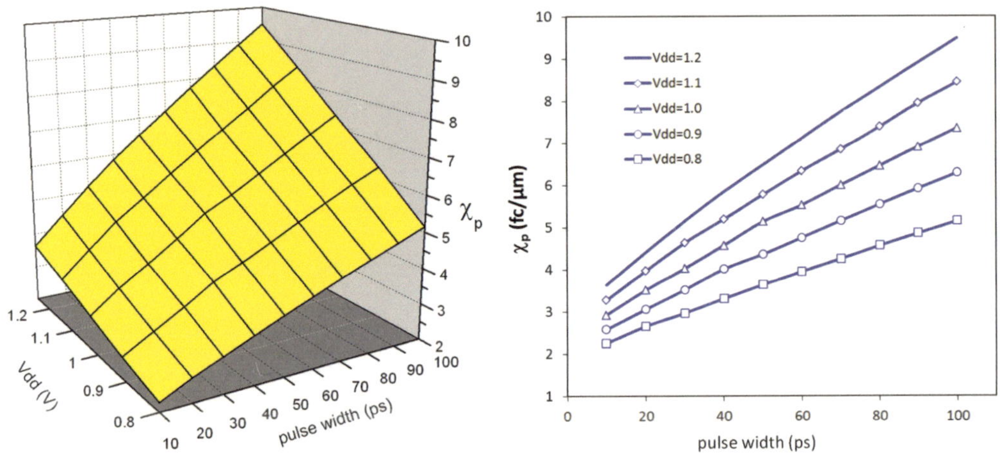

Figure 10. Dependence of $\chi_{p,e}$ with pulse width and supply voltage.

The graph in **Figure 10** shows that reducing both the supply voltage and the pulse width decreases the efficiency, in terms of critical charge, of modulating the pMOS transistors channel width.

From all the results presented in this section, it can be deduced that if the SEU robustness of an SRAM cell is to be increased in a certain percentage, increasing the widths of only the pMOS and leaving the nMOS unmodified is more efficient than any other combination of transistor width modulation. Or, for a given percentage area budget, increasing only pMOS widths maximizes critical charge.

r_p	W_p (μm)	$Q_{crit,e}$ (fC)	$Q_{crit,e}$ increment with respect to minimum cell (%)	Area increment with respect to minimum cell (%)
1.0	0.15	1.72	0	0
1.5	0.23	2.14	24	9
2.0	0.30	2.51	46	17

Table 1. Critical charge and cell area increment for three different values of r_p, and r_n = **1** (W_{min} = 0.15 μm). The supply voltage is nominal.

Table 1 shows the critical charges for a pulse of 30 ps for three values r_p (and $r_n = 1$) at nominal voltage. In addition, it shows the increased area with respect to the minimum sized cell (r_p = 1, $r_n = 1$). Areas are obtained by designing cells with the regular layout features and restrictions described earlier.

Table 1 shows that, for example, for an area increase of 17%, an increment 46% in critical charge is achieved.

To sum up, the transistors channel width modulation technique has shown by simulation to be effective in terms of improving critical charge. For this reason, it was decided to implement and test this technique in a real memory prototype (test chip) described in Section 4.1.

4. Experimental results of the modulation technique

4.1. Test chip description

The transistor width modulation technique was implemented in a custom fabricated SRAM test chip in a 65-nm CMOS commercial technology. Memory cells are six-transistor (6T) cells and were implemented following regular layout design specifications to minimize parameter variations. The regular layout characteristics were described in Section 3, and include the use of straight diffusion regions and regular alignment of word line polysilicon lines.

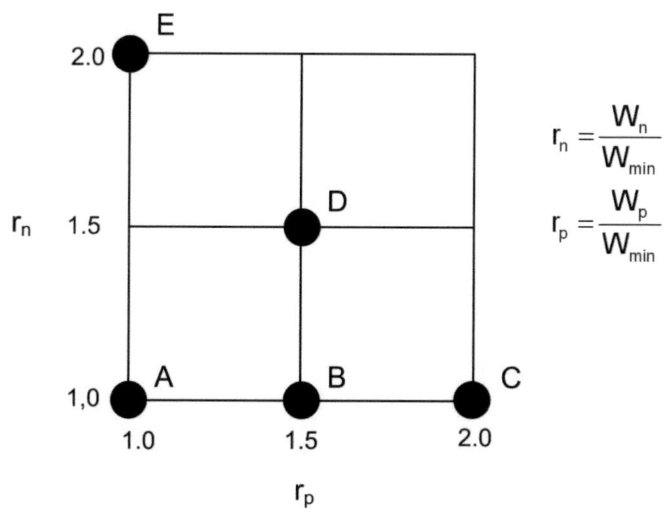

Figure 11. Schematic representation of the five cell types implemented in the test chip.

From all the previously simulated cells, five of them were implemented in the test chip (five different combinations of transistors channel widths). All these combinations satisfy the restrictions imposed for a regular layout. The selected combinations (cell types) of r_n and r_p are schematized in **Figure 11** and detailed in **Table 2**. For each one of the five cell types, a total of

4096 cells were implemented. Finally, the test chip was irradiated following the procedure detailed in Section 4.2 to experimentally test the modulation technique.

Cell type	pMOS width, W_p (μm)	nMOS width, W_n (μm)	Cell height (μm)	Cell width (μm)	Cell area (μm²)	Cell area increment with respect to A (%)
A	0.15	0.15	0.58	1.75	1.01	0
B	0.23	0.15	0.58	1.91	1.10	9
C	0.30	0.15	0.58	2.05	1.18	17
D	0.23	0.23	0.58	2.07	1.19	18
E	0.15	0.30	0.58	2.05	1.18	17

Table 2. Main geometric features of the five cell types implemented in the test chip.

4.2. Experimental irradiation procedure

The objective of the experiment is to obtain the soft error rate (SER) of each one of the five cell types, that is, the number of soft errors (SEUs) for time unit.

The 65-nm CMOS test chip was mounted on a specifically designed PCB and controlled by an FPGA to drive and capture data.

As a radiation source, it was used an Am-241 alpha source with a 5 kBq activity providing alpha particles of 5.5 MeV. The source active area was 7 mm in diameter and was placed atop the unencapsulated chip, and all five cell types were irradiated at the same time. The control FPGA was not irradiated because the objective of the experiment was only to study the behavior of the test chip SRAM cells under radiation conditions.

The test procedure was performed following the subsequent steps:

1. Write all memory cells to a known value.
2. Read all memory cells, and compare to the written values.
3. Initiate the memory radiation.
4. Wait for a sampling time T_s.
5. Read the whole memory and determine the number of cells whose state changed. Go to Step 4.

Steps 4–5 were cycled until the experiment was finished. The overall number of SEUs, N_{TOT}, is given by the addition of the number of SEUs recorded at each sampling period (N_i), i.e.

$$N_{TOT} = \sum_{i=1}^{n} N_i \tag{4}$$

http://dx.doi.org/10.5772/66195

with n being the number of times that the memory is read. The overall time experiment (t_{exp}) is given by $t_{exp} = nT_s$. The SER at each sampling time period (SER_i) is given by $SER_i = N_i/T_s$, while the mean SER of the overall experiment is given by

$$SER = \frac{\sum_{i=1}^{n} SER_i}{n} = \frac{\sum_{i=1}^{n} N_i}{n \cdot T_s} = \frac{N_{TOT}}{t_{exp}} \tag{5}$$

The determination of the sampling period T_s is important, since it must guarantee that the probability of a given cell to experience two or more flips within the same sampling period is negligible, while keeping the overall read time small with respect to the overall hold time (we are interested in computing the memory SER when the memory is not being accessed) [16]. We ran an initial experiment using a small one-minute T_s value and determined an SER order of magnitude of 1 SEU/minute. Based on this, we set a T_s value of 30 min to not increase the memory read rate. The mean estimated SEU error using this T_s value is 1‰.

4.3. Experimental results

The experiment was conducted under the conditions and procedure described in Section 4.2 for a total time of 72 h to accumulate enough SEUs as to obtain a reliable SER result.

The SEU count evolution is shown in **Figure 12**. As expected, results show that the accumulated SEU count with time is linear. An alternative way to calculate SER is by obtaining the slope of the plot of accumulated number of SEU as a function of time.

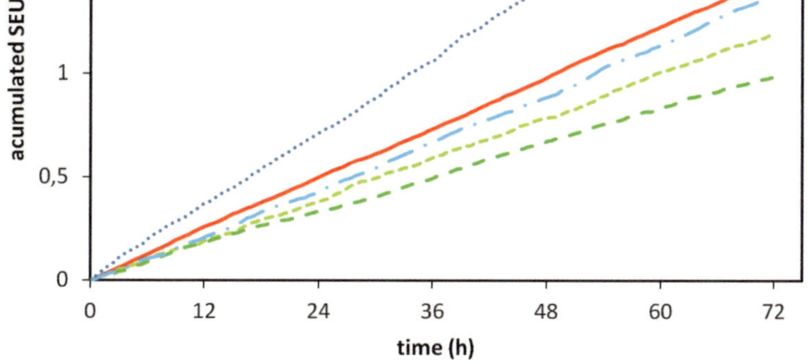

Figure 12. Accumulated SEUs in a 72 h period irradiation for the five cell types.

The first important result from **Figure 12** is that different memory cell types have different SER values (i.e., different slopes). If each SER is computed and represented in a bar plot, **Figure 13** is obtained.

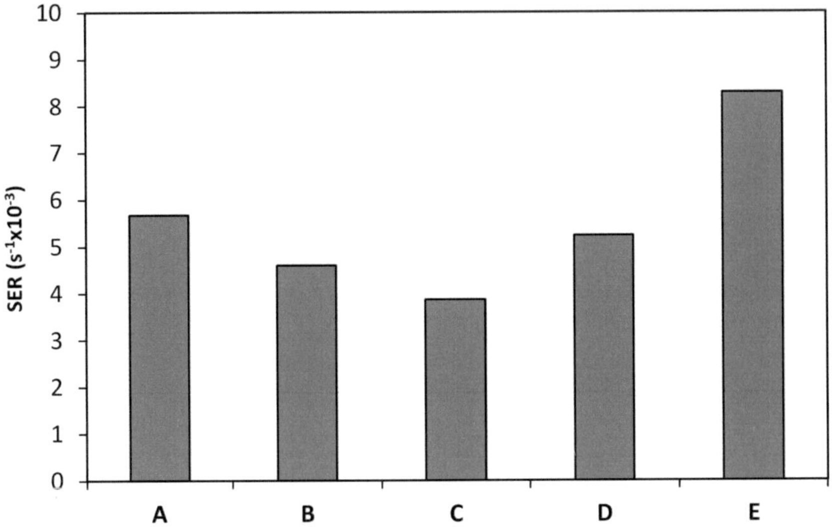

Figure 13. SER of 4096 cells for each one of the five cell types.

In addition, SER values are tabulated in **Table 3** along with critical charge results. Keep in mind that a more robust cell means more critical charge but less SER.

Cell type	SER ($s^{-1} \times 10^{-3}$)	$Q_{crit,e}$ (fC)
C	3.87	2.51
B	4.60	2.14
D	5.24	2.44
A	5.68	1.72
E	8.30	2.26

Table 3. SER and critical charge values for the five different cell types (sorted by SER value).

From **Figure 13** and **Table 3**, it is observed that the stronger cell—from a SER point of view—is the C, followed by B, and that the less robust is E. In addition, if critical charge is also taken into account, the following can be observed:

- The best cell is C; note that this occurs from both critical charge and SER points of view.
- Increasing the pMOS transistors channel widths (cells A, B, and C) causes an increase in critical charge, which directly results into a decrease in SER. That is, cell C is more robust than B, and B more robust than A, from both from critical charge and SER point of view.

- There is no the same direct correlation when cells in which nMOS transistors have been modified are involved. Cells D and E are among the most robust ones in terms of critical charge, and yet are among the ones that show worst SER.
- In Section 3.2, it was justified that increasing pMOS transistor widths was, from a critical charge point of view, the most efficient way to use the additional area. Cells B and C are the ones in which only pMOS transistor width is increased. From these results, it can be concluded that, in terms of SER, increasing only the pMOS transistors width is also the best way to improve SRAM cells robustness.

In short, increasing the pMOS transistors channel width improves critical charge and SER. However, increasing the nMOS transistors channel width improves critical charge, but worsens SER. The reason for this nonsymmetrical behavior must be sought in the fact that increasing critical charge by widening the channel of the transistors has a dual effect on SER:

- It increments cell robustness, because more charge is needed to flip the cell (higher critical charge).
- It lowers cell robustness because a wider transistors channel involves a sensitive area increase, which may also involve an increase in the ability of the cell nodes to collect the charge that has been deposited by an impacting energetic particle.

The key point is that the relative contribution of these two factors (critical charge and area increase) is not the same in the case of widening nMOS and pMOS transistors. Increasing the channel size of pMOS implies an area increase inside the well, while increasing the channel size of nMOS increases the area directly on the substrate. The different ability to collect charge of pMOS (in the well) or nMOS (on the substrate) is the qualitative explanation of the observed relation between SER and critical charge for nMOS and pMOS width modulation. This behavior is quantitatively explained in the following section.

4.4. Analysis of the results

Experimental data show that maintaining minimum nMOS transistors width ($r_n = 1$) while increasing pMOS transistor channel widths improves both critical charge and SER for a 6T memory cell. However, increasing nMOS transistor channel width improves memory cell critical charge, but worsens SER. As it has been mentioned before, this can be qualitatively explained as follows: Increasing transistor width has two competing effects on SER. On the one hand, SEUs are more difficult to occur, because Q_{crit} is raised due to the increase of both the drain capacity and the transistor width, which enhances transistor strength. On the other hand, widening a transistor increases its sensitive area, raising the probability of the cell to collect charge and thus be flipped by the effect of an energetic particle. The relative contribution of these two opposite effects on SER depends on the transistor type (nMOS or pMOS), especially for CMOS bulk technologies with well areas for pMOS transistors [17].

To model these two effects, it is necessary to use an expression that relates SER and critical charge. The following expression [18] will be used:

$$\mathrm{SER} = \kappa\left(A_{\mathrm{diff},n} \cdot e^{\frac{-Q_{\mathrm{crit,e}}}{\eta_{\mathrm{e}}}} + A_{\mathrm{diff},p} \cdot e^{\frac{-Q_{\mathrm{crit,h}}}{\eta_{\mathrm{h}}}} \right) \tag{6}$$

where $A_{\mathrm{diff,n}}$ and $A_{\mathrm{diff,p}}$ are the nMOS and pMOS sensitive drain area. $Q_{\mathrm{crit,e}}$ and $Q_{\mathrm{crit,h}}$ are respectively the critical charges due to the collection of electrons and holes, and κ is a parameter that depends on the radiation flux. Parameters η_{e} and η_{h} represent electron and hole charge collection efficiency. To compute SER, parameters κ, η_{e}, and η_{h} need to be experimentally obtained, as they depend on the environment and on the device precise characteristics. Note that the model includes both critical charges ($Q_{\mathrm{crit,e}}$ and $Q_{\mathrm{crit,h}}$) introduced in Section 3.1.

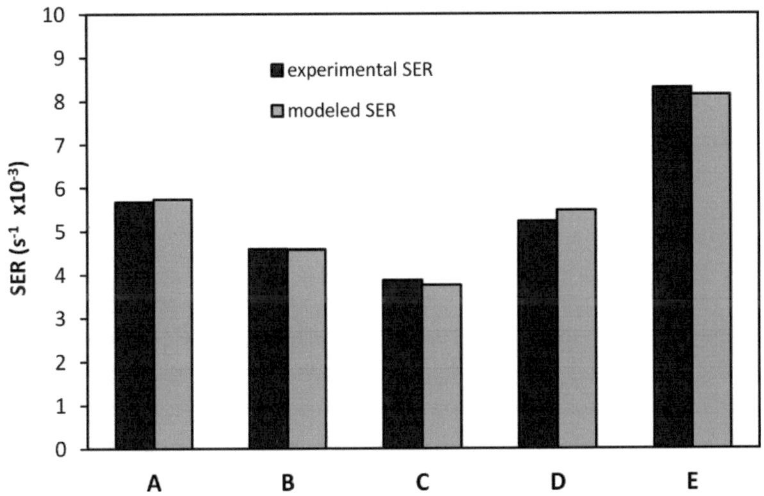

Figure 14. SER (experimental and modeled) of 4096 cells for each one of the five cell types.

In our case, since we obtained SER and critical charge for different cell types, we can fit SER experimental data to the calculated critical charge values and obtain the unknown parameters κ, η_{e}, and η_{h}. Diffusion areas can be expressed as $A_{\mathrm{diff,n}} = W_n \cdot H_{\mathrm{n}}$, and $A_{\mathrm{diff,p}} = W_{\mathrm{p}} \cdot H_{\mathrm{p}}$, being H_{n} and H_{p} the diffusion lengths of the drains of the nMOS and pMOS transistors. The design rules restrictions for symmetrical and regular cell layout impose H_{n} to be slightly longer than H_{p} (in fact we used the minimum possible diffusion length in the pMOS transistor, $H_{\mathrm{p}} = H_{\mathrm{min}}$, while $H_{\mathrm{n}} = K_{\mathrm{diff}} \cdot H_{\mathrm{min}}$ with $K_{\mathrm{diff}} = 1.1$ for the five different cells). Introducing again r_{n} and r_{p} coefficients defined in Eq. (2) we obtain:

$$\begin{aligned} A_{\mathrm{diff,n}} &= r_{\mathrm{n}} \cdot W_{\mathrm{min}} \cdot K_{\mathrm{diff}} \cdot H_{\mathrm{min}} = r_{\mathrm{n}} \cdot K_{\mathrm{diff}} \cdot A_{\mathrm{min,diff}} \\ A_{\mathrm{diff,p}} &= r_{\mathrm{p}} \cdot W_{\mathrm{min}} \cdot H_{\mathrm{min}} = r_{\mathrm{p}} \cdot A_{\mathrm{min,diff}} \end{aligned} \tag{7}$$

where $A_{\mathrm{min,\,diff}} = W_{\mathrm{min}} \cdot H_{\mathrm{min}}$. Therefore, Eq. (7) becomes:

http://dx.doi.org/10.5772/66195

$$SER = \overbrace{\kappa \cdot A_{min,diff}}^{K_A}\left(K_{diff} \cdot r_n \cdot e^{\frac{-Q_{crit,e}}{\eta_e}} + r_p \cdot e^{\frac{-Q_{crit,h}}{\eta_h}}\right) \tag{8}$$

The values of *SER*, $Q_{crit,e}$, $Q_{crit,h}$, r_n, r_p, and K_{diff} in Eq. (8) are known and, therefore, K_A, η_e and η_h remain as fitting parameters, being K_A the product of κ and $A_{min,\,diff}$. The obtained values after the fitting for these parameters are: $K_A = 3.13 \times 10^{-6}\ s^{-1}$, $\eta_e = 2.02$ fC, and $\eta_h = 0.79$ fC.

Figure 14 compares the experimental and fitted SER. As it can be seen, Eq. (8) accurately describes the experimental SER as a function of critical charge and geometrical parameters. In addition, the model properly describes quantitatively the asymmetrical influence of nMOS and pMOS transistor width in terms of SER, which was previously interpreted qualitatively.

The experimentally fitted parameters and the resulting critical charge values from Eq. (8) allow to plot SER as a function of r_n and r_p. The resulting surface is shown in **Figure 15**.

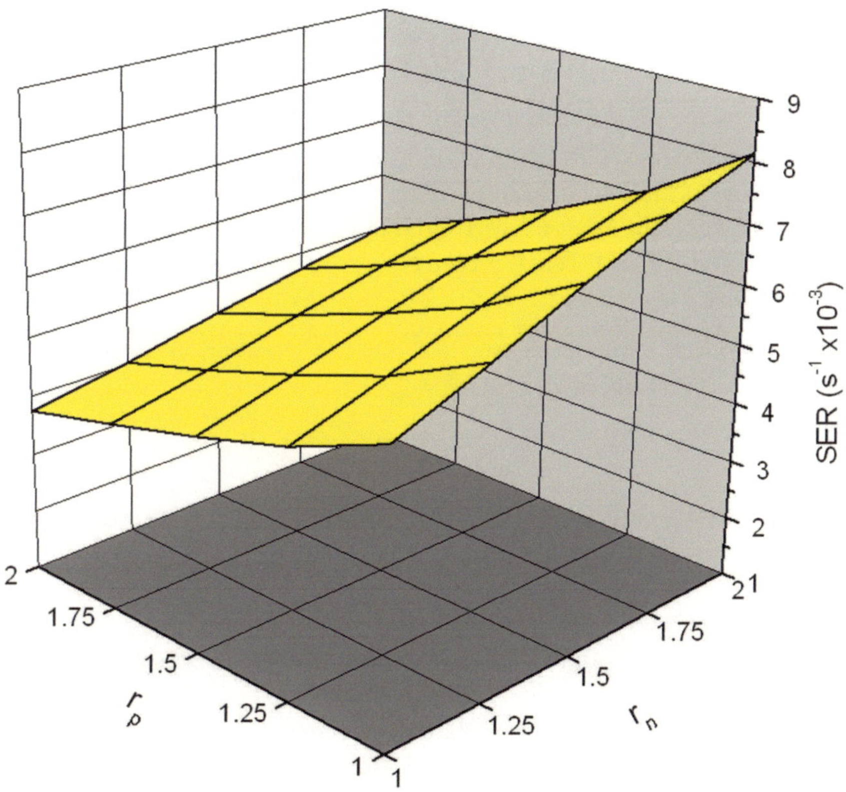

Figure 15. SER as a function of r_n and r_p.

Results of **Figure 15** confirm that increasing r_p leads to a SER reduction, whereas increasing r_n produces an undesired SER increment. This SER surface can be compared to the critical charge surface of **Figure 6**, where critical charge was improved as both r_n and r_p were increased.

If the charge collection efficiency values obtained as fitting parameters are analyzed, it is confirmed that charge collection efficiency for electrons (η_e) is higher than for holes (η_h) [19]. In addition, critical charge for electrons ($Q_{crit/e}$) is smaller than for holes ($Q_{crit/h}$). This electron and hole asymmetry in terms of charge collection efficiency and in terms of critical charge is the root cause of the observed differences of SER dependency with r_n and r_p.

Usual 6T-cells are designed with minimum sized access transistors ($W_{acc} = W_{min}$), minimum sized pMOS ($W_p = W_{min}$), and non-minimum-sized nMOS ($W_n = \mathrm{CR} \cdot W_{min}$). The CR parameter is called cell ratio and is usually greater than 1, being the most frequent values between 1.5 and 2.5 as a trade-off to assure cell stability during write and read operations [3]. Note that this cell with this transistor dimensions does not have straight diffusions. In addition, also note that this cell has the internal latch (cross coupled inverters) equal to the ones in E cell.

From the irradiation experiments, it has been obtained that the C cell shows an SER that is a 46% of the E cell SER, that is, C cell receives less than half the number of SEUs per time unit than E cell. Note that this improvement is achieved only by adequate transistor sizing, because both cells (C and E) have the same area. If instead of considering this two cells, we compare the C cell with respect to a usual cell with CR = 2, then the SER of the C cell is a 57% the SER of the CR = 2 cell.

The effects of the transistor width modulation technique on power consumption and access time are summarized in **Table 4**. For example, it can be observed that C and E cells show similar access times and power consumption levels (although there is an increase of the energy needed to change the logic state of the C cell, it presents lower leakage current than the E cell).

Cell type	Leakage (pW/cell)	Write energy (fJ/cell)	Write time (ns)	Read time (ns)	RSNM (mV)	WSNM (mV)
A	125.5	4.65	0.32	0.28	168	468
B	134.2	5.93	0.33	0.28	178	429
C	144.1	7.10	0.35	0.28	184	346
D	163.8	6.45	0.36	0.27	165	468
E	180.6	5.59	0.36	0.26	149	517

Table 4. Summary of different power, speed and stability figures of the fife different cell types.

Finally, it was also analyzed how the modulation technique affects read and write stability, by computing two well-known parameters: read static noise margin (RSNM) and write static noise margin (WSNM). As it can be seen in **Table 4**, RSNM is not very affected. Despite that, in [20], a technique to recover the RSNM of a 6T cell is analyzed. In addition, WSNM is degraded in some cell types (the ones in which pMOS transistors are increased in size). To overcome that, if needed, there are write assist techniques that could be suitable to improve WSNM [21, 22]. However, all tested cells types are experimentally writable with no write assist technique applied.

http://dx.doi.org/10.5772/66195

5. Conclusions

Due to technology scaling, radiation effects have become a major concern for modern integrated circuits even at ground level. FPGA SRAMS are not an exception, and radiation effects are even maximized, because these circuits are usually designed with transistors sizes close to the minimum allowed by technology. The so-called SEUs are the main radiation issue for SRAMs. SEUs are capable of altering the memory content of SRAM cells without permanently damaging the circuit.

A technique based on transistor width modulation was developed and tested. The technique consists in modifying the cell transistors channel width in a way that is compatible with the so-called regular layouts (i.e. avoiding the formation of bends in the diffusion regions). The main advantage of this layout scheme is that it reduces parameter variation. Nevertheless, it imposes some geometrical restrictions over transistor sizes, so that the modulation technique has to be designed to meet those constraints.

The technique was implemented and tested using two approaches: critical charge and experimental SER. Critical charge is a parameter cheap and easy to obtain, because it can be calculated using electrical simulations. However, as it was shown, it does not give a directly accurate measurement of the robustness of an SRAM cell if transistor areas are modified. Conversely, SER is a better parameter to assess cell robustness. The main drawback of SER is that it can only be directly obtained with experimental measurements, which are expensive and time consuming. After a preliminary analysis, the most interesting transistor size combinations where selected and implemented in a custom-fabricated test chip. The test chip has 4096 cells of each one of the five selected cells types, and all of them where irradiated with alpha particles to experimentally obtain SER.

Results show that some of the cell types are much more robust to radiation than others. In addition, results also reveal that, while a larger critical charge can lead to a better SER, some memory cells with higher critical charge also exhibit worst SER. This behavior was found when increasing nMOS channel widths. This suggests that special care must be taken when comparing SRAM cells with different transistor areas using critical charge as a figure of merit. Despite that, results indicate that SER can be estimated from critical charge with a model if some cell intrinsic cell parameters are known.

Results also show that SER is improved by increasing the pMOS transistors channel width (W_p), and worsened when the nMOS transistors channel width is increased (W_n). For this reason, the best way to design a hardened 6T SRAM cell is by minimizing the nMOS transistors channel width and dedicating all additional area to increase pMOS transistor channel width. In addition, for a 65-nm CMOS commercial technology, SER was reduced to a 57% of the value that conventional nonstructured layout cells exhibit. Due to careful transistor sizing, this radiation robustness improvement was achieved with minor area penalty. However, this hardened cells with wider pMOS transistors, also show a reduction in cell writability. To overcome this issue, write assist techniques can be implemented. Nevertheless, if a trade-off between writability, area, and radiation robustness is achieved by proper transistor sizing, hardened cells remain writable without any further action. Finally, with the modulation technique presented in this chapter, the achieved cell radiation robustness gain is fundamentally an area trade-off, provided that the cell remains writable. For this reason, at design level, radiation robustness can be set as an adjustable parameter in memory compilers.

Acknowledgements

This work has been supported by the European FEDER fund and the Spanish Ministry of Science and Innovation under Grant no. AP2006-03170 and TEC2008-04501 and TEC2011-25017 projects. It has also received funding to support competitive research groups from the Balearic Government (2011–2013), financed jointly by FEDER fund.

In addition, I want to sincerely thank all the members of the Electronics Systems Group (GSE-UIB) of the University of the Balearic Islands who have contributed to this research.

Author details

Gabriel Torrens

Address all correspondence to: gabriel.torrens@uib.edu

University of the Balearic Islands, Palma de Mallorca, Spain

References

[1] Yamauchi, H. Embedded Memories for Nano-Scale VLSIs. Aurburn: Springer Publishing Company; 2009.

[2] Yamauchi, H. Embedded SRAM circuit design technologies for a 45nm and beyond. In: 7th International Conference on ASIC; Guilin. IEEE Press; 2007. p. 1028–1033.

[3] Pavlov, A.; Sachdev, M. CMOS SRAM Circuit Design and Parametric Test in Nano-Scaled Technologies: Process-Aware SRAM Design and Test. Auburn: Springer Publishing Company; 2008.

[4] Normand, E. Single Event Effects in Avionics and on the Ground. International Journal of High Speed Electronics and Systems. 2004;14(2): 285–298.

[5] Baumann, R. C. Soft errors in advanced semiconductor devices-part I: the three radiation sources. IEEE Transactions on Device and Materials Reliability. 2001;1(1):17–22.

[6] S. V. Walstra, C. Dai. Circuit-Level Modeling of Soft Errors in Integrated Circuits. IEEE Transactions on Device and Materials Reliability. 2005;5(3):358–364.

[7] N. Seifert, P. Slankart, M. Kirsh, B. Narasinham, V. Zia, C. Brookseron, A. Vo, S. Mitra, B. Gill, J. Maiz. Radiation-Induced Soft Error Rates of Advanced CMOS Bulk Devices. In: IEEE Int. Reliability physics symposium; 2006. p. 217–224.

[8] P. Jain, V. Zhu.. Judicious Choice of Waveform Parameters and Accurate Estimation of Critical Charge for Logic SER. In: International. Conference on Dependable Systems and Networks.; Edinburgh, UK. IEEE/IFIP; 2007.

[9] T. Heijmen.. Factors that Impact the Critical Charge of Memory Elements. In: IOLTS 2006 Proceedings; Como, Italy. IEEE Computer Society; 2006. p. 6.

[10] M. Nicolaidis. Design for soft error mitigation. IEEE Trans. on Device and Materials Reliability. 2005;5(3):405–418.

[11] Z. Liu, V. Kursun. Characterization of a Novel Nine-Transistor SRAM Cell. IEEE Trans. On VLSI systems. 2008;16(40):488–492.

[12] Ban P. Wong, Anurag Mittal, Yu Cao, and Greg Starr. Nano-CMOS Circuit and Physical Design. Hoboken, New Jersey: John Wiley … Sons; 2005.

[13] K. Osada. Universal-Vdd 0.65-2.0 V 32 kB cache using voltage-adapted timing-generation scheme and a lithographical-symmetric cell. In: IEEE International Solid-State Circuits Conference; 2001. p. 168–169.

[14] F. Hamzaoglu, K. Zhang, Y. Wang, H.J. Ahn, U. Bhattacharya, Z. Chen, Y.-G. Ng, A. Pavlov, K. Smits, M. Bohr. A 3.8 GHz 153 Mb SRAM Design With Dynamic Stability Enhancement and Leakage Reduction in 45 nm High-k Metal Gate CMOS Technology. IEEE Journal of Solid-State Circuits. 2009;44(1):148 - 154.

[15] M. Yamaoka, K. Osada, K. Ishibashi. 0.4-V Logic Library Friendly SRAM Array Using Rectangular Diffusion Cell and Delta-Boosted-Array-Voltage Scheme. IEEE Journal of Solid-State Circuits. 2004;39(6):934–940.

[16] N. Seifert ; P. Slankard ; M. Kirsch ; A. Vo ; S. Mitra ; B. Gill ; J. Maiz. Radiation-Induced Soft Error Rates of Advanced CMOS Bulk Devices. In: IEEE International Reliability Physics Symposium Proceedings; San Jose, CA. IEEE Electron Device Society; 2006. p. 217–225.

[17] D. E. Fulkerson. An Engineering Model for Single-Event Effects and Soft Error Rates in Bulk CMOS. IEEE Transactions on Nuclear Science. 2011;58(2):506–515.

[18] T. Heijmen, P. Roche, G. Gasiot, K.R. Forbes, and D. Giot. A comprehensive study on the soft-error rate of flip-flops from 90-nm production libraries. IEEE Transactions on Device and Materials Reliability. 2007;7(1):84–96.

[19] P. Hazucha, and C. Svensson. Impact of CMOS technology scaling on the atmospheric neutron soft error rate. IEEE Transaction on Nuclear Science. 2000;47(6):2586–2594.

[20] S. Keshavarapu, S. Jain and M. Pattanaik. A New Assist Technique to Enhance the Read and Write Margins of Low Voltage SRAM Cell. In: International Symposium on Electronic System Design (ISED); Kolkata. IEEE Computer Society Conference Publishing Services (CPS); 2012. p. 97–101.

[21] R. Gupta, V. Gadi, H. A. Upendar. Write Assist Scheme to Enhance SRAM Cell Reliability Using Voltage Sensing Technique. In: International Conference on Embedded Systems; 2016. p. 318–322.

[22] S. Keshavarapu, S. Jain and M. Pattanaik. A New Assist Technique to Enhance the Read and Write Margins of Low Voltage SRAM Cell. In: International Symposium on Electronic System Design; 2012. p. 97–101.

Chapter 10

Power Efficient Data-Aware SRAM Cell for SRAM-Based FPGA Architecture

Ajay Kumar Singh

Additional information is available at the end of the chapter

http://dx.doi.org/10.5772/67257

Abstract

The design of low-power SRAM cell becomes a necessity in today's FPGAs, because SRAM is a critical component in FPGA design and consumes a large fraction of the total power. The present chapter provides an overview of various factors responsible for power consumption in FPGA and discusses the design techniques of low-power SRAM-based FPGA at system level, device level, and architecture levels. Finally, the chapter proposes a data-aware dynamic SRAM cell to control the power consumption in the cell. Stack effect has been adopted in the design to reduce the leakage current. The various peripheral circuits like address decoder circuit, write/read enable circuits, and sense amplifier have been modified to implement a power-efficient SRAM-based FPGA.

Keywords: FPGA, ASIC, static power, dynamic power, leakage current, SRAM cell, subthreshold cell, data-aware SRAM cell

1. Introduction

Field programmable gate array (FPGA) is prefabricated integrated circuit (IC), which contains programmable gate matrix to implement logic functions and interconnect resources to connect the logic functions and I/O blocks. These interconnect resources can be electrically programmed by the user to implement any digital circuits and systems. Due to faster time to market, lower cost, and flexibility, FPGA prefers over ASIC (application-specific IC) design although it has disadvantages like larger size, slower speed, and larger power consumption. Due to the flexibility of FPGA, it is possible to partially program any portion of the FPGA depending on the requirement even when the rest of an FPGA is still running. Computer-aided design (CAD) tools and architecture are the two important technologies, which differentiate FPGAs. First memory-based programming FPGAs were introduced in 1986 by Xilinx Inc., San Jose, CA [1].

The programmable term in FPGA only reflects that any new function can be implemented on the chip even after its fabrication. Programmability/reconfigurability of an FPGA is based on an underlying programming technology, which can cause a change in behavior of a prefabricated chip. The main programming technologies used in FPGAs are static random memory (SRAM), flash memory, and antifuse [2–5].

The SRAM-based FPGAs provide ideal prototyping medium and are widely used to integrate FPGAs in an embedded system [6–8] due to the use of standard CMOS technologies, higher performance, and reprogrammability. However, the larger static power consumption in SRAM cell limits the use of SRAM-based FPGAs in portable embedded system compared to flash-based FPGAs [9, 10]. The other concern related to SRAM-based FPGA is its volatile nature. Although the dynamic power management and duty-cycling techniques [11, 12] have been used to save static power during idle mode of FPGA, these techniques are not very effective due to the energy consumption associated with the resulting reconfiguration process. Due to large load capacitance and high access rate, SRAM cells are responsible for consuming significant portion of the total power of the design. Thus, SRAM power consumption is an important consideration for designers to find the balance between the performance and the overall power consumption. The speed of the SRAM cell in FPGA is not a critical factor because it does not affect the operating speed of the circuit implemented in FPGA as mentioned in ref. [13].

In this chapter, we investigate the various factors responsible for power consumption in SRAM-based FPGAs and review the different techniques proposed in the literature to save the power. We will also consider the static and dynamic power in the conventional 6T SRAM cell and its architecture. Various design techniques, presented in the literature, to reduce power consumption in SRAM cell will be reviewed in detail with their merits and demerits. A data-aware power-efficient SRAM cell will be discussed to save power and to optimize the stability.

2. SRAM-based FPGAs

SRAM cells are the basic cells used for SRAM-based FPGA. These cells are scattered throughout the design in form of an array and mainly used to program: (1) the routing interconnects of FPGAs and (2) configurable logic blocks (CLBs) that are used to implement logic functions. SRAM-based programming technology has become the dominant approach for FPGAs because of its reprogrammability and the use of standard CMOS process technology, which results in larger package density and higher speed. Due to the volatile nature of SRAM technology, SRAM-based FPGAs lose their configured data whenever power supply is switched off and need to be reprogrammed every time when the power supply is turned on. Hence, almost every system using SRAM-based FPGAs contains an additional nonvolatile memory such as flash programmable read only memory (PROM) or EEPROM to store the configuration data and load it into the SRAM-based FPGA whenever power is on. In many applications, a complex programmable logic device (CPLD) is used in addition to the external configuration memory to perform the vital functions of the system necessary at power-up. The first static memory-based FPGA (commonly called an SRAM-based FPGA) was proposed by Wahlstrom in 1967 [14]. This architecture is allowed for both logic and interconnection configuration using a stream of configuration bits. From a practical standpoint, an SRAM cell can be programmed indefinite number of times. Dedicated circuitry on the FPGA initializes all the SRAM bits on power up and configures the bits with a user-supplied configuration. No special processing steps are needed in SRAM cells unlike other programming technologies. Although static memory offers the most flexible approach for device programmability, it imposes a significant area penalty per programmable switch compared to ROM implementations.

3. Power consumption in SRAM-based FPGAs

In the recent years, the traditional FPGA research area has shifted from speed and area overhead issues to design of power-efficient FPGAs due to increased applications of FPGA in portable and nonportable devices. In portable devices power saving is required to enhance the battery life time, whereas in nonmobile devices power saving decides the cost, performance, and reliability of the device. The main sources of power consumption in FPGA are static and dynamic power [10, 12, 15, 16].

Static power is consumed when device/system is idle and leakage current flows in the system. The various leakage currents in OFF transistor are subthreshold leakage current, gate-induced drain leakage, junction leakage current, and direct tunneling current [17–19].

Dynamic power consumption is due to the switching activity of the transistors in normal operational mode. The dynamic power consumption depends on the parasitic capacitance, power supply, switching activity, and frequency of operation and mathematically expressed as [20]:

$$P_{dyn} = \eta C_L V_{dd}^2 f \tag{1}$$

where C_L is the load capacitance, V_{dd} is the power supply, f is the frequency of operation, and η is the switching activity.

FPGA design consumes larger static power than the ASIC design due to excessive leakage currents [21–23], which is due to more number of transistors per logic. Other components, which are responsible for larger power consumption, are circuits used to provide flexibility to FPGA, number of configuration bits, lookup-tables (LUTs), and presence of large number of programmable switches.

4. Techniques adopted to reduce power consumption in SRAM-based FPGA

4.1. Leakage power reduction

The important method to control the leakage current in the system is to switch off the transistors, which are not being used at that time. This can be achieved by using the dual threshold voltage transistor FPGA routing design [24–26]. In this technique, high threshold voltage is applied to one subset of multiplexer transistors and low threshold voltage to the rest of the transistors. High threshold voltage controls the leakage current effectively on the cost of performance degradation. This technique increases the complexity at router level. By allowing body-bias effect, the threshold voltage of a multiplexer transistor, which is not a part of the selected path, can be raised [27]. This method increases the fabrication complexity and cost. The leakage current can also be controlled by applying negative bias voltage on the gate of the OFF multiplexer transistor, which results in drastic drop in subthreshold current on the cost of hardware burden [28].

Stack effect is another effective method to reduce the leakage current in any circuit [29–31]. Stack effect means two series connected OFF transistors in the same path. These two OFF transistors offer a high resistive path to the current flow. To utilize this concept in the FPGA design, researchers [32, 33] have introduced an extra configuration SRAM cells (redundant cells) to allow multiple OFF transistors on unselected path. Due to redundant cell approach, the unselected path contains two OFF transistors, which limits the subthreshold current along the unselected path.

Calhoun et al. [34] have proposed the creation of fine-grained "sleep region" to control the leakage current in the system. With this technique, it becomes possible to put unused LUTs and flip-flops to sleep mode independently. Gayasen et al. [35] have proposed coarse-grained sleep strategy. In this technique, the entire region of the FPGA is partitioned into logic blocks so that each region can be put into sleep mode independently whenever it is not used.

Several methods have been proposed by researchers to save the leakage/static power consumption in FPGA design at the architectural level [36–39]. Tran et al. [40] have proposed low-power FPGA architecture based on fine-grained V_{dd} control scheme, called micro-V_{dd}-hopping. They have grouped four CLB into one block to share the V_{dd}. In the micro-V_{dd}-hopping scheme, V_{dd} of each block is varied between high and low V_{dd} to save power consumption without scarifying performance. In their design, they have introduced a level shifter and incorporated zigzag power-gating scheme to control the sneak leakage path problem. They have experimentally observed that the dynamic power can be reduced by 86% when the required speed is half of the highest speed. They have simulated their proposed designed at 90 nm technology and observed that 95% static power saving on the cost of 2% area overhead. In zigzag power gating scheme wake up time is smaller than other gating technique because the INVs and 2-NAND are always in between V_{dd} and V_{ss} during standby mode. Since they have off-off stacking structure, leakage current is suppressed by an order of magnitude even if the overdrive voltage is zero.

Srinivasan et al. [41] have proposed a technique to reduce the leakage current of interconnect fabric. They have put every multiplexer in its least-leakage state by setting its undriven inputs to desired values with a circuit-level modification in the routing multiplexer. The main advantage of this technique is that it has negligible impact on the performance of the design and has small area penalty.

In their research paper, Hasan et al. [42] have reduced the leakage current in the multiplexer-based interconnect matrix by controlling the inputs of unused FPGA routing multiplexers. The simulation results on different sizes and topologies of routing multiplexers show that the minimum leakage vector varies significantly at 22 nm compared to the 65 nm nodes because

http://dx.doi.org/10.5772/67257

of higher gate leakage current and output stage loading effects. Their proposed technique reduces the static power significantly without imposing any area overhead because most of the routing multiplexers are unused in an FPGA.

A directional coarse-grained power-gated FPGA switch box and power gating aware routing algorithm was proposed by Hoo et al. [43] to address the leakage current concern in FPGA. After considering the trade-offs among different PG designs, authors have considered: (1) A novel directional coarse-grained power-gated FPGA switch box. (2) A power-aware routing algorithm to leverage on new PG architecture. In their proposed architecture, multiple buffers in each direction of the switch box are power gated independently of the buffers in the other directions. Due to the homogeneous structure of the switch box, proper sizing of the sleep transistors is not an issue. To maximize the leakage reduction of the coarse-grained PG architecture, they have also adopted the routing algorithm. They have proposed a new cost function for the VPR routing algorithm to support the new routing architecture.

4.2. Dynamic power reduction

Dynamic power is consumed during normal operation when switch toggles. It depends on the frequency of the operation, load capacitance, and square of power supply as clear from Eq. (1). The total dynamic power consumed by a device is given by the sum of the dynamic power of each resource. Due to the programmability of the FPGA, the dynamic power is design dependent. The important contributors for dynamic power are effective parasitic capacitance of the resources, resource utilization, and switching activity of the resources [44]. The effective capacitance of the resources come from parasitic capacitance of interconnect wires and transistors. The dynamic power of the device can be reduced by addressing each of the parameters in Eq. (1) effectively. Various methods have been proposed by researchers to handle the dynamic power consummation [37, 45–47]. The general adopted methods are using clock scheme, reducing toggling activity of the logic, reducing RAM and I/O powers.

Since faster switching logic consumes more dynamic power than the slower switching logic, it is required to partition the clock so that the fast clock should be assigned to those portions of the logic which require a fast clock and slow clock should be assign to those which can be run at a slower speed. This way the switching activity of various logics can be controlled to save the overall dynamic power [9, 10, 15].

Dynamic voltage scaling is another power-saving design technique because supply voltage significantly impacts power efficiency. The power supply scaling technique can be utilized in the design of power-efficient FPGA by considering devices like tunnel-FET, FinFET, etc. [48–51] because these devices can operate at ultra-low voltage.

The dual or multi-V_{dd} techniques [52–54] are other important methods to save the dynamic power. In dual V_{dd} scheme, the noncritical delay circuit is connected with low power supply, whereas delay-critical circuit is powered by high voltage. This concept is also applied in the FPGA design [55–57]. In heterogeneous architecture, some logic blocks are fixed to operate at high power supply and some logic blocks (not limited by speed) are fixed to operate at low voltage. This heterogeneous scheme helps only in small power saving due to the rigidity of the fixed fabric and loss associated with the mandatory use of low-V_{dd} in certain cases. The dual V_{dd} technique cannot be applied to the interconnect wires which is the main source of power consumption. To overcome this problem, Li et al. [58] have proposed V_{dd} programmability technique to reduce power consumption of interconnect wire. They have selectively applied low-V_{dd} to interconnect circuits such as routing and connection switches. The V_{dd} selection for different applications is obtained by programmable dual-V_{dd} technique to both logic blocks and interconnect. On average, they observed a total of 50-55% power is reduction.

Although voltage scaling is the best way to reduce the power consumption in FPGA array, one has to scarify the performance of the circuit. To improve the power efficiency of FPGAs without scarifying performance, Li et al. [59] have explored the different supply voltage (V_{dd}) levels option. According to the authors, a predefined dual-V_{dd} FPGA fabric, in general, cannot achieve better power performance trade-off than the V_{dd} scaling because the predefined dual-V_{dd} fabric is not flexible enough for a variety of applications. To address this issue they have introduced the field programmability for the V_{dd} level by proposing three types of logic blocks: H-block, L-block, and a p-block as shown in **Figure 1**. H-block and L-block are connected to supply voltages VDDH and VDDL, respectively. H-block provides higher speed due to high supply voltage whereas L-block has reduced power consumption at the cost of the increased delay. They have implemented P-block by inserting PMOS transistors (called power switches) between the power supply rails and the logic block. The configuration bits were used to control the switching behavior of these switches so that an appropriate supply voltage can be chosen for the P-block. To avoid the short circuit current, they have introduced a level converter in between VDDH and VDDL.

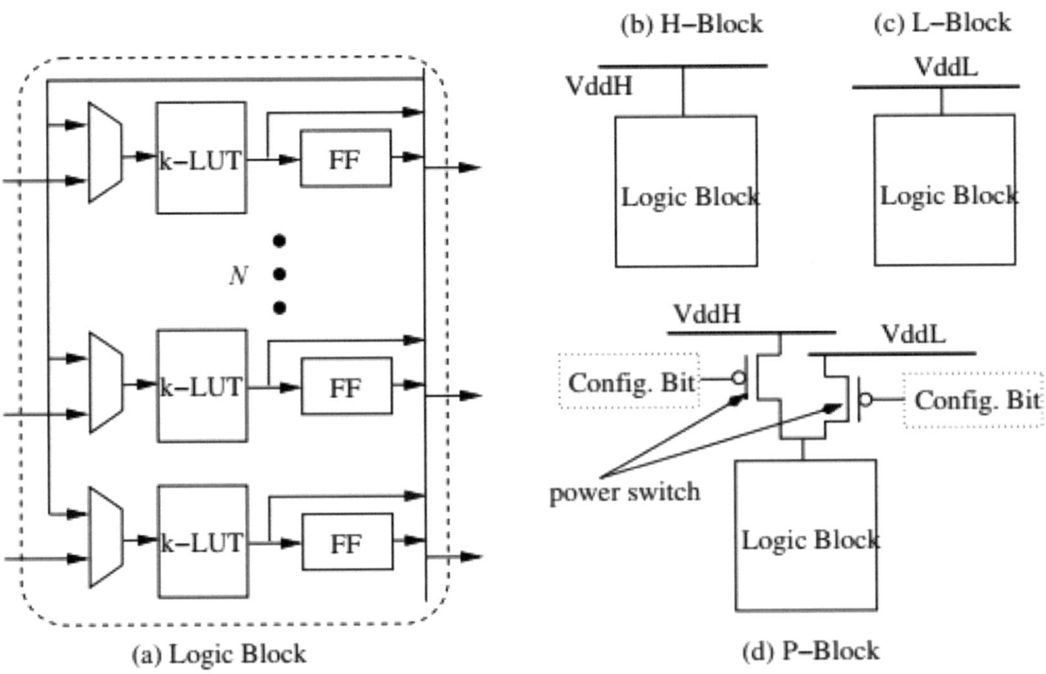

Figure 1. Logic blocks in dual-V_{dd} and V_{dd}-programmable FPGAs [59].

Selective power-down is another method to save power in FPGA. This technique (known as power gating) refers to shut down the power supply of certain portions of a chip which are not performing any task for a long time to save the static power considerably. This can be achieved by implementing a multisupply strategy in which the power grid of some blocks is decorrelated from others in order to allow for selective shutdown. Sleep modes within the FPGA architecture can also be deployed to selectively reduce the power supply of those blocks, which are not in use [60, 61].

Power consumption in interconnect dominates dynamic power in FPGAs [62–64] due to the interconnect structure, which consist of prefabricated wire segments. Each segment is attached with used and unused switches. Wire lengths in FPGAs are generally longer than in ASICs due to the larger area consumed by SRAM cells and circuitry. The larger power consumption in interconnect in FPGA makes it high-level target for power optimization. Anderson et al. [65] have presented a novel FPGA routing switch design to reduce the leakage and dynamic power consumption. The switch can be programmed to operate in any one of the mode: high speed, low speed, or sleep mode. In high-speed mode, power and performance characteristics are similar to those of current FPGA routing switches. Low-power mode offers reduced leakage and dynamic power on the cost of degraded performance. Sleep mode, which is suitable for unused switches, reduces the static power drastically. Three key observations (which hold for majority of Xilinx Spartan-3 commercial FPGA and are specific to FPGA interconnect) were made, namely (1) routing switch inputs are tolerant to "weak-1" signals, (2) there exists sufficient timing slack in typical FPGA designs to allow a considerable fraction of routing switches to be slowed down, without impacting the overall design performance, and (3) most routing switches simply feed other routing switches, authors have proposed the design of new switch as shown in **Figure 2**. The designed switch includes parallel combination of NMOS and PMOS sleep transistors which can operate in three different modes as follows: In high-speed mode, the PMOS is turned ON which results in full rail-to-rail swing of output. The gate terminal of NMOS is left at V_{dd} in high-speed mode. During 0–1 logic transition the virtual V_{dd} may temporarily drop below V_{dd} - VTH, causing the NMOS to leave cut-off and assist with charging the switch's output load. In low-power mode, the PMOS is turned OFF and NMOS is turned ON. The buffer is powered by the reduced voltage, VVD ≈ V_{dd} – VTH.

Clock-gating is an effective and most widely used method to reduce the dynamic power. This technique is based on the principle that only active portion of the system should be connected to the clock tree and others should not be served by the clock tree. A logic circuit must be included in the design for the selection of which portions are clocked and which portions are blocked. This reduces switching activity which results in dynamic power saving. The clock gating can be applied at the chip level as well as at the design level. The gating technique has been successfully used in ASICs, but it is not very effective in SRAM-based FPGAs because a large component of power consumption in FPGA is due to the switching activities of the clock signals along the routing switches. For this reason, researchers investigated the possibility of modifying the way a circuit is mapped on the FPGA array by acting on the synthesis, technology mapping, or placement and routing algorithms [66, 67]. Since clock is distributed in the chip through the global FPGA routing network, the placement of clock loads has a considerable impact on clock wire usage. Clock load placement should be done in such a way that one should get lower clock capacitance, which results in lower dynamic power consumption.

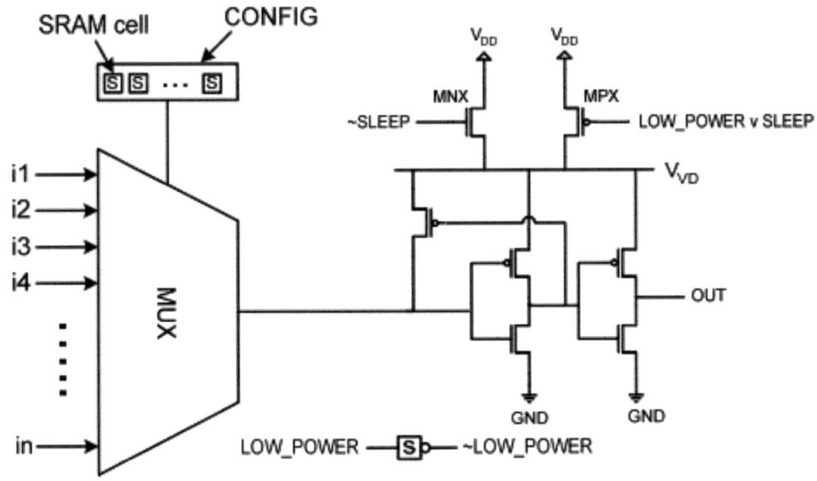

Figure 2. Proposed new programmable low-power FPGA routing switch [65].

Placement and routing (P&R) on the chip also affects the dynamic power consumption because it decides the total parasitic capacitance in the design. To minimize the parasitic capacitance, it is essential to optimize the P&R strategy. It is always advisable to place two connected functional instances closer because it will reduce the interconnect wire-length which in turn can reduce the capacitive loading of the net and lead to dynamic power reduction. The modern FPGA development software typically supports power-driven layout to automatically accomplish this task. Power-driven layout tools examine connection between functional instances for optimization [68–70]. Power-analysis tools are used to further optimize the power saving. Power-analysis tools examine each subcomponent in a design hierarchy to highlight power consumption. Careful examination of this information and subsequent manipulation of the design can result in significant power savings.

Reducing the power supply of I/O can save up to 80% dynamic power. The switching activity of I/O can be controlled by using techniques like time multiplexing, minimum I/O count design portioning [71–73], and reducing I/O drive strength/slew rates. A considerable amount of dynamic power can be saved by adopting differential I/O standards and resistively terminated I/O standards for highest toggling frequency and single-ended I/O standard for low toggling frequency.

Tsang et al. [74] have studied the effectiveness of employing precomputation in reducing dynamic power consumption in commercial off-the-shelf (COTS) FPGAs. Precomputation is a high-level logic optimization technique that lowers power consumption of a design by disabling part of the circuit based on a few relatively simple precomputation conditions. With careful design considerations and increased logic utilization, its associated power consumption can be reduced by disabling much larger part of the design with negligible increase in resource overhead.

http://dx.doi.org/10.5772/67257

In the literature, several techniques/methods are presented in detail to address the issue of dynamic power consumption in FPGA [10, 75–77].

5. SRAM power reduction

The design of low power and high performance SRAM cell becomes a necessity in today's FPGAs because SRAM is a critical component in FPGA design. Although SRAM-based FPGA acquires larger area on the chip but still one of the most useful SRAM-based structure is the lookup table (LUT).

SRAM-based FPGAs such as those manufactured by Xilinx and Altera comprise the largest fraction of the overall market. These FPGAs utilize SRAM for routing and programmability, typically through the use of LUTs and multiplexers. Due to the large number of cells within SRAM FPGA interconnects, a considerable leakage current (of order of milliamps) flows at standby [78]. However, leakage current increases as process geometry shrinks which further exacerbates the power problem. The dynamic power consumption in cell is a serious threat because of large parasitic capacitance (due to longer metallic bitline) which results in larger charging/discharging activity at the bitline. Study on the leakage current and dynamic power in Xilinx Spartan-3 FPGA [79] (**Figure 3**) and Xilinx Virtex-4 [80] (**Figure 4**) show that the major contributor for power consumption in FPGA is configurable SRAM; hence, the new design technique becomes essential to increase the lifetime of the battery. Several techniques have been proposed in the literature [81–85] to address the power consumption problem in SRAM cell. It is worth to disable the SRAM devices that are temporarily unused. This technique will avoid the power consumption by unused components. A system controller can deactivate the device when it is not required in the current operation, or put the device in its sleep mode when that device will not be accessed for an extended period of time. Implementing such a system controller in FPGA reduces the overall switching activity of the system. As discussed by Tuan et al. [86], the data of the configurable SRAM cell alter only when FPGA is configured. FPGA is configured only when power supply is turned on. Therefore, it is necessary to control the leakage current in the cell during idle phase to save the overall power.

Wang et al. [87] have proposed the design of an ultra-low voltage 9T SRAM cell. Their designed cell consists of a 6T SRAM part (for write operation) and a dedicated read port. The read port comprises three NMOS transistors for realizing equalized bitline leakage and improving bitline sensing margin in a single-ended read bitline (RBL). The write access paths and the data storage latch are implemented with HVT devices for leakage reduction while the read port employs LVT devices for better performance. Their test chip shows an improvement of 40% in energy efficiency with the minimum energy per operation of 2.07 pJ at 0.4 V. This design increases the fabrication complexity due to the use of LVT and HVT transistors.

Although much research has been done in order to design a power-efficient SRAM circuit, still interest in power-efficient cell design at the architecture level continues to increase due to the occupation of considerable fraction of total area on chip by configurable SRAM cells and circuitry in the FPGA design. Ye et al. [13] have observed that more than 40% of the total FPGA's logic block area is occupied by SRAM cells. Such huge area overhead results in larger wire length, which leads in larger parasitic capacitance at load. This increased capacitance increases the dynamic power consumption. The most widely used and well accepted SRAM cell is 6T cell [88] (as shown in **Figure 5**) due to its symmetric structure and larger data storage capacity. The cell has two cross-coupled inverters which form latch to keep the programmed data intact. Two pass transistors are used to transfer the data from bitline to cell node (write operation) or cell node to bitline (read operation). The actual control of the FPGA is handled by the Q and Qbar outputs. The main drawbacks of the conventional 6T cell are: poor stability, large power consumption, and degraded performance.

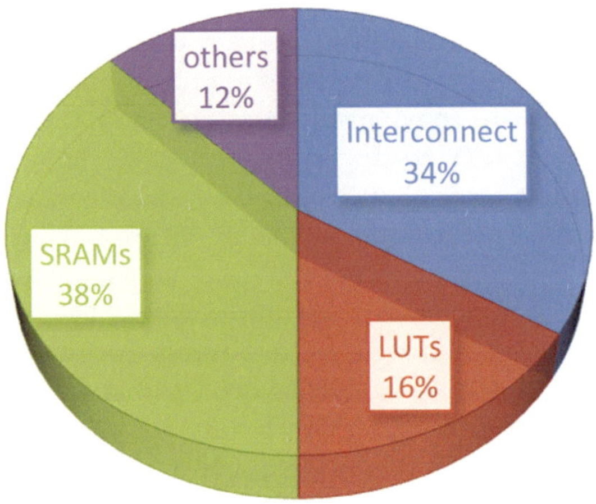

Figure 3. Leakage power breakdown in Xilinx Spartan [79].

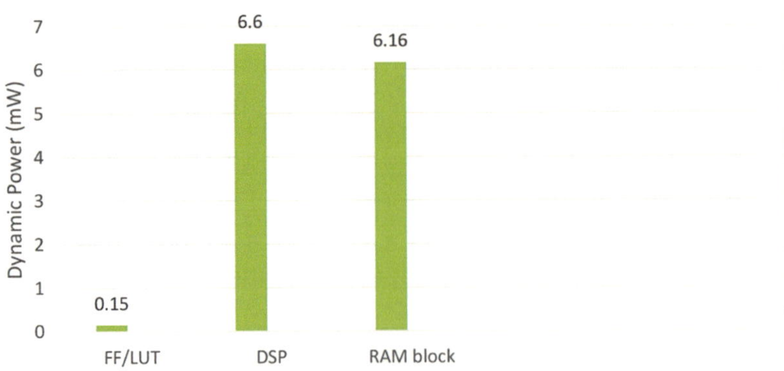

Figure 4. Dynamic power breakdown in Xilinx Virtex-4 [80].

http://dx.doi.org/10.5772/67257

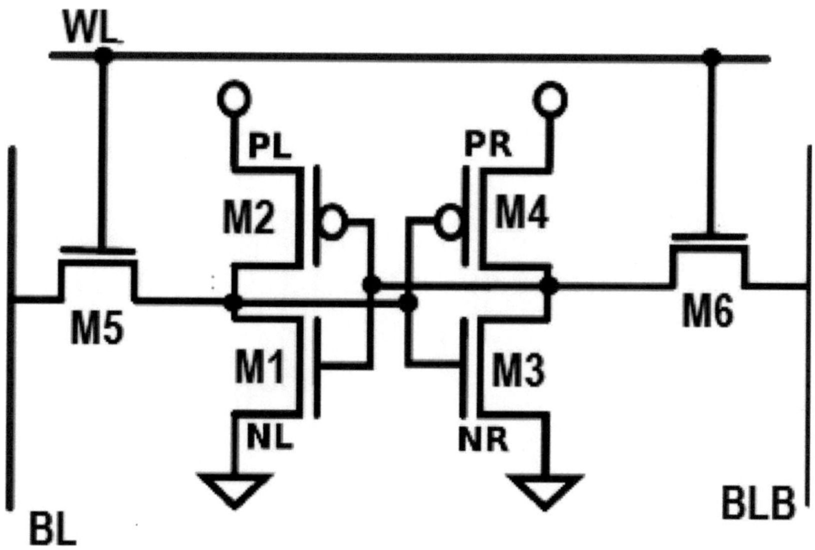

Figure 5. Architecture of the conventional 6T SRAM cell [88].

5.1. Subthreshold SRAM cell

Subthreshold operation is achieved when the device is allowed to operate at power supply (V_{dd}) lower than its threshold voltage. Using this concept, researchers [89–94] have proposed the subthreshold SRAM cells to reduce the overall power consumption in the cell. Teman et al. [95] have designed a robust, low-voltage SRAM bit cell with reduced 5 transistors compared to the standard 6T circuit. Their designed cell can operate at voltage as low as 400 mV in a commercial 40 nm CMOS process. At this supply voltage, the proposed bit cell provides 6σ stability and an average static power reduction of 21× compared to the 6T cell. The main drawback of the circuit is its extra processing complexity due to HVT and SVT transistors.

Calhoun et al. [90] have proposed 10T subthreshold bit cell (**Figure 6**). Transistors M1 through M6 forms conventional 6T cell except that the source of M3 and M6 tie to a virtual supply voltage rail (VVDD). The proposed cell has distinct read and write ports to improve the stability of the cell. Eliminating the read SNM problem allows this bitcell to operate at half of the V_{dd} of a 6T cell while retaining the same 6σ stability. Transistors M7–M10 are used to remove the read SNM problem by buffering the stored data during read operation. M10 is mainly included in the cell to control the leakage current. Their experimental results show that the proposed cell saves 2.5× and 3.8× leakage power at V_{dd} = 0.6 V and V_{dd} = 0.4 V at room temperature. This saving is more aggressive (60×) when power supply is scaled down to 0.3 V.

A design of 10T SRAM is proposed by Jiangzheng et al. [96] by employing voltage lowering techniques to effectively control the leakage current in the cell after allowing cell to operate in subthreshold region. The proposed circuit generates a subthreshold read pulse for transferring the data out of the SRAM. The floating write bitlines minimizes write bitline leakage on the cost of degraded stability. Short read bitlines improve read speed and suppress read power on the cost of area overhead.

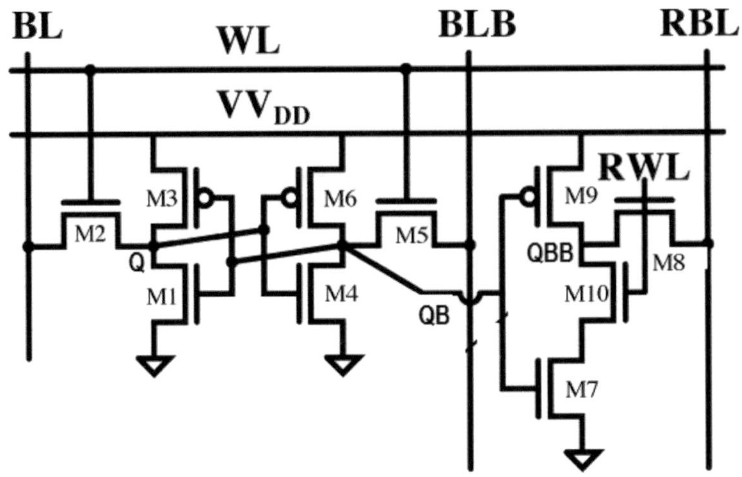

Figure 6. Architecture of 10-T subthreshold bitcell [90].

Kushwah et al. [97] have proposed a single-ended dynamic feedback control 8T static RAM (SRAM) cell to enhance the static noise margin (SNM) for ultralow power supply. It achieves write SNM of 1.4× and 1.28× as that of isoarea 6T and read-decoupled 8T (RD-8T), respectively at 300 mV. The standard deviation of write SNM for 8T cell is reduced to 0.4× and 0.56× as that for 6T and RD-8T, respectively. The proposed 8T consumes about 0.6× less write power and 0.48× less read power than 6T cell.

5.2. Data-aware power-efficient SRAM cell

The main drawbacks of subthreshold cells are poor stability and degraded performance. Besides the cell leakage, the bitline leakage is another dominating factor for power consumption. The overall bitline power consumption is data dependent. Many data-aware cells have been reported in the literature to control the bitline power consumption [98–102]. Chiu et al. [103] have proposed 8T single-ended subthreshold SRAM with cross-point data-aware write operation. In the circuit write operation is performed by traditional write circuit as in 6T cell, whereas 2T stacked read buffer is used for read operation. Due to stack read circuit, leakage current is controlled and stability is improved. The data-aware cross-point write operation improves the writeability. The main drawback of the circuit is large voltage swing on bitline during write operation.

A 130 mV SRAM with expanded write and read margins for subthreshold applications was proposed by Chang et al. [104] to reduce the voltage swing on the respective bitlines during write operation. They have used two separate signals SCR and SCL to perform write operation. The proper selected value of these two signals controls the write power consumption after reducing the discharging activity at the bitline. The isolated read circuit improves the stability of the cell on the cost of large parasitic capacitance and resource burden due to two extra signals.

Singh et al. [105] have designed a data aware dynamic 9T SRAM cell to reduce the bitline power consumption. The dynamic nature of the cell flips the data faster at the bitline so that the average discharging activity is reduced. The cell contains nine transistors with isolated read and writes circuits. The write operation is performed using write signal WS. The value of write signal is chosen based on the write operation. The simulation results predicted the 47% lower write power consumption compared to the 6T. They also observed that power saving varies from 42.45 to 61.3% when no peripheral devices are included in the array during hold mode because of lower leakage current from write bitlines and lower discharging activity at RBL. The cell imposes hardware and wiring burden due to extra signal.

The bit-interleaving-enabled 8T SRAM architecture is proposed by Wen et al. [106]. The proposed cell features shared data-aware write structure and utterly eliminates the half-select disturbance. In their proposed design, shared write and separated read behaviors are implemented by activating horizontal cells and vertical bitlines instead of enabling blocks. They also proposed a reference-based sense amplifier (SA) to coordinate the column-selection array to further optimize the area efficiency. The proposed SRAM operates at a frequency of 125 kHz and consumes a total power of 5.1 μW.

5.3. Data-dependent-write-assist dynamic (DDWAD) SRAM cell

Recently, we have designed a power-efficient SRAM cell [107] by utilizing dynamic data aware concept for write operation and stack effect to control the read leakage current. The architecture of the cell is shown in **Figure 7(a)**. The designed cell has distinct read and write ports with single bitline to improve the overall stability of the cell. To flip the data at the storage node faster without waiting bitline BL to charge/discharge completely we have introduced a write signal WS and broken the latch of the cell (since WL = high). To control the leakage current in read circuit during write operation and hold mode, stack technique is (three series connected OFF transistors in read path) used on the cost of increased delay. The write signal (WS) has been generated according to the data to be stored at Q and Qbar with the help of circuit as shown in **Figure 7(b)** [107]. During read and hold mode, WS maintains its previous value and latch nature of the cell is restored to keep the stored data intact. The proposed cell and other cells were simulated at layout level using Cadence 6.1 CMOS design rules for 65 nm technology. The large write power saving (**Figure 8**) is due to no discharging activity at the bitline BL due to high resistive path (NM1 Turns OFF because WS = 0 (write 1 operation)). Similarly, for WS = high, OFF transistor PM1 does not allow any current to flow between V_{dd} and ground. This causes low voltage at the storage node Q. In both write operations, a small voltage drops at BL results in considerable dynamic power saving. Due to OFF transistors NM4 and NM6 (since RWL = 0 during write operation) in the read path, the leakage current through RBL is restricted.

Due to the forbidden discharging of precharged RBL during read 1 operation and stack effect in read path, a considerable power saving is achieved compared to the conventional 6T cell (**Figure 9**). In hold mode, WS maintains its value due to internal latch. The static power consumption in the proposed cell is lower than the 6T cell and other proposed cells in the literature irrespective of the power supply (**Figure 10**). The lower static power in the proposed cell [107] is due to lower leakage current through write bitline BL and stack effect in read circuit. During simulation, we observed that the proposed cell shows a nominal variation in static power consumption with temperature, which reflects the robustness of the cell against temperature. The data at the storage nodes maintained strongly at their respective values for power supply range of 300 mV ≤ V_{ddmin} ≤ 400 mV. The proposed cell shows larger immunity toward the statistical variation due to signal WS as discussed in our published paper in detail [107].

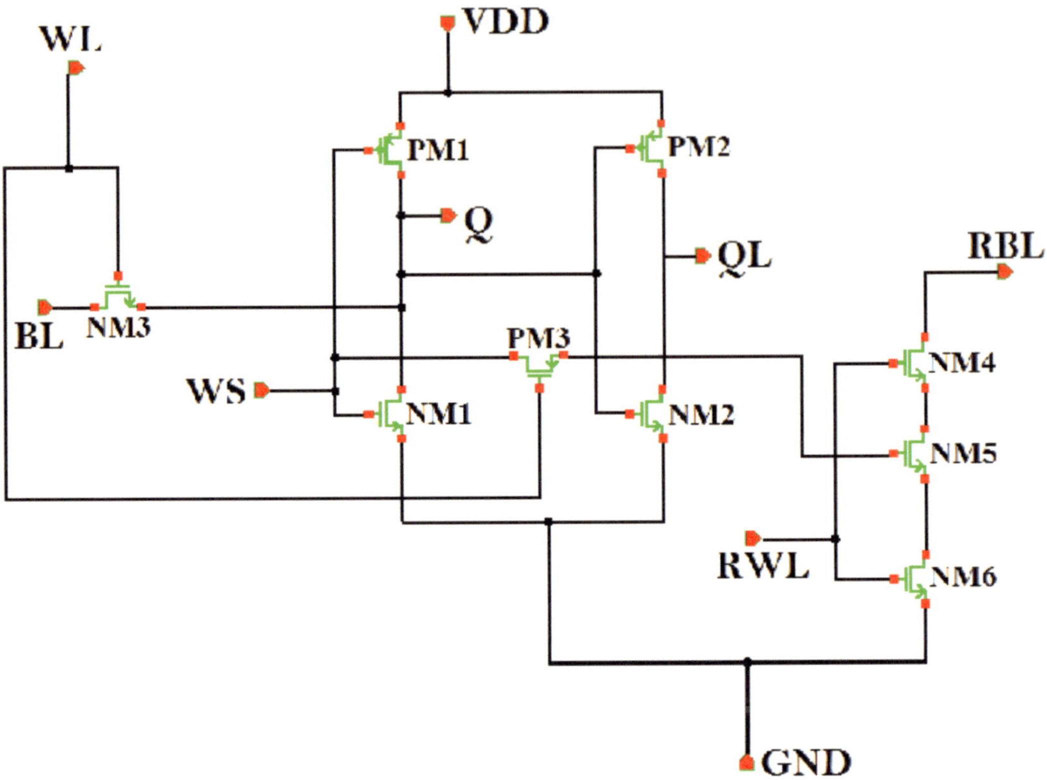

(a)

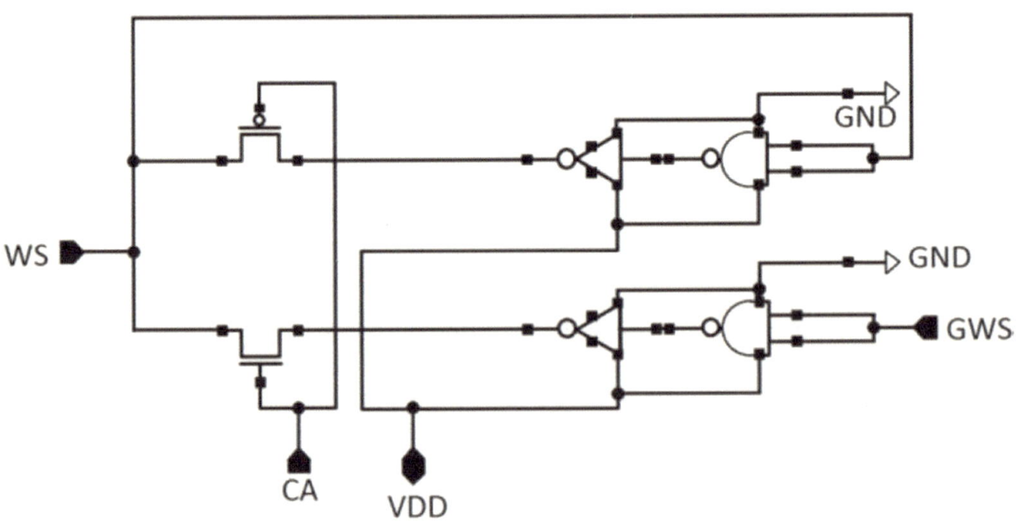

(b)

Figure 7. (a) Architecture of DDWAD SRM cell. (b) Circuit to generate appropriate WS signal depending on write operation [107].

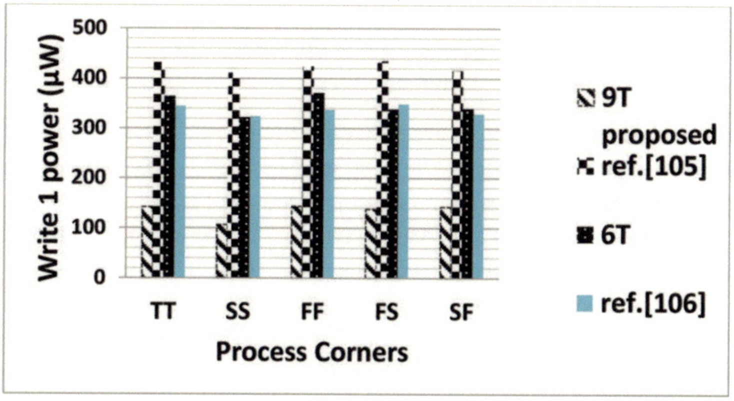

Figure 8. Total power consumption in data aware cell [107].

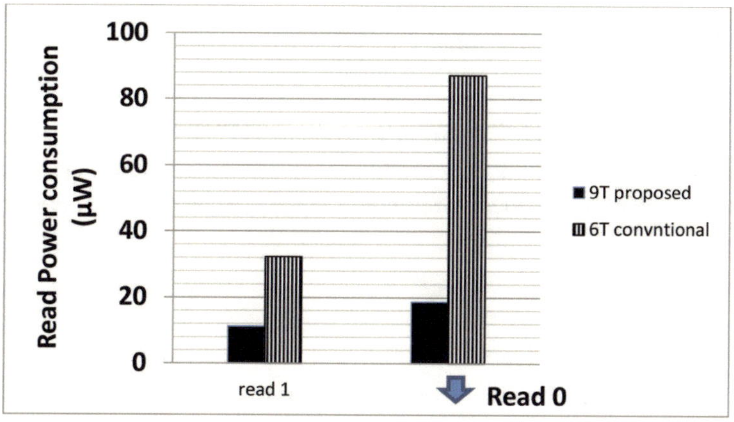

Figure 9. Read power consumption [107].

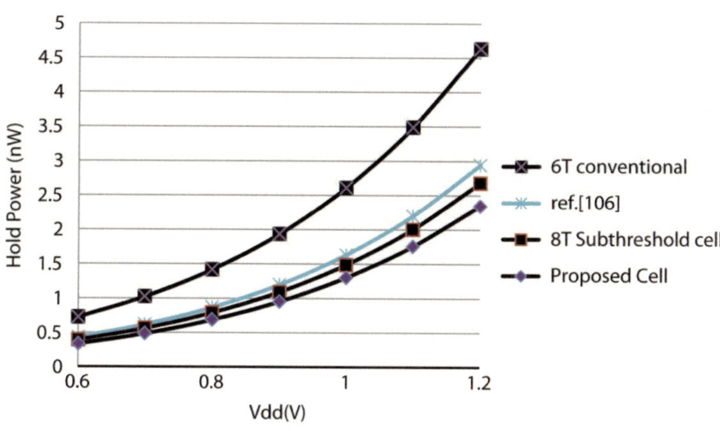

Figure 10. Hold leakage power at various power supplies [107].

Although the proposed cell imposes area overhead compared to the conventional 6T cell, it is not a serious threat in FPGA implementation because of lower leakage current through bitline, more number of cells can be connected on a single bitline in the array.

In SRAM-based FPGA memory accesses are performed with a designed clock and series of interface circuits like row/column decoder, write/read enabled circuit, etc. These peripheral circuits consume a considerable power in the chip. To implement an array using the proposed cell, we have adopted the hierarchical design approach in which instead of giving individual signals (WS, WL, and RWL) to each cell, global signal circuits are used [108]. The main advantage of using the hierarchical design is the use of shorter wires within local blocks, which reduces parasitic capacitances. In this approach, at one time only one block address can be activated which saves considerable power. Each global signal is connected to corresponding local signal through NMOS pass transistor to save the area. The column-based approach is adopted in which signal WS is routed parallel to write bitline BL. To avoid the column half selected disturbance in the array due to toggle of the signal WS during write operation, we proposed a circuit as shown in **Figure 7(b)** [107].

5.4. Proposed decoder circuits and sense amplifier

The most important signals that affect the power dissipation in SRAM memory are the address lines, read and write enable circuits, block select, and sense amplifier. To address these concerns, we have designed new architectures for these circuits to reduce the power consumptions. The detail about these circuits is available in our published work [108, 109].

The proposed column decoder circuit is shown in **Figure 11** [108], where C_{Lj} represents the address of the columns to be selected (*j* is an integer number). The architecture of the other decoder circuits is explained in Ref. [108]. Since the proposed decoder is implemented without using NAND gates as in the conventional decoder, the number of transistors is reduced to 546 compared to 1939 in the conventional decoder [108]. The reduced number of transistor results in lower parasitic capacitance, which leads to approximately 76% power saving [108]. The proposed WL driver consumes lower power compared to other designs due to the compactness of the circuit.

As we know most of the current will be dissipated in the SRAM cell by sense amplifier. To address this issue we have also designed a single-ended sense amplifier [109]. The proposed SA (sense amplifier) reduces the power consumption by controlling the leakage current during evaluation/precharge mode. The circuit can be used even at higher temperature with minimum power consumption. The working of the circuit is explained in detail in Ref. [109].

Table 1 gives the comparison of read power consumption in various sense amplifiers. The main reason for lower power consumption in the proposed circuit is due to lower average current during evaluation mode, small voltage drops on RBL, and lower leakage current compared to other circuits [110, 111]. During hold mode, power consumption in the proposed circuit is lower than the other circuits [110, 111] due to gating effect.

We have implemented 32Kb SRAM array using the proposed cell and proposed decoder circuits/sense amplifier. The simulation results were compared with ref. [112] array. The results were encouraging in terms of power consumption as seen in **Figures 12** and **13**, respectively. The lower hold power obtained in the implemented cache is due to write signal WS and stack effect (read path).

http://dx.doi.org/10.5772/67257

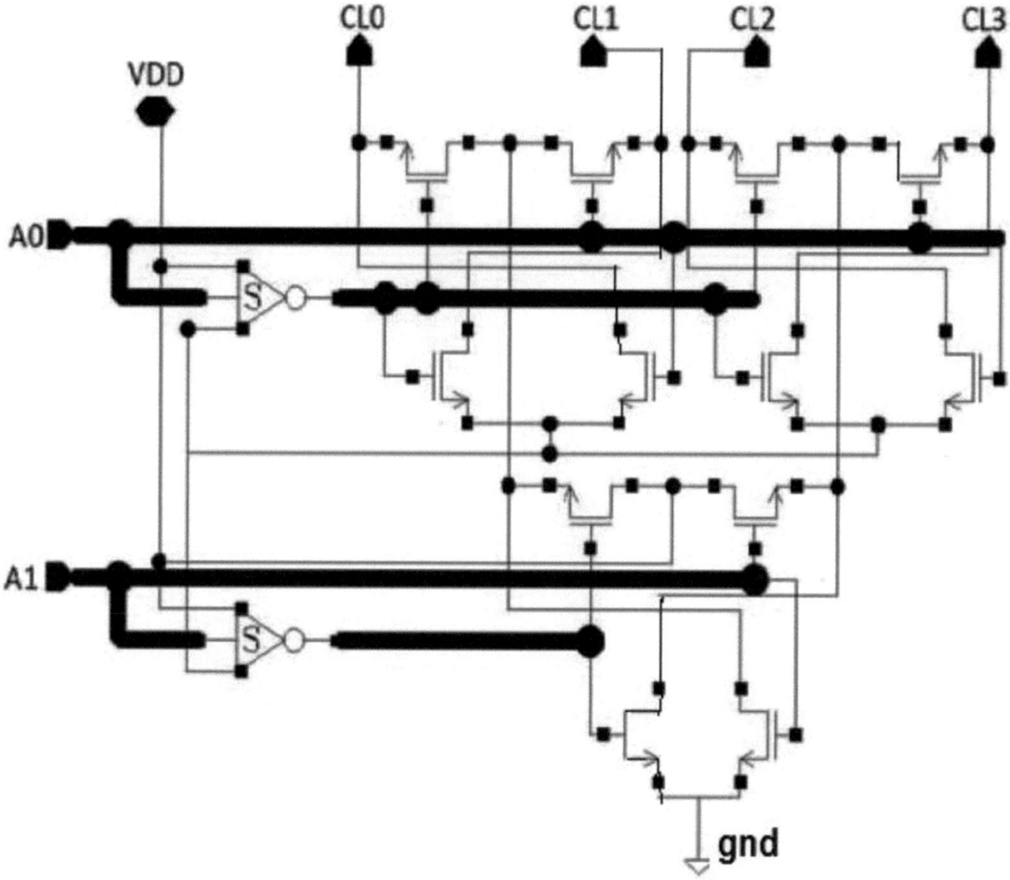

Figure 11. Proposed decoder [108].

Type of circuit	Read power consumption (μW)	
	Read 0	Read 1
Proposed	7.768	8.699
Ref. [110]	26.674	59.856
Ref. [111]	77.840	18.795

Table 1. Read power consumption in various SA [109].

The overall reduction in dynamic and static power in the proposed cell, decoder, and sense amplifier make them an ideal choice for the implementation of power-efficient and reliable SRAM-based FPGA.

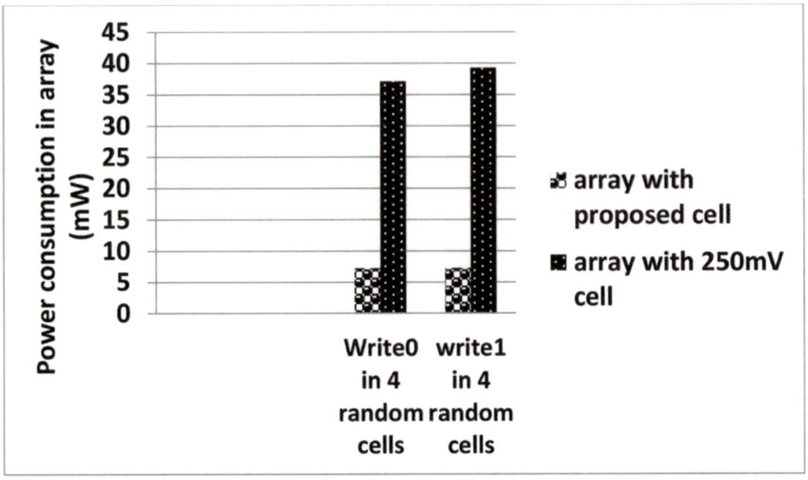

Figure 12. Write power consumption in 32 kb SRAM array.

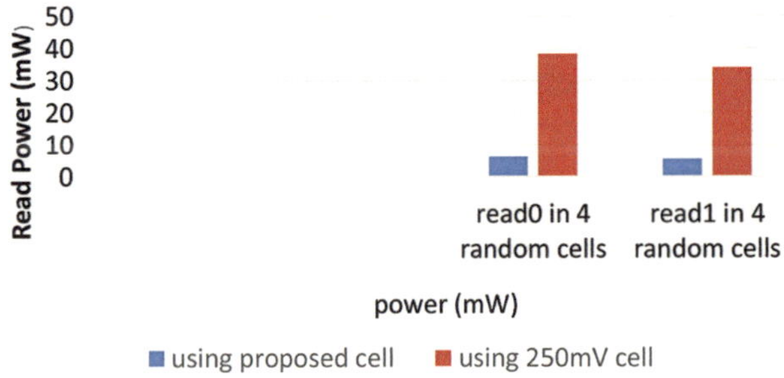

Figure 13. Read power consumption in 32 kb array.

6. Conclusion

The various issues related with the power consumption in FPGA have been discussed in detail with solutions/techniques as presented in the literature. Power gating/clock gating, dual threshold/multithreshold voltage, programmable V_{dd}, etc. are the important and well-accepted methods to control the static and dynamic power consumption in the SRAM-based FPGA. SRAM is the basic component used in the implementation of SRAM-based FPGA and occupies larger area in the chip and consumes considerable amount of static/dynamic power. The power consumption in the cell can be reduced by reducing the bitline length, designing compact peripheral circuits, or improving the cell at the architecture level. Researchers have proposed subthreshold SRAM cell to reduce the power consumption but it degrades the reliability of the cell. To address dynamic power and static power consumption in the cell, a data aware cell is proposed with isolated write and read ports. Both operations are performed on single bitline. Power-efficient peripheral circuits like write/read decoder, address decoder circuit, and sense amplifier were also presented in the chapter to realize the SRAM

http://dx.doi.org/10.5772/67257

array. The proposed cell and implemented array consume lower overall power due to lower discharging activity at BL and leakage current control due to stack effect. The area overhead in the proposed cell is not a serious threat in the implementation of array because of lower bitline leakage more number of cells can be connected on the same bitline.

Author details

Ajay Kumar Singh

Address all correspondence to: aks_1993@yahoo.co.uk

Faculty of Engineering and Technology, Multimedia University, Melaka, Malaysia

References

[1] W.S. Carter, K. Duong, R.H. Freeman, H.C. Hsieh, J.Y. Ja, J.E. Mahoney, L.T. Ngo, and S.L. Sze, "A user programmable reconfigurable logic array," in IEEE 1986 Custom Integrated Circuits Conferences, pp. 233–235, 1986.

[2] A. Gupta, V. Aggarwal, R. Patel, P. Chalasani, D. Chu, P. Seeni, P. Liu, J. Wu, and G. Kaat, "A user configurable gate array using CMOSEPROM technology," in Proceedings of Custom Integrated Circuits Conferences, pp. 31.7.1–31.7.4, 1990.

[3] J. Birkner et al., "A very-high-speed field-programmable gate array using metal-to-metal antifuse programmable elements," Microelectronics Journal, vol. 23, pp. 561–568, 1992.

[4] D. Tavana, W. Yee, S. Young, and B. Fawcett, "Logic block and routing considerations for a new SRAM-based FPGA architecture," in Proceedings of Custom Integrated Circuits Conferences, pp. 511–514, 1995.

[5] R. Patel et al., "A 90.7 MHz-2.5 million transistors CMOS PLD with JTAG boundary scan and in-system programmability," in Proceedings of Custom Integrated Circuits Conferences, pp. 507–510, 1995.

[6] P. Chow, S.O. Seo, J. Rose, K. Chung, G. P'aez-Monz'on, and I. Rahardja, "The design of an SRAM-based field-programmable gate array—part I: Architecture," IEEE Transanctions on Very Large Scale Integration (VLSI) SYSTEMS, vol. 7, no. 2, pp. 191–197, 1999.

[7] P. Graham, M. Caffrey, J. Zimmerman, D.E. Johnson, P. Sundararajan, and C. Patterson, "Consequences and categories of SRAM FPGA configuration SEUs," Proceedings of the Military and Aerospace Applications of Programmable Logic Devices (MAPLD), Washington DC, September 2003.

[8] C. Bolchini, A. Miele, and C. Sandionigi, "A novel design methodology for implementing reliability - aware system on SRAM based FPGAs", IEEE Transactions on Computers, vol. 60, no. 12, pp. 1744–1758, 2011.

[9] J. Lamoureux and W. Luk, "An overview of low-power techniques for field-programmable gate arrays", Proceedings of IEEE NASA/ESA Conference on Adaptive Hardware and Systems, pp. 338–345, 2008.

[10] P. Singh and S.K. Vishvakarma, "Device/circuit/architectural techniques for ultra-low power FPGA design," Microelectronics and Solid-State Electronics, vol. 2, no. 2A, pp. 1–15, 2013.

[11] I. Brynjolfson and Z. Zilic, "Dynamic clock management for low-power applications in FPGAs", Proceedings of IEEE Custom Integrated Circuits Conference, pp. 139–142, 2000.

[12] K. Shahzad and B. Oelmann, "Investigation of energy consumption of an SRAM-based FPGA for duty-cycle applications", in ParaFPGA2013, Parallel Computing with FPGAs, Munich, Germany, 10–13 September 2013.

[13] A. Ye, J. Rose, and D. Lewis, "Using multi-bit logic blocks and automated packing to improve field-programmable data path circuits", in IEEE International Conference on Field-Programmable Technology, pp. 129–136, Brisbane, Australia, 2004.

[14] S.E. Wahlstrom, "Programmable Logic arrays — cheaper by the millions," Electronics, vol. 40, pp. 90–95, 1967.

[15] N. Grover and M.K. Soni, "Reduction of power consumption in FPGAs – an overview," Information Engineering and Electronic Business, vol. 5, pp. 50–69, 2012.

[16] J. Tarrillo and F.L. Kastensmidt, "Estimating power consumption of multiple modular redundant designs in SRAM-based FPGAs for high dependable applications," in 24th International Workshop on Power and Timing Modeling, Optimization and Simulation (PATMOS), 2014.

[17] K. Roy, S. Mukhopadhyay, and H. Mahmoodi-Meimand, "Leakage current mechanisms and leakage reduction techniques in deep-submicrometer CMOS circuits," IEEE Proceeding, vol. 91, no. 2, pp. 305–327, 2003.

[18] P.F. Butzen and R.P. Ribas, "Leakage current in sub-micrometer CMOS gates," University of Federal do Rio Grande do Sul, 2005.

[19] Y.-B. Kim, "Challenges for nanoscale MOSFETs and emerging nanoelectronics," Transaction on Electrical and Electronic Materials," vol. 11, no. 3, pp. 93–105, 2010.

[20] N.H.E. Weste, D. Harris, and A. Banerjee, "CMOS VLSI Design a Circuits and Systems Perspective," 4th Edition, Addison-Wesley, ISBN 10: 0-321-54774-8, ISBN 13: 978-0-321-54774-3, UK, 2005.

[21] D. Rittman, "Structured ASIC design: A new design paradigm beyond ASIC, FPGA and SoC", 2004. http://www.tayden.com/publications/Structured %20ASIC%20Design.pdf.

[22] S.M.H. Ho, "Structured ASIC: Methodology and comparison," Proceedings of 2010 International Conference Field-Programmable Technology (FPT), pp. 377–380, 2010.

[23] Y. Cai, K. Mai, and O. Mutlu, "Comparative evaluation of FPGA and ASIC implementations of buffer less and buffered routing algorithms for on-chip networks," in

http://dx.doi.org/10.5772/67257

Proceedings of the International Symposium on Quality Electronic Design (ISQED), pp. 475–484, 2015.

[24] F. Li, Y. Lin, L. He, and J. Cong, "Low-power FPGA using pre-defined dual-Vdd/dual-Vt fabrics", in Proceedings of ACM/SIGDA International Symposium on Field Programmable Gate Arrays, pp. 42–50, 2004.

[25] A. Kumar and M. Anis, "Dual-threshold CAD framework for subthreshold leakage power aware FPGAs," in IEEE Transactions of Computer-Aided Design of Integrated Circuits and Systems, vol. 26, no. 1, pp. 53–66, 2007.

[26] R. Jaramillo-Ramirez and M. Anis, "A dual-threshold FPGA routing design for subthreshold leakage reduction," in 2007 IEEE International Symposium on Circuits and Systems, New Orleans USA, pp. 3724–3727, 27–30 May 2007.

[27] S. Bae, R. Krishnan, and N. Vijaykrishnan, "A novel low area overhead body bias FPGA architecture for low power applications," IEEE Computer Society Annual Symposium on VLSI, pp. 193–198, 2009.

[28] J.H. Anderson, and F.N. Najm, "Active leakage power optimization for FPGAs," IEEE Transactions on Computer-Aided Design of Integrated Circuits and Systems, vol. 25, no. 3, pp. 423–437, 2006.

[29] S. Narendra, S. Borkar, V. De, D. Antoniadis, and A. Chandrakasan, "Scaling of stack effect and its application for leakage reduction", in Proceedings of International Symposium on Low Power Electronic Design (ISLPED), pp. 195–200, 2001.

[30] F. Fallah and M. Pedram, "Standby and active leakage current control and minimization in CMOS VLSI circuits", IEICE Transactions on Electronics (Special Section on Low-Power LSI and Low-Power IP), vol. E88-C, no. 4, pp. 509–519, 2005.

[31] P.E. Gaillardon, E. Beigne, S. Lesecq, and G. De Micheli, "A survey on low-power techniques with emerging technologies: From devices to systems", ACM Journal on Emerging Technologies in Computing Systems, vo. 12, no.2, 2015, pp. 12.1–12.26.

[32] A.A.M. Bsoul and S.J.E. Wilton, "An FPGA architecture supporting dynamically controlled power gating," in International Conference on Field-Programmable Technology, ser. FPT'10, pp. 1–8, 2010.

[33] M.K.J. Hussein and M. Hart, "Lowering power at 28 nm with Xilinx 7 series devices," Xilinx, White Paper, WP389 (v1.2), 2013.

[34] B. Calhoun, F. Honore, and A. Chandrakasan, "Design methodology for fine-grained leakage control in MTCMOS," in Proceedings of IEEE International Symposium on Low Power Electronics and Design (ISLPED), 2003.

[35] A. Gayasen, Y. Tsai, N. Vijaykrishnan, M. Kandemir, M. Irwin, and T. Tuan, "Reducing leakage energy in FPGAs using region-constrained placement", Proc. ACM/SIGDA Int. Symp. Field Programmable Gate Arrays, pp. 51–58, 2004.

[36] V. George and J. Rabaey, "Low-Energy FPGAs: Architecture and Design," Kluwer Publication, New York, 2001.

[37] K. Poon, A. Yan, and S.J.E. Wilton, "A flexible power model for FPGAs", in Proceedings of Int. Conf. Field Programmable Logic and Applications, pp. 312–321, 2002.

[38] J. Lach, J. Brandon, and K. Skadron. "A general post-processing approach to leakage current reduction in SRAM-based FPGAs." In International Conference on Computer Design, 2004.

[39] R. Ahmed, "Towards High-Level Leakage Power Reduction Techniques for FPGAs," PhD Thesis, College of Graduate Studies (Electrical Engineering), University of British Columbia (Okanagan), 2015.

[40] C.Q. Tran, H. Kawaguchi, and T. Sakurai, "The 95% leakage reduced FPGA using zig-zag power-gating, Dual-VTH/VDD and Micro VDD hopping," in 2005 Asian Solid-State Circuits Conference, Hsinchu, pp. 149–152, 2005.

[41] S. Srinivasan, A. Gayasen, and T. Tuan, "Leakage control in FPGA routing fabric", in Proceedings of Asia South Pacific Design Automation Conference, pp. 661–664, 2005.

[42] M. Hasan, A.K. Kureshi, and T. Arslan. "Leakage reduction in FPGA routing multiplexers," in 2009 IEEE International Symposium on Circuits and Systems, Taipei, pp. 129–1132, 24–27 May 2009.

[43] C.H. Hoo, Y. Ha, and A. Kumar, "A directional coarse-grained power gated FPGA switch box and power gating aware routing algorithm", in Proceedings of 23rd International Conference on Field Programmable Logic and Applications (FPL), pp. 1–4, 2013.

[44] J. Anderson and F. Najm, "Power estimation techniques for fpgas," IEEE Transactions on Very Large Scale Integration (VLSI) Systems, vol. 12, no. 10, 2004.

[45] F. Li, Y. Lin, L. He, D. Chen, and J. Cong, "Power Modeling and Characteristics of Field Programmable Gate Arrays, "IEEE Transactions on Computer-Aided Design Integrated Circuits and Systems, vol. 24, no. 11, pp. 1712–1724, 2005.

[46] J.H. Anderson, "Power optimization and prediction techniques for FPGAs", Department of Electrical and Computer Engineering, Univeristy of Toronto, 2005.

[47] J.R. Templin and J.R. Hamle, "A new power-aware FPGA design metrics," Journal of Cryptographic Engineering, vol. 5, no. 1, pp. 1–11, 2015.

[48] R. Mukundrajan, "Tunnel FET based field programmable gate arrays", PhD Thesis, The Graduate School, College of Engineering, The Pennsylvania State University, USA, 2011.

[49] M. Abusltan and S.P. Khatri, "A comparison of FinFET based FPGA LUT design," in Proceeding GLSVLSI'14, 24th Edition of Great Lakes Symposium on VLSI, pp. 353–358, 2014.

[50] M.M. El-Din, H. Mostafa, H.A.H. Fahmy, Y. Ismail, and H. Abdelhamid, "Performance evaluation of FinFET-based FPGA cluster under threshold voltage variation," in

13th International Conference on New Circuits and Systems Conference (NEWCAS), Grenoble, pp. 1–4, 7–10 June 2015.

[51] A. Davidson, "A new FPGA architecture and leading-edge FinFET process technology promise to meet next generation system requirements," High-End FPGA Products, San Jose, CA, June 2015.

[52] W. Hung, Y. Xie, N. Vijaykrishnan, M. Kandemir, M.J. Irwin, and Y. Tsai, "Total power optimization through simultaneously multiple-VDD multiple-VTH assignment and device sizing with stack forcing," ISLPED'04, Newport Beach, California, USA, August 9–11, 2004.

[53] H.S. Deogun, R. Senger, D. Sylvester, R. Brown, and K. Nowka, "A dual-VDD boosted pulsed bus technique for low power and low leakage operation," in ISLPED'06 Proceeding of the 2006 International Symposium on Low Power Electronics and Design, pp. 73–78, 2006.

[54] K. Kim and V.D. Agrawal, "Ultra low energy CMOS logic using below-threshold dual-voltage supply," Journal of Low Power Electronics, vol. 7, pp. 1–11, 2011.

[55] A. Gayasen, K. Lee, N. Vijaykrishnan, M. Kandemir, M. Irwin, and T. Tuan, "A dual-Vdd low power FPGA architecture", in Proceedings of International Conference on Field Programmable Logic and Applications, pp. 145–157, 2004.

[56] F. Li, Y. Lin, H. Lei, and J. Cong, "Low-power FPGA using pre-defined dual-Vdd/Dual-Vt FABRICS", in FPGA'04, Monterey, California, USA, 22–24February 2004.

[57] R. Mukherjee and S. Ogrenci, "Mimic evaluation of dual VDD fabrics for low power FPGAs," in Proceedings of Asia and South Pacific Design Automation Conference, pp. 1240–1243, 2005.

[58] F. Li, Y. Lin, and L. He, "Vdd programmability to reduce FPGA interconnect power", in Proceedings of International Conference on Computer-Aided Design, pp. 760–765, 2004.

[59] F. Li, Y. Lin, and L. He, "Field programmability of supply voltages for FPGA power reduction", IEEE Transactions on Computer-Aided Design of Integrated Circuits and Systems, vol. 26, no. 4, pp. 752–764, 2007.

[60] Y. Meng, T. Sherwood, and R. Kastner, "Leakage power reduction of embedded memories on FPGAs through location assignment", in DAC 2006, July 24–28, San Francisco, California, USA, 2006.

[61] I. Ashraf, F. Boccardi, and L. Ho, "Alcatel-Lucent, SLEEP mode techniques for small cell deployments", in IEEE Communications Magazine, pp. 72–79, August 2011.

[62] M. Lin and A. El Gamal, "A low-power field-programmable gate array routing fabric", IEEE Transactions on Very Large Scale Integration (VLSI) Systems, vol. 17, no. 10, pp. 1481–1494, 2009.

[63] S.D. Pable and M. Hasan, "Performance analysis of FPGA interconnect fabric for ultra-low power applications", in ICCCS'11, Rourkela, Odisha, India, 12–14 February 2011.

[64] K. Siozios, F. Pavlidis, and D. Soudris, "A novel framework for exploring 3-D FPGAs with heterogeneous interconnect fabric", ACM Transactions on Reconfigurable Technology and Systems, vol. 5, no. 1, article 4, pp. 4:1–4:23, 2012.

[65] J. Anderson and F. Najm, "A novel low-power FPGA routing switch", in Proceedings of IEEE Custom Integrated Circuits Conference, pp. 719–722, 2004.

[66] T. Gao, K.C. Chen, J. Cong, Y. Ding, and C.L. Liu, "Placement and placement driven technology mapping for FPGA synthesis", in Proceedings of IEEE International ASIC Conference, pp. 87–91, 1993.

[67] E. Bozorgzadeh, S.O. Memik, X. Yang, and M. Sarrafzadeh, "Routability-driven packing: Metrics and algorithms for cluster-based FPGAs," Journal of Circuits, Systems and Computers, vol. 13, no. 1, pp. 77–100, 2004.

[68] M. Xu and F.J. Kurdahi, "Layout-driven high level synthesis for FPGA based architectures", in Proceeding of the Conference on Design, Automation and Test in Europe, DATE'98, pp. 446–450, 1998.

[69] D.P. Singh and S.D. Brown, "Incremental placement for layout-driven optimizations on FPGAs", Proc. Int. Conf. Comput.-Aided Des., pp. 752–759, 2002.

[70] V. Betz and J. Rose, "Circuit design, transistor sizing and wire layout of FPGA interconnect", IEEE 1999 Custom Integrated Circuits Conference, pp. 171–174, 1999.

[71] N. Kapre, N. Mehta, M. deLorimier, and R. Rubin, "Packet switched vs. time multiplexed FPGA overlay networks", in IEEE Symposium on Field-Programmable Custom Computing Machines (FCCM 2006), 24–26 April 2006.

[72] I. Kuon, R. Tessier, and J. Rose, "FPGA architecture: Survey and challenges", Foundations and Trends in Electronic Design Automation, vol. 2, no. 2, pp. 135–253, 2007.

[73] R. Seelam, "I/O design flexibility with the FPGA mezzanine card (FMC)", White Paper, WP315 (v1.0), pp. 1–7, 19 August 2009.

[74] C.C. Tsang and H.K.-H. So, "Reducing dynamic power consumption in FPGAs using precomputation", Proceedings of International Conference on Field Programmable Technology (FPT 2009), December 2009.

[75] J. Lamoureux, G. Lemieux, and S. Wilton, "GlitchLess: Dynamic power minimization in FPGAs through edge alignment and glitch filtering", IEEE Transactions on Very Large Scale Integrated Systems, vol. 16, no. 11, pp. 1521–1534, 2008.

[76] C. Ravishankar, J.H. Anderson, and A. Kennings,"FPGA power reduction by guarded evaluation considering logic architecture", IEEE Transactions on Computer-Aided Design of Integrated Circuits and Systems, vol. 31, no. 9, pp. 1305–1318, 2012.

[77] K. Subraniyam, "Proven power reduction with Xilinx ultrascale FPGAs", White Paper, WP466, vol. 1.1, pp. 1–13, 15 October 2015.

[78] C.Q. Tran, H. Kawaguchi, and T. Sakurai, "More than two orders of magnitude leakage current reduction in look-up table for FPGA's", IEEE International Symposium on Circuits and Systems, vol. 5, pp. 4701–4704, 23–26 May 2005.

http://dx.doi.org/10.5772/67257

[79] T. Tuan and B. Lai, "Leakage power analysis of a 90 nm FPGA", in IEEE Custom Integrated Circuits Conference, pp. 57–60, San Jose, CA, 2003.

[80] D. Curd, "Power consumption in 65nm FPGAs", White Paper: Virtex-5 FPGAs, WP 246, vol. 1.2, pp. 1–12, February 2007.

[81] V. Rozic, W. Dehaene, and I. Verbaushede, "Design solutions for securing SRAM cell against power analysis", in Symposium on Hardware-Oriented Security and Trust (HOST), pp. 122–127, 3–4 June 2012.

[82] J. Lach, J. Brandon, and K. Skadron, "A general post-processing approach to leakage current reduction in SRAM-based FPGAs", in Proceedings of the IEEE International Conference on Computer Design (ICCD'04), pp. 144–150, 11–13 October 2004.

[83] M. Qazi, M.E. Sinangil, and A.P. Chandrakasan, "Challenges and directions for low-voltage SRAM", in IEEE Design & Test of Computers, pp. 32–43, January/February 2011.

[84] Z. Liu and V. Kursun, "Characterization of a novel nine-transistor SRAM cell", IEEE Transactions on Very Large Scale Integration (VLSI) Systems, vol. 16, no. 4, pp. 488–492, 2008.

[85] K. Blomster and J.G. Delgado-Frias, "Reducing power and delay in memory cells using virtual source transistors," in 48th IEEE International Midwest Symposium on Circuits and Systems, pp. 299–302, 2005.

[86] T. Tuan, S. Kao, A. Rahman, S. Das, and S. Trimberger, "A 90 nm low-power FPGA for battery-powered applications", 14th International Conference on Field Programmable Gate Arrays, FPGA'06 Proceedings, pp. 3–11, 2006.

[87] B. Wang, T.Q. Nguyen, A.T. Do, J. Zhou, M. Je, and T. Tae-Hyoung Kim, "Design of an Ultra-low Voltage 9T SRAM With Equalized Bitline Leakage and CAM-Assisted Energy Efficiency Improvement", IEEE Transactions on Circuits and Systems—I: Regular Papers, vol. 62, no. 2, pp. 441–448, 2015.

[88] L. Zhang, C.-H. Chang, Z.H. Kong, and C.Q. Liu, "Statistical analysis and design of 6T SRAM cell for physical unclonable function with dual application modes", in IEEE International Symposium on Circuits and Systems (ISCAS), Lisbon, pp. 1410–1413, 24–27 May 2015.

[89] B.H. Calhoun and A.P. Chandrakasan, "Static noise margin variation for sub-threshold SRAM in 65-nm CMOS", IEEE Journal of Solid-State Circuits, vol. 41, no. 7, pp. 1673–1679, 2006.

[90] B.H. Calhoun and A.P. Chandrakasan, "A 256-kb 65-nm sub-threshold SRAM DESIGN for ultra-low-voltage operation", IEEE Journal of Solid-State Circuits, vol. 42, no. 3, pp. 680–688, 2007.

[91] M.-T. Chang and W. Hwang, "A fully-differential subthreshold SRAM cell with auto-compensation," in IEEE Asia Pacific Conference on Circuits and Systems, APCCAS, Macao, pp. 1771–1774, 30 Nov–3 Dec. 2008.

[92] M.-H. Tu, J.-Y. Lin, M.-C. Tsai, S.-J. Jou, and C.-T. Chuang, "Single-ended subthreshold SRAM with asymmetrical write/read-assist", IEEE Transactions on Circuits and Systems-I: Regular Papers, vol. 57, no. 12, pp. 3039–3047, 2010.

[93] L. Ming, C. Hong, L. Changmeng, and W. Zhihua, "An ultra-low-power 1 kb subthreshold SRAM in the 180 nm CMOS process", Journal of Semiconductors, vol. 31, no. 6, pp. 065013-1–065013-4, 2010.

[94] B. Amelifard, F. Fallah, and M. Pedram, "Reducing the sub-threshold and gate-tunneling leakage of SRAM cells using dual-Vt and dual-Tox assignment", in Proceedings of DATE, pp. 1–6, 2006.

[95] A. Teman, A. Mordakhay, J. Mezhibovsky, and A. Fish, "A 40 nm sub-threshold 5T SRAM bit cell with improved read and write stability", IEEE Transactions on Circuits and Systems II: Express Briefs, vol. 59, no. 12, pp. 873–877, 2012.

[96] C. Jiangzheng, Z. Sumin, Y. Jia, S. Xinchao, C. Liming, and H. Yong, "A 320 mV, 6 kb subthreshold 10T SRAM employing voltage lowering techniques", Journal of Semiconductors, vol. 36, no. 6, pp. 065007-1–065007-6, 2015.

[97] C.B. Kushwah and S.K. Vishvakarma, "A single-ended with dynamic feedback control 8T subthreshold SRAM cell", IEEE Transactions on Very Large Scal Integration (VLSI) Systems, vol. 24, no. 1, pp. 373–377, 2016.

[98] M.-F. Chang, J.-J. Wu, K.-T. Chen, and H. Yamauchi, "A differential data aware power-supplied (D^2AP) 8T SRAM cell with expanded write/read stabilities for lower VDDmin applications", Symposium on VLSI Circuits, Kyoto, Japan, pp.156–157, 16–18 June 2009.

[99] N. Gong, S. Jiang, A. Challapalli, S. Fernandes, and R. Sridhar, "Ultra-low voltage split-data-aware embedded SRAM for mobile video applications", IEEE Transactions on Circuits and Systems-II: Express Briefs, vol. 59, no. 12, pp. 883–887, 2012.

[100] C.M.R. Prabhu and A.K. Singh, "Novel eight transistor SRAM cell for write power consumption," IEICE Electronics Express (ELEX), vol. 7, no. 16, pp. 1175–1181, 2010.

[101] C.M.R. Prabhu and A.K. Singh, "Low-power fast (LPF) SRAM cell for write/read operation," IEICE Electronics Express, vol. 6, no. 18, pp. 1473–1478, 2011.

[102] Y.-W. Lin, H.-I. Yang, M.-C. Hsia, Y.-W. Lin, C.-H. Chen, C.-T. Chuang, W. Hwang, N.-C. Lien, K.-D. Lee, W.-C. Shih, Y.-P. Wu, W.-T. Lee, and C.-C. Hsu, "A 55nm 0.5V 128Kb cross-point 8T SRAM with data-aware dynamic supply write-assist", in Proceedings of IEEE International SoC Conference (SOCC), pp. 218–223, 12–14 September 2012.

[103] Y.-W. Chiu, J.-Y. Lin, M.-H. Tu, S.-J. Jou, and C.-T. Chuang, "8T single-ended subthreshold SRAM with cross-point data-aware write operation," in Proceedings of IEEE ISLPED, August 2011.

[104] M.-F. Chang, S.-W. Chang, P.-W. Chou, and W.-C. Wu, "A 130 mV SRAM with expanded write and read margins for subthreshold applications", IEEE Journal of Solid-State Circuits, vol. 46, no. 2, pp. 520–529, 2011.

[105] A.K. Singh, M.M. Seong, and C.M.R. Prabhu, "A data aware (DA) 9T SRAM cell for low power consumption and improved stability". International Journal of Circuit Theory and Applications, vol. 42, no. 9, pp. 956–966, September 2014.

[106] L. Wen, X. Cheng, K. Zhou, S. Tian, and X. Zeng," Bit-interleaving-enabled 8T SRAM with shared data-aware write and reference-based sense amplifier", IEEE Transactions on Circuits and Systems—II: Express Briefs, vol. 63, no. 7, pp. 643–647, 2016.

[107] A.K. Singh, M.-S. Saadatzi, and C. Venkataseshaiah, "Design of a single-ended energy efficient data-dependent-write-assist dynamic (DDWAD) SRAM cell for improved stability and reliability", Accepted for the publication in Analog Integrated Circuits and Signal Processing. C. Analog Integr Circ Sig Process (2016).

[108] A.K. Singh, M.-S. Saadatzi, and C. Venkataseshaiah, "Design of peripheral circuits for the implementation of memory array using data-aware (DA) SRAM cell in 65 nm CMOS technology for low power consumption", Journal of Low Power Electronics, vol. 12, pp. 1–12, 2016.

[109] A.K. Singh, M.M. Seong, and C.M.R. Prabhu, "Low power and high performance single-ended sense amplifier", Journal of Circuits, Systems, and Computers (Published by World Scientific), vol. 22, no. 7, pp. 1350062-1–1350062-12, 2013.

[110] L. Jiang, W. Xueqiang, W. Qin, W. Dong, Z. Zhigang, P. Liyang, and L. Ming, "A low voltage, sense amplifier for high-performance embedded fash memory", Journal of Semiconductor, vol. 31, pp. 1–5, 2010.

[111] H.-I. Yang, M.-H. Chang, S.-Y. Lai, H.S.-F. Wang, and W. Hwang, "A low-power low swing single-ended multi-port SRAM", in International Symposium VLSI Design, Automation and Testing 2007, VLS-DAT 2007, Hsinchu, pp. 1–4, May 2007.

[112] A. Teman, L. Pergament, O. Cohen, and A. Fish, "A 250 mV 8 kb 40 nm ultra-low power 9T supply feedback SRAM (SF-SRAM)", IEEE Journal of Solid State of Circuits, vol. 46, no. 11, pp. 2713–2726, October 2011.

Chapter 11

High-Speed Deterministic-Latency Serial IO

Raffaele Giordano, Vincenzo Izzo and
Alberto Aloisio

Additional information is available at the end of the chapter

http://dx.doi.org/10.5772/67012

Abstract

In digital systems, serial IO at speeds in the range from 1 to 20 Gbps is realized by means of dedicated transceivers, named serializer-deserializers (SerDeses). In general, due to their internal architecture, the data transfer delay, or the latency, may vary after a reset of the device. On the other hand, some applications, such as high-speed transfer protocols for analog-to-digital and digital-to-analog converters, trigger and data acquisition systems, clock distribution, synchronization and control of radio equipment need this delay to be constant at each reset. In this chapter, we focus on a serial IO architecture based on configurable transceivers embedded in field-programmable gate arrays (FPGAs). We will show how it is possible to achieve deterministic-latency operation in a line-code-independent way. As a case study, we will consider a synchronous 2.5-Gbps serial link based on an 8b10b line code.

Keywords: FPGA, serial links, line coding, latency, high-speed data transfers

1. Introduction

Deterministic-latency serial IO is highly desirable in applications such as high-speed transfer protocols for analog-to-digital and digital-to-analog converters (ADCs and DACs), trigger and data acquisition systems, clock distribution, synchronization and control of radio equipment. Unfortunately, deterministic-latency operation of serial transceivers generally requires dedicated circuitry and it is not generally supported by most of the devices on the market. Let us discuss a few examples.

Since their appearance on the market, the performance of high-speed DACs and ADCs in terms of sample rate and bit range improved continuously. This in turn generated the need for faster digital interfaces, evolving from traditional parallel single-ended buses in CMOS technology running at few hundred megabits per second to low voltage differential signalling (LVDS) operating up to 1 Gbps, also leading to higher power consumption. The CMOS and LVDS interfaces required laying out multiple traces on the printed circuit board (PCB) from the converter to the data processor and they imposed the usage of several pins on integrated circuits. In 2006, the Joint Electron Device Engineering Council (JEDEC) proposed a serial IO protocol designed for interfacing data processors to ADCs and DACs. In the latest revision of the standard (JESD204B [1]), line rates can reach up to 12.5 Gbps per serial lane and specific signals are introduced to synchronize the read out of multiple converters in the same system.

The Precision Time Protocol (PTP) is defined by the IEEE 1588 [2] standard, which defines specifications for synchronizing clocks in a networked system to a reference, precise clock (the grandmaster clock). The precision of the synchronization can be better than 1 μs, depending on the timing determinism of the network and on the asymmetries in up and down link delays. The synchronization is based on the exchange of messages between the sub-systems pertaining to the grandmaster clock and the slave clocks for estimating the clock offset and correcting it. The protocol assumes the up and down link delays to be equal, which is true only down to a certain resolution. The timing offset between the clocks grows linearly with the delay difference. Minimizing this difference, also by achieving deterministic-latency transmission at each system reset, allows the system to improve the accuracy of the synchronization [3].

One of the main goals in radio transmission systems is to keep the technological evolution of radio equipment (RE) and radio equipment control (REC) independent. Protocols, such as the Common Public Radio Interface (CPRI) [4], have been designed for this purpose. The protocol explicitly sets constraints on the latency and on the round trip delay of the data paths between RE and REC. All the delay requirements may be satisfied by means of deterministic-latency transmission between RE and REC, even for the multi-hop paths foreseen by the protocol.

Deterministic-latency links [5–8] find also application in data acquisition systems of nuclear and sub-nuclear physics experiments, specifically in the trigger sub-systems, where it is crucial to preserve the timing information associated with the transferred signals.

As we have exemplified, several application domains exist, very different from each other, which need fixed-latency data transmissions. In general, different applications adopt different protocols; therefore, in this chapter, we discuss the basic requirements for a fixed-latency link architecture and how to customize it in order to support any line coding or serial protocol. We highlight dependency of major blocks from the protocol and we discuss the aspects to be taken care of during the implementation phase. As a case study, we consider the GTP SerDes embedded in Xilinx Virtex-5 Field Programmable Gate Arrays (FPGAs), but the methodology can be easily exported to other SerDes devices.

In the following sections, we present concisely the GTP transceiver's architecture and we briefly outline the possible causes of variable latency in a serial link. We discuss the relationship between the logical alignment and the latency of the link and we present a general protocol-independent link architecture. Eventually, we show two specific implementations based on the 8b10b protocol and we highlight which features in a SerDes are keys for achieving fixed latency.

http://dx.doi.org/10.5772/67012

2. A configurable SerDes: the GTP transceiver

The SerDes used for discussing deterministic-latency concepts in this chapter is the configurable GTP transceiver [9] of the Xilinx Virtex 5 [10] FPGA family. GTPs are available as configurable hardware blocks. Each block hosts two transmitter (Tx) and receiver (Rx) pairs. The architecture of one pair is schematically represented in **Figure 1**. Some components, such as a phase locked loop, the dynamic reconfiguration port and the reset logic, are shared by the Tx and Rx. The GTP does not work with fixed latency in configurations based on its internal resources. The user has to develop a configuration based on an external logic controlling the alignment, which forces the SerDes to have a deterministic latency through its data path. We will now concisely present the features of the GTP essential to fixed-latency operation. The reader willing to get deeper insight might find more details in the device user's guide.

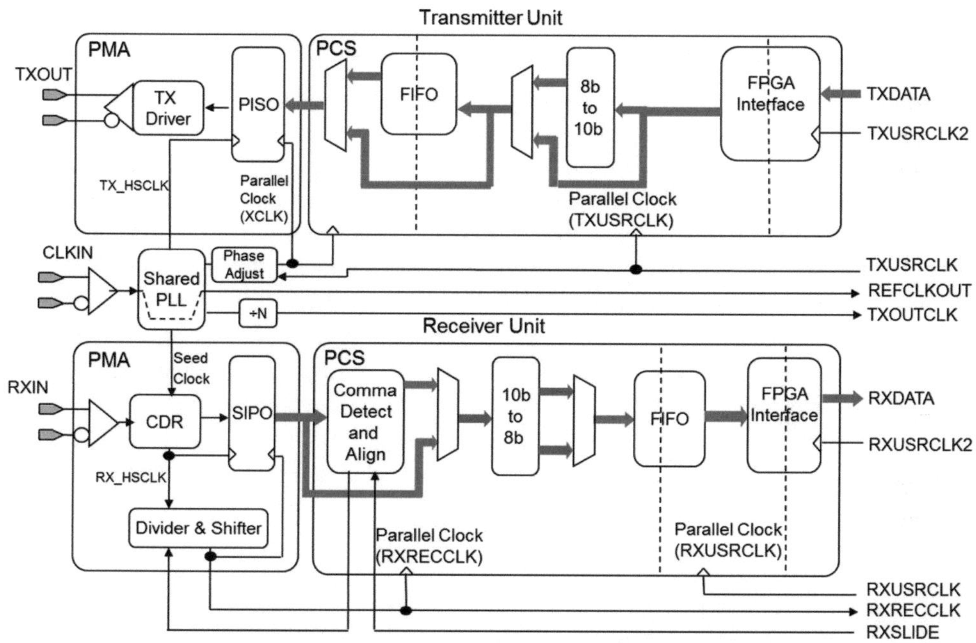

Figure 1. Simplified block diagram of the GTP transceiver. Half of the configurable hardware block is shown.

In order to transfer data to the FPGA fabric, the GTP offers a parallel IO interface which can be configured to be 8-, 10-, 16- or 20-bit wide. The lower two sizes are referred to as single-width, the higher ones as the double-width. The so-called physical medium attachment (PMA) sub-layer performs the actual data serialization and deserialization, whereas the physical coding sub-layer (PCS) processes parallel data. A reference clock (CLKIN) is routed to the shared PLL, which generates the high-speed clock for the serializer (TX_HSCLK), the parallel-side clock (XCLK) for the parallel input to serial out (PISO) block and a seed clock for the clock and data recovery circuit (CDR).

At the transmitter side, data flow from the fabric clocked by TXUSRCLK2 through the FPGA interface. When the FPGA interface is configured with a 16- or 20-bit size (double width modes), data are multiplexed 2:1 into 8- or 10-bit words and retimed on the TXUSRCLK clock, which in this case runs at double the rate of TXUSRCLK2. When the FPGA interface is configured for single-width operation, data are passed through without any processing and the two TXUSRCLK and TXUSRCLK2 coincide. A dedicated encoder can be activated for 8b10b-based protocols, while an elastic buffer (i.e. a first in first out memory) is included to cross the clock domain boundary from TXUSRCLK and XCLK reliably. In some applications, it may happen that XCLK and TXUSRCLK are derived from the same clock, therefore, they toggle at the same average frequency, with a constant phase difference. In this case, the elastic buffer can be bypassed and a dedicated circuitry is used to ensure a safe transfer of data from the TXUSRCLK clock domain to XCLK. The PISO block serializes data and outputs them synchronously with TX_HSCLK. It is worth mentioning that the PLL produces also another clock (TXOUTCLK), which can be routed to a clock buffer in the fabric and used as a TXUSRCLK. Unfortunately, due to architectural constraints of the GTP, this signal cannot be used when the elastic buffer is not in use.

At the receiver side, the CDR extracts the receiver high-speed clock (RX_HSCLK) from the stream and recovers the serial data. A dedicated prescaler divides RX_HSCLK down to generate the RXRECCLK, namely the recovered clock for clocking data out from the parallel output block and for the PCS operation. Since it is synchronous with the parallel data in the PCS, this clock can also be forwarded to the fabric and it can be used to synchronize the logic processing the deserialized data. An interesting and very useful block is the "Comma Detector and Aligner" which can search for special symbols in the serial stream and align the symbol boundary to them automatically, saving the designer to perform this operation in the fabric. The rest of the blocks in the receiver's PCS are symmetrical to the transmitter ones, they perform elastic buffering toward the RXUSRCLK clock domain and data demultiplexing when needed (FPGA interface). The RXUSRCLK2 signal synchronizes the data from the FPGA interface into the fabric. For single-width operation modes, RXUSRCLK and RXUSRCLK2 are the same signal, while for double-width modes they are edge-aligned but RXUSRCLK2 toggles at half the frequency with respect to RXUSRCLK.

3. Variation of the latency in a SerDes device

A SerDes may show latency variations related to its serial and/or parallel sub-components. In the serializer, the transmission clock that strobes the parallel data into the device is multiplied to provide the high-speed serial clock for the PISO. On the other hand, in the deserializer, the high-speed serial clock is recovered from the stream and divided back to obtain the clock for the parallel data. These clocks are used to clock, respectively, the serial and parallel side of the SIPO. However, the clock division leads to an uncertainty of the phase of the recovered clock. The phase of the recovered clock may vary in integer multiples of the unit intervals (UIs) and it causes a consequent variation of the delay of the data strobed by the clock.

Let us imagine to multiply a signal *clk* in frequency by a factor M and let us call clk_M, the result of this operation (**Figure 2**). We can label each clk_M edge with an integer number from 0 to M - 1. If T_{clk} is the clock period of clk, the *i*th edge of clk_M will be shifted by a delay $\frac{i}{M}T_{clk}$ with

http://dx.doi.org/10.5772/67012

respect to the rising edge of *clk*. Let us now suppose to use a counter to divide clk_M in frequency back by a factor M, there are now M possible phases for the result, which are represented by the clk_i signals (with $i = 0$ to $M-1$). The obtained signal depends on which edge of the clk_M signal marks the 0 in the counter. Data crossing the clock domain from *clk* to one of the clk_i will do it with a latency related to their relative phase. After a reset or a power cycle of the system, the resulting clk_i signal might vary and the data latency with it. The system designer has to foresee a dedicated logic to remove this variation and to generate always the same clk_i signal and consequently the same data delay.

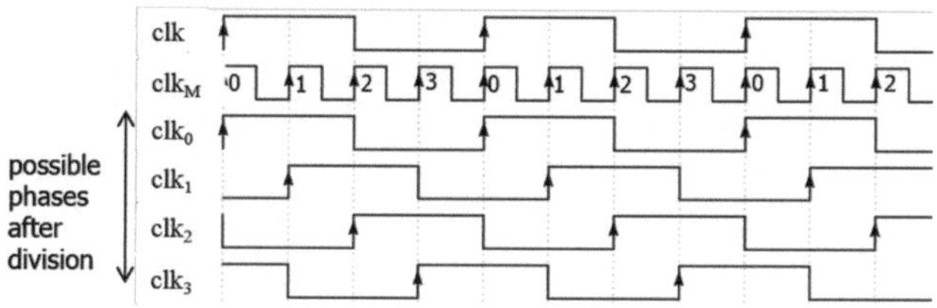

Figure 2. Clock multiplication and subsequent division (case M=4).

In the parallel sub-components of the SerDes, elastic buffers might induce variations of latency. Even if a buffer is written to and read from at the same frequency, after each reset or power cycle its latency depends on the difference between the internal write and read pointers. This difference in turns depends on the functionality of the logic accessing the buffer and it induces variations in terms of integer multiples of the clock period used for reading and writing. A dedicated logic has to be added in order to match the number of written words before they start being read at each reset or power up.

Therefore, the overall latency variation ΔL between two resets or power cycles of the device is as follows:

$$\Delta L = nT_{ser} + mT_{par} \quad (1)$$

where T_{ser} is the high-speed serial clock period, T_{par} is the parallel clock period, n and m are integers and their ranges depend on the SerDes device and how it is configured and operated.

4. Word alignment mechanisms

The recognition of the symbol boundary in the serial stream and the consequent alignment operation has a direct impact on the latency of a deserializer. The Xilinx GTP transceiver enables user logic implemented in the fabric to control the alignment. The RXSLIDE signal can be used to make the device shift the symbol boundary by one bit. Two modes are supported for performing this operation: a first one realized in the PMA, which shifts the recovered clock phase and combines it with the logical shifting of the data ("PMA mode") and a second one which only shifts the data logically ("PCS mode") while leaving the phase of the recovered clock unchanged. As discussed in Section 3, in order to remove latency variation, a dedicated mechanism for selecting always the same recovered clock phase must be added. Therefore, the PMA with its clock phase shifting capability allows the design to partially remove the phase variation.

Since there is not an official documentation about the internal architecture of the PMA, after some experimental tests, we have built a conceptual model of how the GTP performs the phase shift in PMA mode (**Figure 3**). The RX_HSCLK is recovered by the CDR running at half the line rate (i.e. $T_{RX\_HSCLK} = 2$ UIs) and both its edges are used for sampling the serial stream. The serial input shift register of the SIPO is therefore double data rate (DDR) and synchronous with RX_HSCLK. A register in the SIPO synchronizes the output from the shift register on the recovered clock (RXRECCLK). The "Clock Divider and Shifter" divides RX_HSCLK down by 5 and routes it to a 5-bit shift-register, which provides five copies of the input signal, where nth copy is delayed by $n \cdot T_{RX\_HSCLK}$ with $n = 0$ to 4, therefore, spanning a full RXRECCLK period. The RXSLIDE signal from the user logic drives a modulo-10 counter, which in turn drives the selector of a multiplexer. According to the number of RXSLIDE pulses received, a specific phase for RXRECCLK is chosen. At the interface between the serial and parallel sections of the SIPO, this phase determines which bit of the parallel register latches the least significant bit of the shift register. In other words, changing the selection of the multiplexer modifies the alignment of the parallel data with respect to the serial data. Due to the double data rate operation, this mechanism shifts the recovered clock phase relatively to the stream in steps of 2 UIs. In order to produce a correct logical shift for any number of required bit slides, the devices use the least significant bit (lsb) of the counter to drive a barrel shifter. If the number of required slides is odd, the lsb is set and the barrel shifter shifts the data by one bit, otherwise it leaves the data unchanged (**Figure 4**). Data are therefore always correctly shifted, but for each of the five possible recovered clock phases, there are two possible alignments of the data, one for odd bit slides and one for even slides. Operation in PMA mode requires external logic to drive the RXSLIDE signal, while in PCS mode the internal comma detector and aligner can be used. We remark that, by themselves, none of the native alignment modes (PCS or PMA) operate with fixed latency. We will see in the coming sections how a smart configuration of the GTP plus a dedicated logic implemented in the fabric allows the firmware designer to achieve fixed-latency operation.

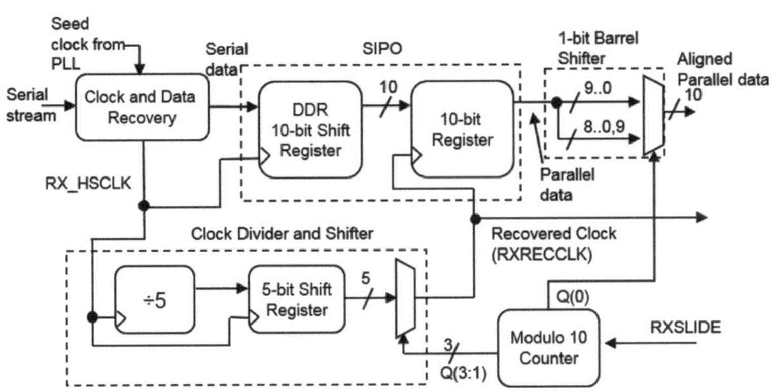

Figure 3. Conceptual block diagram of the shift architecture used in PMA mode.

http://dx.doi.org/10.5772/67012

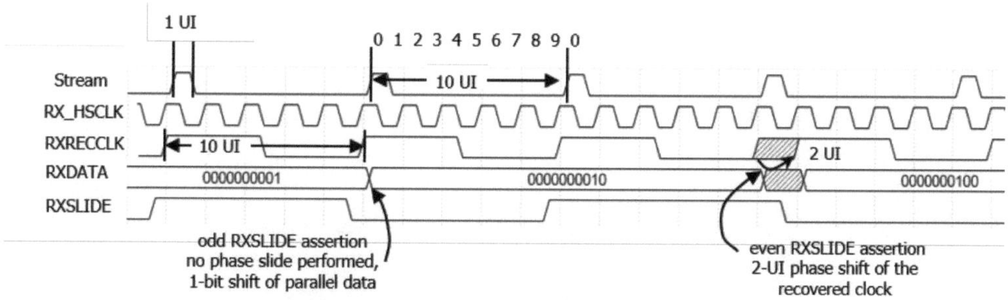

Figure 4. Recovered clock phase adjustment by means of bit slips.

5. Encoding-independent, fixed-latency operation

This section discusses key points that the designer should keep in mind in the implementation of fixed latency operation, independently of line coding and communication protocol. We are going to discuss a block diagram of the architecture shown in **Figure 5**. In order to make the discussion more practical, let us suppose the GTP runs at 2.5 Gbps with a 10-bit interface to the fabric.

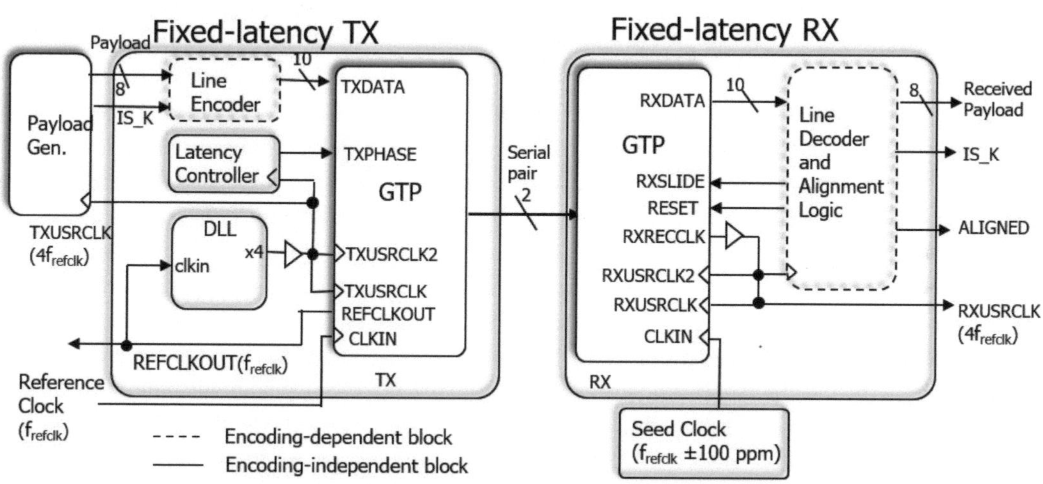

Figure 5. Customizable link architecture based on the GTP transceiver. Protocol-dependent blocks are outlined with dashed lines, protocol-independent blocks with solid lines.

Since this is single data width configuration, the GTP User's Guide prescribes the transmit clocks (TXUSRCLK and TXUSRCLK2) must be tied together and as well as the receive clocks (RXUSRCLK and RXUSRCLK2). We use a delay locked loop (DLL) for deriving the transmit clocks (toggling at 250 MHz) from the reference clock (toggling at 62.5 MHz).

Let us focus on the transmitter node. The serial coding is provided by the line encoder, which operates on data words before they enter the GTP. The latency controller adjusts the latency through the GTP transmitter to be constant at each power up or reset. The encoder is implemented by means of fabric resources in order to show how to achieve fixed latency data transfers with any coding, not only the internally supported 8b10b. A vast majority of commercial protocols allow the user to send data and control symbols. In this example, the IS_K input of the line encoder determines whether the data word will be encoded as a data or control symbol. The parallel clock for the PISO in the transmitter (XCLK) is generated by multiplying of the reference clock and then by dividing it back to obtain the desired frequency. As we discussed in Section 3, at each power up or reset, its phase can be different with respect to previous power ups or resets. The latency controller exploits a dedicated phase alignment circuit internal to the GTP, which trims the phase of XCLK to the one of the reference clock. The procedure is based on GTP features and it can be implemented by following guidelines provided in the documentation. The payload generator, not really part of the link, is explicitly included in the block diagram in order to show that data are synchronous with the transmit clock, rather than with the reference clock. It is also possible to generalize this architecture in order to transmit data synchronously with the reference clock, as we will show in Section 7. It is important to remark that, on the transmitter, there is not dependence between latency control and data encoding. There is no exchange of information between the line encoder and the latency controller, which ensures fixed latency operation by itself.

Let us now discuss the receiver node. There is a single block performing line decoding and logic alignment and it is implemented in the fabric. On the contrary of what happens for the transmitter, now the decoding and alignment are interdependent. In fact, alignment requires processing deserialized data, which in general might need decoding. The GTP embeds an 8b10b line decoder and an alignment logic which operate with variable latency. Therefore, we reimplement in the fabric the decode and alignment logic. A clock source with a frequency within 100 ppm with respect to the transmitter reference clock is needed as a seed for the correct lock-up of the CDR. As we discussed in Section 3, at each CDR lock-up, the recovered clock edge might have 10 possible phases. We configure the GTP in PMA slide mode, therefore, the alignment is controlled by asserting the RXSLIDE signal. For each possible recovered clock phase with respect to the stream, there are two possible logical alignments, one requiring an odd number of bit slides and one requiring an even number. Since we require a bi-unique relationship between the number of slides and the recovered clock phase, we reject CDR locks pertaining to one of the two possibilities, for instance, we reject those requiring odd slides. Which possibility we decide to reject is immaterial, but the alignment logic has to perform this rejection. It is important to remark that although the recovered clock shifting feature of the GTP is useful for achieving fixed-latency operation, it is not necessary. Other strategies can be implemented for SerDes devices which do not support that, as we will discuss at the end of Section 6.

Some serial protocols (e.g. SONET [11]) need to decoding for assessing the correctness of the alignment. The alignment logic checks received data according to protocol-specific criteria and if the check is failed, it changes the symbol alignment. When the check is passed, the correct alignment is found. On the other hand, a serial line code might not need data decoding for finding the correct alignment. For instance, the 8B10B code uses special bit sequences, called

http://dx.doi.org/10.5772/67012

commas, which cannot be obtained by concatenating two symbols of the code. Finding a comma in the encoded stream allows the receiver to determine the word boundaries and performs the alignment, without the need for data decode.

For a given line code, once the correct number of bit slides needed to achieve the correct logic alignment is determined, the pertaining logic must check whether the SerDes can implement that sliding by changing the recovered clock phase.

If that is not possible, the alignment logic can force the CDR to lose the lock (for instance, by resetting the CDR or by resetting the whole SerDes) and wait for a relock.

If it is possible, the alignment logic can use appropriate features of the SerDes, such as the RXSLIDE signal for the GTP, to shift the recovered clock phase. When this procedure is complete, data are correctly aligned to the symbol boundary and the recovered clock edge has a known phase relationship with respect to the stream. This technique is referred as the *roulette* approach and it will be further discussed in the following sections.

6. An 8B10B serial link implementation

This section shows how to include, in our architecture, one of the most used coding schemes for serial data: the 8B10B encoding. At the transmitter end, the encoder has to be configured to use the 8B10B coding, according to the architecture described in Ref. [12] or in the original 8b10b patent [13]. At the receiver side, we need to include three different elements for designing the correct decoder and alignment logic: a 10b to 8b decoder, a comma detector and an aligner (**Figure 6**). At the output of the decoder, besides the 8-bit decoded data, also a flag is provided, that, when active, indicates that the received word is a control character (IS_K). The Comma Detector module looks at the deserialized data and searches for specific 10-bit symbols. When the Comma Detector finds an expected symbol, its bit offset with respect to the word boundary is sent out (on the 4-bit bus "Bit-offset") and a "Found" flag is activated. The bit shifter of the GTP is driven by the Aligner block. When the "Found" flag is asserted a number of RXSLIDE pulses corresponding to the bit-offset are generated by the Aligner block, otherwise a reset to the GTP is produced. It is worth noting that according to the 8B10B coding, some symbols can be represented with two different 10-bit words, where one word is the complement of the other. The 8B10B encoder chooses one word or its complement by minimizing the so-called "running disparity," i.e. the difference between the number of 1s and 0s sent on the serial channel. The Comma Detector (**Figure 7**) can search in the incoming data for two independent 10-bit symbols. The searched symbols are described in the hardware description language (HDL) source code as two parameters, which can be modified before the synthesis and implementation of the design. The two symbols were programmed in order to be the two different possible versions of the 8B10B coded word to be found (indicating them as Comma+ and its complement Comma-). In our design, the serial stream is sliced into 10-bit words, for this reason, part of a comma word can be in a 10-bit word and the remaining part can be in an adjacent word. Thus, the search procedure has to look into the stream and to examine each symbol and the first 9 bits from the next symbol. Following this consideration, we designed a 2-level pipeline in the Comma Detector that we used to combine each incoming 10-bit word (DATAIN bus) with the 9 adjacent bits from the next 10-bit word, so to build an overall 19-bit word (WORD bus). All the 10 possible portions of 10 adjacent bits of the 19-bit WORD bus (WORD(9+i:i) with i = 0, 1···9) are compared with the Comma- and Comma+ symbols, by means of an array of comparators. When the comparator finds a match with the WORD(9+i:i) segment, the comparator also asserts the corresponding iFound(i) signal. The "Binary Encoder" block collects all the iFound signals and then produces a 4-bit binary code (Bit-offset), obtained by encoding the index i for the asserted iFound(i) signal. The "Binary Encoder" block also asserts the "Found" output, when at least one iFound(i) signal is asserted. A closer look into the Aligner block reveals that it consists of a finite state machine (FSM) (**Figure 8**), which continuously checks the outputs of the Comma Detector; the Aligner logic is also made of a register and a counter (not shown for simplicity). When the FSM is in the "Idle" state, it is continuously waiting for a comma: when a comma is found, the FSM captures the data on the Bit-offset input into a special register, which keeps the data on the internal bus "iBit-offset." Then, the FSM performs the "roulette approach" algorithm, in particular,

1. When the data on the iBit-offset bus is zero, the FSM asserts the "Aligned" flag and returns back into the "Idle" state.

2. When the data on the iBit-offset bus is non-zero and odd, the machine performs a full reset of the GTP and then waits for the CDR to lock again.

3. When the data on the iBit-offset bus is non-zero and even, the FSM generates a sequence of pulses on the RXSLIDE output, where each pulse requests a bit slide to the GTP. According to the GTP specifications, each RXSLIDE pulse must stay active for one clock cycle and, between two consecutive pulses, a minimum interval of two clock cycles is required. A specific counter (Pulses bus) is used to store the amount of sent pulses. When the "Pulses bus" value reaches the same value latched on the iBit-offset bus, the production of the RXSLIDE pulses is stopped, the "Aligned" flag is activated and the FSM returns back to the "Idle" state.

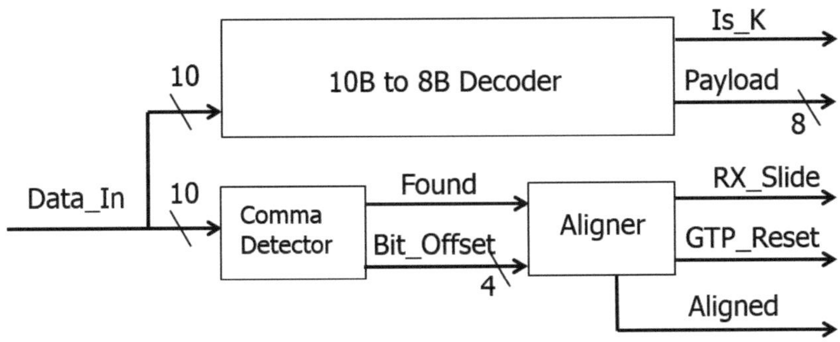

Figure 6. Internals of the line decoder and alignment logic.

http://dx.doi.org/10.5772/67012

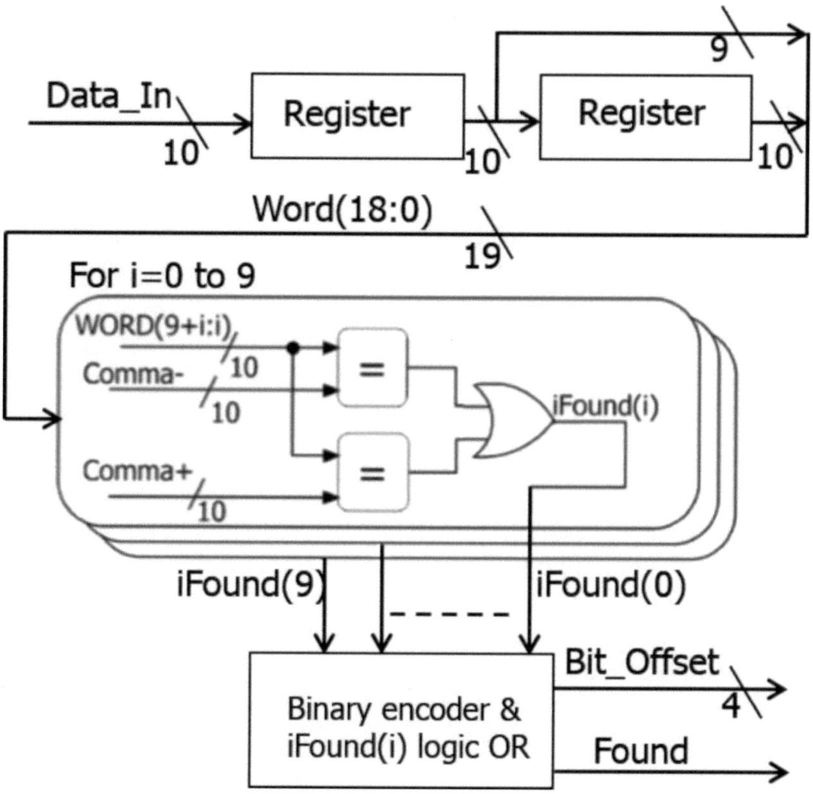

Figure 7. Simplified block diagram of the Comma Detector.

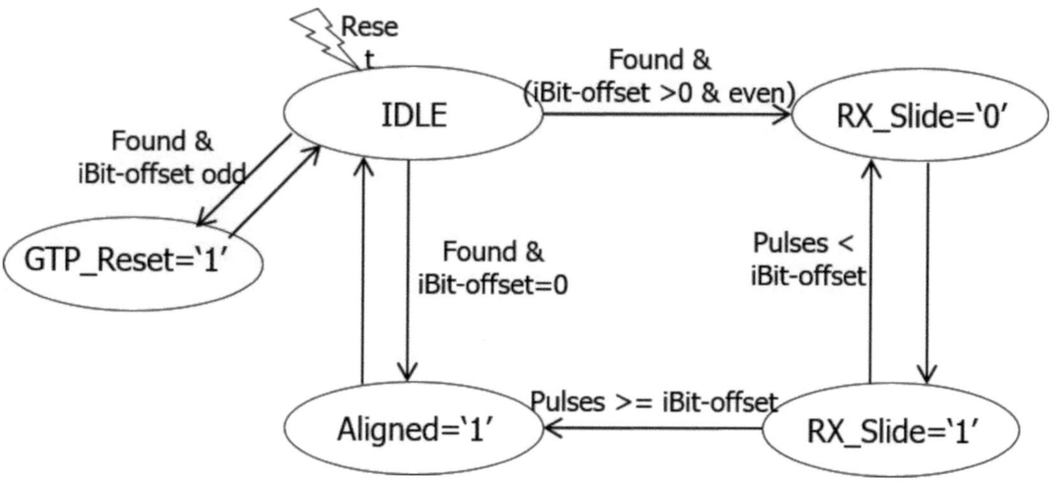

Figure 8. Simplified bubble diagram of the Aligner.

By looking at the algorithm, when there is a non-convenient value of the bit offsets (e.g. the odd ones), the CDR can be just reset, in order to wait for a relock on a most advantageous bit offset (e.g. an even one); thus, the alignment technique based on the "roulette approach" just described can be used to bypass the limitation on the GTP, that is capable only to shift by 2-UI steps. This approach brings to a lock time of the link that doubles, on average, as the lock of the CDR is rejected the 50% of the times.

When the bit offset is odd, solutions can be adopted, instead of resetting the CDR. For instance, the recovered clock phase can be shifted by 1 UI using a programmable delay (e.g. by means of a DLL, a PLL or an open-loop fine-grained programmable delay [14]). Also this method has a drawback, as it requires a higher complexity in the circuitry surrounding the GTP and it might introduce a higher clock jitter and a possible phase-skew between the alignment obtained with odd bit offset and the alignment obtained with even bit offsets, as the different delay elements are used in the two different cases.

A design based on the roulette approach can be deployed in many applications, where the deserializer architecture does not offer phase-shifting capabilities of the recovered clock. The aligner should simply monitor that the received comma has a certain bit offset (e.g. zero) and, in this case, it should perform a reset of the CDR until the required bit offset is detected. The roulette approach greatly simplifies the logic of the aligner block (**Figure 9**) and easily helps to obtain a recovered clock with a fixed-phase, without the need to perform a phase shift. As already noted, a disadvantage of the approach is the increase in the average lock time. As an example, the average lock-time is increased by a factor 10 (as the bit offset of a comma has the required value 1 time out of 10). When used in bidirectional links, an increase in the number of commas to be sent before the lock of the link is reached may help to soften this effect. For instance, the JESD204B protocol foresees an initial transmission of commas to be interrupted only after the receiver has locked and the increase of lock time would be minimal in this case. Anyway, using the roulette approach always requires a trade-off between the complexity of the aligner logic and the average lock time.

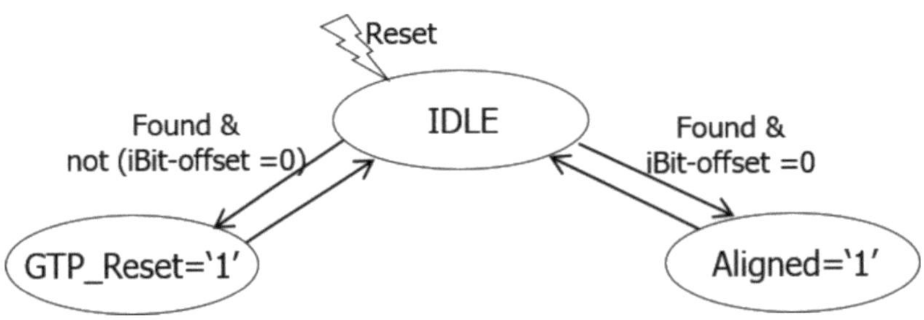

Figure 9. Simplified bubble diagram of the aligner implementing a pure roulette approach.

6.1. Frequency performance and resource occupancy

Tables 1 and **2** show the resource occupancy for the presented fixed-latency architecture, with details of resources for both the transmitter and the receiver design. For each block constituting the design, the usage of FPGA primitives is shown, separated by type; moreover, the used percentage of the design resources is shown, for a medium-size Virtex-5 FPGA.

Module\primitives	Slices	Slice Regs	LUTs	BUFG	DCM_ADV	GTP_DUAL
Latency controller	15	27	8	0	0	0
8B10B encoder	4	6	2	0	0	0
DLL	1	0	1	2	1	0
GTP	3	9	0	1	0	1
Total	23	42	11	3	1	1
Available in a V5LX50T	7200	28,800	28,800	32	12	6
% Used in a V5LX50T	0.3	0.1	0.04	9	8	17

Table 1. Resource occupation of the fixed-latency transmitter.

Module\primitives	Slices	Slice Regs	LUTs	BUFG	DCM_ADV	GTP_DUAL
Aligner	7	13	14	0	0	0
Comma detector	15	34	39	0	0	0
10B8B decoder	4	2	8	0	0	0
GTP	3	9	0	2	0	1
Total	29	58	61	2	0	1
Available in a V5LX50T	7200	28,800	28,800	32	12	6
% Used in a V5LX50T	0.4	0.2	0.2	6	0	17

Table 2. Resource occupation of the fixed-latency receiver.

A small logic foot-print (in terms of slice occupancy) is used for both the transmitter and receiver, respectively, requiring 23 and 29 slices, which correspond to 0.3% and 0.4% of a V5LX50T. On the transmitter side, the DLL block requires one digital clock manager primitive (DCM_ADV) and three clock buffers (BUFGs): one is used for driving the reference clock, one is used for the transmit clock and one for the DLL input clock. On the receiver side, only two buffers are needed (one for the reference clock and one for the receive clock) as the clock requirements are simpler.

Given such a small use of logic in the fabric, in most cases, there is no effort needed to reduce or optimize it. However, the designer should make an additional effort in order to have the system working with transmit clock and reference clock having the same frequency, so to avoid needing a DLL and its additional buffer. Such a simplified architecture lowers the occupation of fabric resources and also reduces the power consumption.

The transmitter has a maximum clock frequency (for the fabric resources) of about 370 MHz, which is essentially defined by the encoder logic, as reported by static timing analysis. The receiver has a maximum frequency of 330 MHz, in this case limited by the comma detector.

Thus, the presented architecture is able to operate up to the maximum transfer rate supported by the GTP (3.125 Gbps) and there is no need for increasing the frequency performance in the fabric.

7. Fixed-latency, packet-based transmission

In the previously described design, the data clock (which is used to transmit and receive the data) operates at 250 MHz, which is a frequency that should be easily reached when the clock is limited on a single board; however, such frequency could be too high for the propagation of the clock into a larger or more complex system, e.g. a crate or a board network. Moreover, in some applications (e.g. when using a 64b/66b encoding), an 8-bit parallel word size could be too short and might require a complex additional circuitry. In order to overcome these limitations, we show how to enlarge our architecture, so to build a packet made of several data words, but still having a fixed latency on the link. In this case, we need special care in the clock division and in the word de-multiplexing blocks, in order to obtain our objective. In this section, we describe a link with the same data-rate of 2.5 Gbps (as the previous one) but transmitting 32-bit words at 62.5 MHz, i.e. larger transmitted words. We remark that in Section 5, we presented an iso-synchronous architecture, i.e. the clock for the data source is produced by the link itself. In many applications, the designer could have a system clock which drives a data source and would prefer to use that system clock to transmit the data over the link. In these architectures, the reference clock for the PLL of the transmitter and the transmission clock for the parallel data are the same clock driving the payload. In **Figure 10**, the "fixed-latency tx" or FLT block (and the analogous block "fixed- latency rx" or FLR) represent a GTP transmitter (and the analogous GTP receiver) when opportunely configured and equipped with the logic need for implementing fixed latency operations.

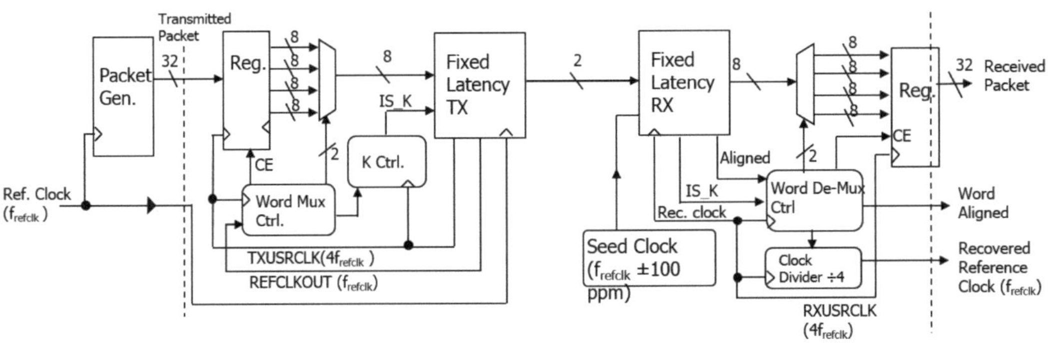

Figure 10. Synchronous implementation of our link architecture.

At the transmitter end, the input data are synchronous with the 62.5 MHz and the reference clock is latched into an input register synchronous with a 250-MHz data clock (TXUSRCLK) generated by the "Fixed Latency Tx." The clock-enable pin of the input register is driven at a 62.5 MHz rate, by the "Word Multiplexer Controller" (WMC) block. The design suggests to use a multiplexer in order to split the incoming 32-bit words into 8-bit words, which are then serialized by the FLT. The "Word Multiplexer Controller" block is also used to drive the select signals of the multiplexer, so that the same byte of the incoming 32-bit word is serialized first, whenever, the circuit is powered-on. In order to perform this task, the WMC samples the edge of the reference clock (REFCLKOUT) with the TXUSRCLK clock and sends the first byte after the edge is detected. Another circuitry is needed to tag the first byte in the word, in order to allow the receiver to correctly recognize and align the 8-bit words. In order to do this, at each every power-up, the "K Controller" module is in charge to assert the IS_K input for the FLT, which then sends a control character on the first byte of the word. Even if everything seems coherent from the logic point of view, there are still some timing issues in the transmitter part of the architecture. Indeed, the two clock signals TXUSRCLK and REFCLKOUT could be edge-misaligned, due to their different paths in the FPGA's layout, even if they are phase-locked. As an example, the "Word Multiplexer Controller" samples the REFCLKOUT signal as a data on the TXUSRCLK edge: thus, the designer needs to carefully verify that the REFCLKOUT signal meets the timing constraints needed by TXUSRCLK. In order to overcome such an alignment issue, the designer can adequately program a DLL inside the FLT, thus finding an appropriate phase of TXUSRCLK with respect to REFCLKOUT. Moreover, the WMC also has to pulse the clock enable signal of the input register with a specific phase, with respect to REFCLKOUT, in order to prevent timing violations during the capture of the input payload. It should also be noted that some of the modules around the FLT (specifically the multiplexer controller, the multiplexer itself and the input register) could be replaced by a dual-port FIFO, with a 32-bit input port and a 8-bit output port. In this case, the incoming 32-bit input words could enter the FIFO synchronously with the edge of the reference clock, while the 8-bit output words could exit the FIFO synchronously with the TXUSRCLK edge. The reader could easily argue that such dual-port FIFO could represent the solution by-design for all the described timing issues; one could also note that Xilinx FPGAs provide embedded dual-port FIFOs as hardware components, thus making the implementation of a dual-port FIFO very easy. However, the use of a FIFO has the following two main drawbacks: first, including a FIFO into the design would also increase the latency, which could not be affordable in some applications, while the architecture previously described keeps the latency as low as possible; second, the latency could not be constant at each power-up and this would require to carefully handle the read operations from the FIFO and the write operations to the FIFO (see the end of Section 3).

At the receiver end, the "Fixed Latency Rx" must be fed with a seed clock for the Clock & Data Recovery circuit (CDR) with a frequency offset below 100 ppm of the reference clock of the transmitter. At its output, the FLR drives a fixed phase 250 MHz recovered clock, to be used by the surrounding logic. The FLR module provides the 8-bit deserialized words to a de-multiplexer and drives some control signals to a specific control logic: these control signals indicate that the received character is a control character (IS_K) and that the link is byte-aligned (Aligned). The comma character is used by the "Word De-Mux Controller" (WDC) in order to handle the de-multiplexer 2-bit selection signal and to drive the clock enable for the output register (driven at a 62.5 MHz rate) with the correct latency. Furthermore, when receiving a comma, the WDC resets a clock divider (by 4) that is used to produce a 62.5 MHz recovered reference clock. The 62.5 MHz clock is then used to transfer the 32-bit payload from the de-multiplexer to the following logic. The WDC has to generate the clock divider reset in such a way to set the recovered reference clock edge in the centre of the data eye of the 32-bit payload provided by the output register. The full synchronous Tx+Rx architecture gives, as a side effect, the possibility to use at the receiver a phase-locked copy of the reference clock of the transmitter, which is a very effective and profitable feature to be used in distributed systems such as TDAQ systems of high energy physics (HEP) experiments. In TDAQ applications of HEP experiments, there is very often the need to distribute a common clock signal to all the elements of the TDAQ system, with a predictable phase and a minimum jitter. These TDAQ systems often rely on serial links, which are already deployed for data transmission. The same serial links, therefore, are a very appealing medium also for delivering the clock to every destination, without the necessity for a separate clock distribution network, thus making the TDAQ system architecture simpler to be implemented and easier to be maintained. Regarding TDAQ system, applications of fixed latency serial links, some measurements can be found in the literature [15], in particular, the measurements performed on 2.5 Gbps links show that it is possible to distribute a clock signal with a rms jitter of about 20 ps. We would like to stress that the reference clock recovered at the receiver of the described architecture cannot be easily handled to achieve a synchronous retransmission with the same GTP, but it requires to pay attention in order to make it work correctly. In fact, by looking at the hardware resources inside a GTP, the reader can easily see that the internal PLL is shared between the transmitters and receivers in the same SerDes; moreover, the PLL is already locked to the seed clock. Thus, the usage of another GTP is mandatory. Alternatively, the designer might change the phase and frequency of the reference clock in order to match phase and frequency of the recovered clock smoothly enough, so that the lock of the link is neither lost at the transmitter nor at the receiver. Furthermore, it could be necessary to filter the recovered clock in order to satisfy the jitter specifications for the GTP reference clock. Such disadvantage is not present in the newer FPGA families, such as the Virtex-6 or the seven series, as these devices are equipped with transceivers that provide separate PLLs for transmitter and for receiver.

8. Key features for fixed latency

In this section, the key features of the GTP for achieving fixed latency are described, together with helpful suggestions to be used in the porting of our results to other SerDeses.

At the transmitter end of the link, there must be a predictable phase relationship between the parallel clock (which drives the PISO) and the external parallel data clock. In many transmitters, the usage of the features for serial channel bonding can be used for satisfying this condition. In fact, channel bonding is widely employed for multi-lane serial buses, such as PCI Express [16], RapidIO [17] and Infiniband [18, 19], requiring the same latency over every bonded data path. For instance, the phase alignment circuit of the GTP, which was described and exploited to lock the latency of the transmitter, has been designed for applications related to channel bonding.

At the receiver end, there must be a predictable phase relationship of the recovered clock with respect to the byte boundary in the incoming serial stream. However, the proposed

http://dx.doi.org/10.5772/67012

architecture could be implemented only if the receiver provides a direct method to establish the phase offset or if the receiver offers some other feature to be used to calculate the phase offset indirectly. As an example, when the receiver device can output encoded and un-aligned parallel words, the phase offset could be determined outside the receiver. Anyway, as the phase offset might change (and usually changes) at each power-up of the receiver, the designer should also add an external logic, which is able to determine the offset (by looking at the data alignment) and to reset the receiver, if the calculated offset is not the desired one: this is described as roulette approach. The external logic should not reset the device, in the case that the desired phase offset is present and thus the link achieves the lock with the same latency it had (and will have) in the other successful locks. The basic idea of the roulette approach is that no alignment of the data is explicitly performed. The receiver is forced by-design to accept only the locks achieved with data already correctly aligned and it simply rejects the locks achieved with data not correctly aligned. This approach makes the additional receive logic simpler than other designs, as there is no need for a comma detector or a word aligner. The only required module is a decoder that verifies that the received data is valid. Having a simple logic has not only benefits in saving resources, but it is desirable in applications to be used in environments with radiation [20–23], in order to lower the chance of single event upsets (SEUs) and single event latch-ups (SELs). Due to the fact that many resets could be performed at the receiver, before the desired alignment is achieved, the roulette approach has an obvious drawback in the increase of the average lock time that can be shown to be proportional to the number of bits in the parallel symbol. For this reason, when planning to use the *roulette* approach, the designer should carefully evaluate a trade-off between the average lock time and the simple receive logic. Another possibility is the use of a device that is not able to output the raw data but can automatically align and decode the data to the byte boundary and possibly can store the phase-offset between the recovered clock and the stream into an internal register. This behaviour has the same effect of externally finding the phase-offset. In the architecture described in this chapter, based on a GTP, we combined the approaches and the strategies described above. Indeed, we used an external logic in the FPGA fabric, designed to inspect the encoded data and to find the phase offset between the recovered clock and the byte boundary. But we also used the roulette approach, as we can shift the phase of the recovered clock only by even numbers of UIs. In case an odd bit shift is required, due to a specific unfortunate phase offset, the logic in the fabric provides a reset of the device and keeps waiting for a phase offset requiring an even bit shift, after achieving a new lock.

9. Conclusions

Commercially available high-speed SerDes devices are usually designed for data transfers at variable latency. This is because fixed-latency operations require dedicated circuitry and they are often not needed in most of datacom and telecom applications. However, fixed-latency serial IO is useful, or even mandatory, in various application, such as high-speed transfer protocols for analog-to-digital and digital-to-analog converters, trigger and data acquisition systems, clock distribution, synchronization and control of radio equipment.

In this chapter, we have shown how to implement fixed-latency serial IO, essentially by opportunely configuring SerDeses (in particular, the SerDes devices embedded in commercially available Xilinx FPGAs) and by adequately adding a specific control logic to such devices. The proposed architecture is able to operate with fixed latency and it is capable to recover the clock from the serial stream with a predictable phase, which does not change after a power-cycle or a reset of the link. We presented a 2.5-Gbps 8B10B link which is able to serialize 8-bit words, as a detailed example of implementation. We also described the procedure for extending the example architecture in order to transfer packets made of several data words and to synchronously transfer data with an external clock. The presented architecture is also code-independent, i.e. it can be used with any data encoding, provided a special care to the various issues described.

Acknowledgements

This work is part of the ROAL project (CINECA Grant no. RBSI14JOUV) funded by the Scientific Independence of Young Researchers (SIR) 2014 program of the Italian Ministry of Education, University and Research (MIUR).

Author details

Raffaele Giordano[1]*, Vincenzo Izzo[2] and Alberto Aloisio[1]

*Address all correspondence to: rgiordano@na.infn.it

1 Università Degli Studi Di Napoli "Federico II" and INFN Sezione Di Napoli, Napoli, Italy

2 INFN Sezione di Napoli, Napoli, Italy

References

[1] Jedec Solid State Technology Association, JEDEC Standard, "Serial Interface for Data Converters," JESD204B.01

[2] IEEE Standard 1588, *IEEE Standard for a Precision Clock Synchronization Protocol for Networked Measurement and Control Systems*, 2008.

[3] M. Lipiński, T. Włostowski, J. Serrano and P. Alvarez, "White rabbit: a PTP application for robust sub-nanosecond synchronization," In: *2011 International IEEE Symposium on Precision Clock Synchronization for Measurement Control and Communication (ISPCS)*, Munich, 2011, pp. 25–30. doi:10.1109/ISPCS.2011.6070148

[4] *CPRI Specification V4.1 (2009-02-18)*, 2009, pp. 21–22 [On-line]. Available: http://www.cpri.info/downloads/CPRI_v_4_1_2009-02-18.pdf

[5] R. Giordano and A. Aloisio, "Fixed-latency, multi-gigabit serial links with Xilinx FPGAs," *IEEE Trans. Nucl. Sci.*, vol. 58, no. 1, Feb. 2010, pp. 194–201. doi:10.1109/TNS.2010.2101083

[6] R. Giordano and A. Aloisio, "Protocol-independent, fixed-latency links with FPGA-embedded SerDeses," JINST,7,P05004, 2012. doi:10.1088/1748-0221/7/05/P05004

[7] J. Wang et al., "FPGA implementation of a fixed latency scheme in a signal packet router for the upgrade of ATLAS forward muon trigger electronics," *IEEE Trans. Nucl. Sci.*, vol. 62, no. 5, Oct. 2015, pp. 2194–2201. doi:10.1109/TNS.2015.2477089

[8] R. Giordano, V. Izzo, S. Perrella and A. Aloisio, "A JESD204B-compliant architecture for remote and deterministic-latency operation," In: *2016 IEEE-NPSS Real Time Conference (RT)*, Padua, 2016, pp. 1–2. doi:10.1109/RTC.2016.7543080

[9] "Virtex-5 FPGA RocketIO GTP Transceiver User Guide," Xilinx, UG196, v1.7, 2008 [On-line]. Available: http://www.xilinx.com/support/documentation/user_guides/ug196.pdf

[10] "Virtex-5 FPGA User Guide," Xilinx, UG190, v4.3, 2008 [On-line]. Available: http://www.xilinx.com/support/documentation/user_guides/ug190.pdf

[11] M. Yan, "SONET/SDH Essentials WHITE PAPER," 2008 Exar Corporation [On-line]. Available: https://www.exar.com/uploadedfiles/home/sonet-sdh-essentials_022508.pdf

[12] A.X. Widmer and P.A. Franaszek, "A DC-balanced, partitioned-block, 8B/10B transmission code," *IBM J. Res. Dev.*, vol. 27, no. 5, 1983, p. 440

[13] A.X. Widmer and P.A. Franaszek, "Byte oriented DC balanced (0,4) 8B/10B partitioned block transmission code," U.S. Patent 4 486 739, Dec. 4, 1984

[14] R. Giordano, A. Aloisio, V. Izzo et al., "High-resolution synthesizable digitally-controlled delay lines," *IEEE Trans. Nucl. Sci.*, vol. 62, no. 6, Dec. 2015, pp. 3163–3171. doi:10.1109/TNS.2015.2497539

[15] A. Aloisio, F. Cevenini, R. Giordano and V. Izzo, "Characterizing Jitter performance of multi gigabit FPGA-embedded serial transceivers," *IEEE Trans. Nucl. Sci.*, vol. 57, no. 2, Apr. 2010, pp. 451–455

[16] PCI Express, "PCI Express Base Specification Revision," 3.0 Nov. 10, 2010 [On-line]. Available: http://composter.com.ua/documents/PCI_Express_Base_Specification_Revision_3.0.pdf

[17] RapidIO, "RapidIO Interconnect Specification," 9/2014. [On-line]. Available:http://www.rapidio.org/wp-content/uploads/2014/10/RapidIO-3.1-Specification.pdf

[18] InfiniBand Trade Association, "InfiniBand Architecture Specification Volume 1," Mar. 3, 2015 [On-line]. Available: https://cw.infinibandta.org/document/dl/7859

[19] InfiniBand Trade Association, "InfiniBand Architecture Specification Volume 2," Nov. 6, 2012 [On-line]. Available: https://cw.infinibandta.org/document/dl/7141

[20] A. Aloisio and R. Giordano, "Testing radiation tolerance of SerDeses for serial links of the SuperB experiment," In: *Proceedings of the 2011 IEEE Nuclear Science Symposium, Medical Imaging Conference*, Valencia, Oct. 23–29, 2011

[21] A. Aloisio and R. Giordano, "Testing radiation tolerance of electronics for the SuperB experiment," In: *Proceedings of the 13th ICATPP Conference on Astroparticle, Particle, Space Physics and Detectors for Physics Applications*, Como, Oct. 3–7, 2011

[22] P. Branchini et al., "Intensive irradiation study on monitored drift tubes chambers," *IEEE Trans. Nucl. Sci.*, vol. 54, no 3, Part 2, 2007, pp. 648–653

[23] P. Branchini et al., "ATLAS MDT chamber behaviour after neutron irradiation and in a high rate background," *Nucl. Instrum. Meth. A*, 581, 2007, pp. 171–174